INTERNATIONAL ASSOCIATION
OF FIRE CHIEFS

A Leadership Guide for Combination Fire Departments

Fred C. Windisch, EFO, CFO

Fred C. Crosby

JONES AND BARTLETT PUBLISHERS
Sudbury, Massachusetts
BOSTON TORONTO LONDON SINGAPORE

Jones and Bartlett Publishers

World Headquarters
Jones and Bartlett Publishers
40 Tall Pine Drive
Sudbury, MA 01776
978-443-5000
info@bpub.com
www.jbpub.com

Jones and Bartlett Publishers Canada
6339 Ormindale Way
Mississauga, Ontario L5V 1J2
Canada

Jones and Bartlett Publishers
International
Barb House, Barb Mews
London W6 7PA
United Kingdom

International Association of Fire Chiefs

4025 Fair Ridge Drive
Fairfax, VA 22033
www.IAFC.org

Jones and Bartlett's books and products are available through most bookstores and online booksellers. To contact Jones and Bartlett Publishers directly, call 800-832-0034, fax 978-443-8000, or visit our website www.jbpub.com.

Substantial discounts on bulk quantities of Jones and Bartlett's publications are available to corporations, professional associations, and other qualified organizations. For details and specific discount information, contact the special sales department at Jones and Bartlett via the above contact information or send an email to specialsales@jbpub.com.

Production Credits

Chief Executive Officer: Clayton Jones
Chief Operating Officer: Don W. Jones, Jr.
President, Higher Education and Professional Publishing: Robert W. Holland, Jr.
V.P., Sales and Marketing: William J. Kane
V.P., Design and Production: Anne Spencer
V.P., Manufacturing and Inventory Control: Therese Connell
Director of Marketing: Alisha Weisman
Publisher–Public Safety Group: Kimberly Brophy
Senior Acquisition Editor: William Larkin
Editor: Jennifer Kling
Senior Production Editor: Susan Schultz
Manufacturing and Inventory Coordinator: Amy Bacus
Composition: Modern Graphics
Text Design: Anne Spencer
Cover Design: Tim Dziewit
Printing and Binding: Malloy
Cover Printing: Malloy

The International Association of Fire Chiefs (IAFC) and the publisher have made every effort to ensure that contributors to *A Leadership Guide for Combination Fire Departments* are knowledgeable authorities in their fields. Readers are nevertheless advised that the statements and opinions are provided as guidelines and should not be construed as official IAFC policy. The IAFC and the publisher disclaim any liability or responsibility for the consequences of any action taken in reliance on these statements or opinions.

Library of Congress Cataloging-in-Publication Data

Windisch, Fred C.
A leadership guide for combination fire departments / Fred C. Windisch, Fred C. Crosby.
p. cm.
At head of title: International Association of Fire Chiefs
Includes index.
ISBN 978-0-7637-3381-0 (casebound)
1. Fire departments—Management—Case studies. 2. Fire departments—Personnel management—Case studies. I. Crosby, Fred C. II. International Association of Fire Chiefs. III. Title. IV. Title: International Association of Fire Chiefs.
TH9158.W565 2008
363.37068'4—dc22

2007026592

6048
Printed in the United States of America
11 10 09 08 07 10 9 8 7 6 5 4 3 2 1

TABLE OF CONTENTS

FOREWORD

James B. Harmes

Many things have changed since I began my career in the fire service 30 years ago. The tactics, technology, scope of the job, and the expectations of the public have all evolved. Gone are the days of riding to calls on the tailboard, open cab rigs, and just providing basic first aid to victims. Today, we have advanced apparatus, complete coverage firefighting gear ensembles, built-in thermal imaging in SCBA masks, and the integration of advanced life support EMS into the fire service.

With these positive changes, the requirements for training, staffing, and customer service continue to grow. While this impacts every fire department from Boston to San Francisco, it impacts volunteer departments in growing communities even more so. Volunteer departments face increasing requirements while the number of community members able to volunteer at that level decreases.

Today, many communities are looking toward combination fire departments to pay for a critical minimum level of select services while still maintaining the qualified and dedicated volunteers to provide the bulk of fire services. The combination model provides a level of flexibility around service and staffing levels while maintaining a realistic budget outlook for communities with limited resources and expanding needs. This is a situation facing more and more communities throughout the United States.

When I became chief 22 years ago, there was no guidebook on how to run a combination fire department. Like many of my peers, I learned by trying, adapting, and trying again until I got it right. *A Leadership Guide for Combination Fire Departments* provides a comprehensive outline of considerations and variables that surround taking a volunteer fire department and transitioning into a combination department. From goal setting, working with your local elected officials, training both a captive and not-so-captive audience to the same standards, the new challenges of dealing with paid employees, and how to lead the entire department to success, this book provides a framework for creating a new, or improving an existing combination fire department.

As more and more communities move toward a combination department, a guide like this is invaluable for you—the leaders of tomorrow's fire service—as we move forward in providing quality services to the public.

Chief James B. Harmes, CFO
IAFC President 2006–2007
Grand Blanc, Michigan

Timothy S. Wall

The fire service is moving ever forward at a quick and steady pace. There has been a steady movement throughout the fire service in the United States toward a combination fire department. Fire chiefs today face many challenges that affect the daily operations of these types of fire departments. *A Leadership Guide for Combination Fire Departments* is a road map with incentives and guidance that will help enhance your education on today's combination fire service. It will help you to build a successful strategic plan for your department, its members, and your community.

Volunteer Chief Timothy S. Wall
North Farms Volunteer Fire Department
Wallingford, Connecticut
Chair, Volunteer & Combination Officers Section of the IAFC

Philip C. Stittleburg

The fire service is changing. While this is not new news, the rate of change is unprecedented. Today's fire chief must be alert to the factors that signal the need for a change in the organizational structure of the fire department entrusted to his or her leadership. *A Leadership Guide for Combination Fire Departments* is an excellent resource for the fire chief challenged by ever increasing demands for service and a rapidly evolving mission. It provides advice on recognizing the need for a shift from a volunteer department, how to make the change, and how to manage that most challenging of fire service organizations: the combination department. Leading diverse members who are expected to perform the same services, but approach the job from different motivations, requires a rare set of skills. This book can help identify those necessary skills, as well as provide resources, examples, and sample forms.

Chief Philip C. Stittleburg
LaFarge Fire Department
LaFarge, Wisconsin
Chair, National Volunteer Fire Council
Chair, Wisconsin State Firefighters Association

PREFACE

This book has been a journey that really started over 10 years ago when many of our fire service brothers and sisters began asking questions about the best way to run a combination fire department. The term *combination* is relatively new, just like most of the agencies that are evolving from pure volunteer to part paid. The intent of this book is to identify the perfect ways to be a perfect leader in a combination system.

In reality, we know there is no one right way to do things, but we have attempted to bring together various subject matter experts and blend their concepts and experiences into a book that should be read from beginning to end. Every community has different needs and available resources, but we have tried to distill the content into some pretty good methods that will assist in system design. The challenges are many, the solutions are few.

We believe that the mission, vision, and values are first and foremost and that strategic planning is imperative. These themes resurface throughout the book because every contributor understands that the building process begins with the foundation. Ultimately, it is important that leaders never forget that the citizen comes first, and if every decision were based on citizen requirements, then the rest of the process would be to fill in the blanks.

Fred C. Windisch, EFO, CFO
Fred C. Crosby
August 2007

ACKNOWLEDGMENTS

A hearty thank you goes to our contributing authors. At the beginning of this project we really didn't know how we were going to get through the pain and agony of it all, but as usual, this network of wonderful and dedicated individuals stepped up to the plate and hit a home run!

Such an undertaking could never be completed without the aid of many people. While we offer this as a guide, we would be remiss if we did not acknowledge that everything we are and everything we hope to be is because of the people who have taught and mentored us along the way. This project presented the opportunity to work with the "best in the business," so rather than taking any credit, we simply say that it has been an honor to work with the best minds in the combination fire service today

We also humbly thank our families for their patience, love, and support, that allows us to follow a path that is really a vocation rather than an occupation. Without our wives, Theresa Windisch and Mary Theresa Crosby, we would be nothing. Theresa Windisch deserves a valor award for putting up with me for 38 years and dealing with my trials and tribulations as we traversed our lives. And, to our son Scott, we are very, very proud of you and your accomplishments. We love you and (grandson) Cody and appreciate how you have become a great citizen and emergency responder—helping others is a great thing, and what better legacy could Theresa and I leave? To my mom and recently departed dad: Thank you for giving me the opportunities and foundation to excel. To my friends and colleagues: Thank you for helping me all the time.

This book is dedicated to those individuals who have chosen this line of work and who want to do the very best for their respective communities.

Fred C. Windisch, EFO, CFO
Fred C. Crosby
August 2007

ABOUT THE AUTHORS

Rick Haase, CFO, CFPS, CEM

Rick Hasse is currently the emergency response coordinator/fire chief of ConocoPhillips Wood River Refinery in Roxana, Illinois. Rick also serves as the volunteer chief of the Staunton Fire Protection District, Staunton, Illinois.

Haase holds associate's degrees in industrial electronics and fire science technology from Lewis and Clark Community College and a bachelor's degree in advanced fire administration from Western Illinois University. He holds numerous state fire service and EMS certifications and is a field staff instructor with the University of Illinois Fire Service Institute.

He is currently the vice-chairman of the IAFC Industrial Fire and Safety Section and is also active in other IAFC sections as well as other state and national emergency response organizations. Rick was honored as the 2001 Illinois Volunteer Fire Chief of the Year and as the 2004 Fire Chief Magazine National Volunteer Fire Chief of Year. He has been in the fire service for 25 years.

Wes A. Cole, AASFSCT

Wes Cole is the fire chief of the Northwest Fire Department in Houston, Texas. Chief Cole began his career in 1976 as a volunteer with the Little York VFD in Houston, Texas and joined the Northwest VFD in 1982. At Northwest he has served in every capacity possible and became the volunteer chief in 1991. In 2000 he transitioned to the first full-time career chief of what it now an integrated combination department. He holds an associate's degree in fire science technology from Cy-Fair College where he serves as an adjunct instructor. He holds certifications as a Master Firefighter with the State Fireman's and Fire Marshal's Association of Texas as well as an advanced certification from the Texas Commission on Fire Protection. He has served as president of the Harris County Firefighters Association, has been an active member of the International Association of Fire Chiefs since 1992 and is an active member of the VCOS. He is also a member of the Texas Fire Chiefs Association.

Richard A. Marinucci, CFO

Richard Marinucci has served as fire chief of the Farmington Hills Fire Department in Farmington Hills Michigan since 1984. Chief Marinucci was president of the International Association of Fire Chiefs in 1997\1998. He was the first chair of the Commission on Chief Fire Officer Designation. In 1999 he served as acting chief operating officer of the United States Fire Administration and received the Outstanding Public Service Award.

He has bachelor's degrees from Western Michigan University, Madonna University, and the University of Cincinnati, where he was the first graduate of the Open Learning Fire Service Program (summa cum laude) and named a distinguished alumnus in 1995. He has been an adjunct faculty member of Madonna University, Eastern Michigan University, and Oakland Community College. He has presented programs for the International Association of Fire Chiefs, Fire Department Instructors Conference, Maryland Fire/Rescue Staff and Command, and Fire Chiefs Association of Japan.

Bruce Mygatt

Chief Mygatt has been the fire chief at Boulder Rural Fire Protection District in Boulder County Colorado since 1996. He has spent his entire fire service career with Boulder Rural starting as a volunteer fire fighter in 1972. During his volunteer career he was instrumental in creating the department's initial training division, held the position of training chief, and worked his way up through the officer ranks and served as the volunteer fire chief. In 1996 he became the first career employee of the district on its way to becoming a combination department.

Chief Mygatt is a past president of the Boulder County Firefighters Association, was very active in establishing the new Boulder County Fire Chiefs

Association, modeled after Colorado State Fire Chiefs Association, and is the President of the Boulder County Regional Fire Training Centers. He is also a member of the Colorado State Fire Chiefs Association Executive Board, the group leader of the CSFCA-Combination Section, a member of the Metro Chiefs Section, a Board Member of the Colorado Emergency Services Association, and a founding member of the Colorado Fire Chaplains Association.

Bruce has a bachelor's degree from Montana State University and maintains his Fire Officer III certification from the State of Colorado. He has lived with his wife, Karen, in Boulder since 1967 close to their two sons, two nice daughters-in-law, and three wonderful grandchildren.

Chief Mygatt can be contacted at 303-530-9575 or by e-mail at Bruce.Mygatt@brfd.org.

John R. Leahy, Jr.

Chief Leahy has served the fire service for over 51 years. Over the span of his career, he has served as: Pittsburgh Bureau of Fire, fire fighter through fire chief; Largo Fire Rescue (FL), fire fighter through deputy chief; Seminole Fire Rescue (FL) as fire chief; Pinellas Suncoast Fire Rescue District (FL), training officer through fire chief; National Fire Academy Board of Visitors; member IAFC, Metro Chiefs and Volunteer and Combination Officers Section Board of Directors; International Society of Fire Service Instructors member, Section Chair through Chairman of the Board; International Association of Fire Fighters, member Local 1 through vice-president; and volunteer since 1958 in Greensburg, Pennsylvania through present. He is the author of numerous articles and past editor of *Metro Network* and *Alaska Firefighter*, as well as editorial advisor to *National Fire & Rescue Magazine*. Leahy mentored both a blood family and a fire professional family. He also authored the *Contract with America*, and the *Firefighter's Bill of Rights*. He is the regional manager of National Fire Sprinkler and is a committed advocate of training for life.

Dan Eggleston, EFO, CFO

Chief Dan Eggleston began his fire service career in 1978 as a volunteer with Chesterfield Fire and EMS. Dan has been a career and volunteer chief officer for over 17 years and is currently the career fire rescue chief with the Albemarle County Department of Fire Rescue in Charlottesville, Virginia, a 600-member combination department that serves a 723-square-mile urban/rural area.

Chief Eggleston has an AAS degree in engineering, a bachelor of science degree in business, and a master of science degree in emergency management from the University of Richmond. He is a graduate of the National Fire Academy's Executive Fire Officer Program and received his Chief Fire Officer Designation from the Commission on Fire Accreditation International.

Chief Eggleston is a Commonwealth of Virginia Department of Fire Programs adjunct fire instructor and is an active member of the International Association of Fire Chiefs and the Virginia Fire Chiefs Association. He has published many articles and research papers relating to fire and EMS administration and emergency management and has lectured and consulted throughout the country.

Richard B. Gasaway, MBA, EFO, CFO

Chief Richard B. Gasaway has served as a career fire chief in combination fire departments since 1990. His emergency services career spans 29 years, serving six organizations in West Virginia, Ohio, and Minnesota. He has served as the chief of the Roseville Fire Department in Roseville, Minnesota since 1999.

Chief Gasaway holds a bachelor of science degree in business administration from West Virginia University, an MBA from the University of Dayton, and is in the final step of completing his PhD from Capella University where his dissertation research is focused on fireground command decision making under stress. He is also a graduate of the National Fire Academy's Executive Fire Officer program.

In 2001 Chief Gasaway became the 30th chief fire officer worldwide to be accredited by the Center for Public Safety Excellence. In 2003 he was appointed by President George W. Bush to serve on the Public Safety Officer Medal of Valor Review Board. The board reviews and recommends recipients to receive the highest civilian award for heroism.

Chief Gasaway lives in Roseville, Minnesota with Carol, his wife of 24 years, and four children.

David B. Fulmer, MPA, EFO, CFO

Chief Fulmer began his fire service career in 1983 with the Pleasant Hill Volunteer Fire Department located in Steubenville, Ohio. Since then Chief Fulmer has served in varying capacities with organizations in Illinois, Wisconsin, and currently serves as the fire chief of the Miami Township Division of Fire/EMS in Miamisburg, Ohio. This combination department is comprised of career and part-time fire fighters that provide service to a 22-square-mile jurisdiction from four fire stations.

Chief Fulmer holds an associate's degree in fire protection, a bachelor's degree in technical education, a master of public administration degree, and is currently pursuing his PhD in leadership and organizational change. Chief Fulmer is a 1998 graduate of the National Fire Academy's Executive Fire Officer program and is a chief fire officer designee through the Commission on Fire Accreditation International.

Chief Fulmer resides in Miami Township with his wife, Amber, and their two children, Ian and Kendal. Chief Fulmer can be reached at fulmerdbf@aol.com.

CHAPTER 1

Why a Combination System?

Rick Haase

Case Study

It is the last Thursday of the month, and the fire chiefs of Johnson County convene for the Johnson County Fire Chiefs Association monthly meeting. After the meeting, a number of the volunteer fire chiefs from the smaller jurisdictions of Johnson County engage in a discussion about the trials and tribulations of their respective fire departments.

Chief Siebold of the Murrayville Township Volunteer Fire Department voices his concerns about the continued increase in emergency medical service (EMS) calls in his district because of the recent construction of a large retirement village. Chief Siebold is worried because his volunteer fire fighters and emergency medical technicians (EMTs) are stretched to the limit with the increasing EMS calls. The new highway opening next month will greatly increase the traffic flow through his district. In addition, six new subdivisions are going to greatly increase the ever-growing population. Chief Siebold expresses his doubts as to whether his volunteers can continue to keep up with the ever-increasing service demands.

Across the table, Chief Cooper of the Longview Volunteer Fire Department expresses his concerns about the dwindling number of daytime volunteers. Although he has a rather large overall staff of 50 active volunteer fire fighters, many of the volunteers leave Longview each morning to commute to work. Chief Cooper has an excellent volunteer turnout, but before 6:00 p.m. and after 6:00 a.m., it is sometimes difficult to get two fully staffed engines out the station door.

Chief Mosser of the Johnsonville Heights Volunteer Fire Department expresses his concerns with the ever-increasing nonresponse workload his volunteers are experiencing. The rapidly growing suburb of Johnsonville Heights has a large number of new strip malls and an expanding school system. His volunteers work nonstop to complete inspections on the newly developed strip mall and to keep up with the fire drills and public education requests from the local school system.

All three of the chiefs continue to express their concerns about meeting the volunteer staffing needs posed by their unique situations. Assistant Chief Quigley of the Johnsonville Fire Department pulls up a seat at the table. He tells the story of how the Johnsonville Fire Department evolved from an all-volunteer staff of 45 personnel to the current combination staff of 75 personnel. Immediately the conversation turns to a lively question-and-answer session on the pros and cons of volunteer versus combination fire departments.

Introduction

Throughout history, fire service organizations have adapted to meet the needs of the community. Fire departments that were once all volunteer often become combination departments or paid departments. There is one major question that any fire department considering the shift from a volunteer department to a combination department (or paid department) has: what is the deciding factor that shifts the volunteer fire department to a paid fire department? Unfortunately, there is no single answer.

There are a multitude of reasons why different fire departments opt to move from a volunteer mode to a combination of volunteers and paid personnel. In some cases, the reasoning is based on an increase in service delivery requirements from the community. In other cases, it is because of the lack of resources. In many cases, it is a combination of many factors that leads to the decision to move to a combination service delivery system.

Overall Call Volume

Overall call volume is a major factor in the transition from a volunteer fire department to a combination fire department. When a volunteer fire department sees a continual increase in responses for an extended period of time, it will look for ways to meet the extra service demands. A continued increase in calls for service can be attributed to multiple factors including rapid population growth, addition of new or different levels of service, and increased hazards within the response district.

A continued growth in the number of responses stretches the resources of a volunteer fire department for several reasons. First and foremost, the sheer time spent responding to increased call volume is a major burden on the volunteer personnel. Other issues include increased paperwork, maintenance operations, and minimum training standards and requirements, as well as additional incident support services (e.g., fire investigation, customer service operations, assurance of apparatus, and equipment readiness).

Major increases in the number of responses burdens the volunteer staff. Time spent away from regular jobs poses issues with employers. Time spent away from the family increases stress on both the volunteers and their families. Juggling the stresses of incident responses, other department operations, regular employment, and family can push volunteers to reconsider volunteering. If some volunteers decide to discontinue their membership because of increased call volume, this adds to the problems of the volunteers who remain with the fire department.

A continuing increase in call volume is a major burden on any fire department—volunteer, combination, or full career—but the burden on an all-volunteer fire department can be extreme. A progressive fire department heeds the warning sign of a continuous increase in the number of responses and develops appropriate plans to meet the concerns of the fire department and the community. The evolution from a volunteer to a combination fire department is a common plan. In general terms, when the number of responses exceeds approximately 750 within a calendar year, then it is necessary to evaluate when to begin the transition process of hiring a fire chief, administrator, or other full-time employees. There is no hard and fast rule on this change, but common sense indicates that an average of two responses per day may indicate a workload that the volunteer fire chief cannot maintain without help. Another indicator is a budget that exceeds $500,000 per year.

Daytime Staffing Shortages

For a long time, daytime staffing shortages have been a major deciding factor for many volunteer fire departments in changing over to a combination fire department. In most volunteer fire departments, many volunteers work regular jobs during normal daytime hours (typically between 7:00 a.m. to 4:00 p.m.). Add the fact that many of the volunteers have 1- to 2-hour daily commutes, and you can understand why many fire departments have a problem providing required staffing from 6:00 a.m. to 6:00 p.m. Monday through Friday.

In today's commuter world, the realization that many volunteers work outside of the response district is only one of the many problems that volunteer departments have to cope with. With personnel working outside of the response district, the fire department's ability to respond during the daytime hours is compromised. Volunteers who have daily commutes are actually only available for daytime duty during their vacation days and perhaps as replacements during a major long-term incident.

Many volunteer fire departments are fortunate to have a core of volunteers who are employed within the response district and can respond to incidents during the day. For years, this was the primary method that many volunteer fire departments used to provide daytime staffing. With the increased employment demands, the ability for volunteers to respond while working is more difficult.

Alternatives that some volunteer fire departments have used in an attempt to increase daytime staffing include recruitment of nontraditional volunteers (e.g., homemakers, retirees, students), shift workers, and home business owners. Some volunteer fire departments have even used "mandatory volunteer" staffing. This consists of city public works personnel and even police officers acting as volunteer fire fighters.

To meet the shortfalls in daytime staffing, combination fire departments with paid daytime duty crews are a popular solution throughout the country. These paid daytime crews provide sufficient staffing for small-scale response operations such as EMS assists, carbon monoxide investigations, and utility responses. These crews also can perform required support operations such as equipment inspections and maintenance, documentation, and public education, thus relieving volunteers of these time-consuming tasks.

Although the use of the daytime paid crews lifts a significant burden off of volunteers, combination departments have found that additional volunteer staffing is still needed during daytime periods to provide sufficient staffing for major incidents. Additionally, many combination departments still provide a large portion of their evening staffing (6:00 p.m. to 6:00 a.m.) with all volunteer resources.

Types of Responses

The types of responses that a fire department is burdened with affects the decision to move towards a combination fire department. For many years, fire departments were primarily tasked with basic firefighting operations—structure fires, vehicle fires, and brush fires. During the last 40 years, the role of the fire service has moved from basic firefighting to a myriad of tasks: from EMS response to technical rescue to hazardous materials incidents to utility service support. Naturally, the increase in the type of responses has increased the total number of responses for the fire departments.

Many fire departments are seeing a major change in the types of incidents that they are responding to. The number of structure fires are decreasing for many fire departments while the number of service calls (e.g., utility emergencies, carbon monoxide investigations, smoke investigations) and alarms sounding (fire detector and smoke detector activations) are increasing. This increase in nonfire incidents increases the burden on many volunteer fire departments to the point where they need to shift to a combination fire department in order to meet the needs of the community.

The nonfire incidents are tying up volunteer fire departments for several reasons. First, the overall call load for these types of incidents increases, especially in suburban areas. Also, in today's workplace, it can be difficult for volunteers to justify leaving the workplace to answer a non-life-threatening call. Many of these types of incidents can be time consuming (e.g., standing by an electrical emergency waiting for utility companies to arrive, searching a large structure for the source of a detector activation). For this reason, many volunteers find it difficult to justify responding to these incidents because these incidents could keep them from their paying jobs for an extended period of time.

An initial response of on-duty paid crew can effectively handle many of these nonfire and nonurgent incidents. This greatly reduces the impact on the volunteer staff—yet the volunteers are still available to respond to significant incidents if needed.

EMS Response

The increased role of the fire service in the delivery of emergency medical services (EMS) is why many volunteer fire departments have opted to change to a combination fire department. Whether a fire department provides first response services, transport services, or a combination of both, the delivery of EMS services has a major effect on the fire service from a staffing perspective.

Whether it is first response services or transport services, EMS response can be a time-consuming operation for a fire department. First and foremost, there is the sheer volume of EMS calls. Most fire departments that provide EMS response see at least 50% of their responses related to EMS incidents. Because the personnel resources required for EMS response places a major burden on the fire department, many fire departments use paid personnel to provide EMS services. Also the use of paid personnel to provide these EMS services drastically reduces the impact on volunteer personnel, making them more available to assist during major incidents and periods of extremely heavy call volumes.

There are several other reasons why combination staffing is a natural fit for fire departments providing EMS response. EMS transport services are time consuming, especially if the fire department provides nonemergency transfer services. This type of time-consuming operation is difficult to provide with an all-volunteer staff. Also there is the issue of the support services required to provide EMS services. The extra paperwork, billing services, restocking of equipment, and maintenance of equipment adds another set of time-intensive burdens that is difficult to provide with an all-volunteer staff. Finally, there is the issue of the required training for EMS responders. Naturally, the higher the level of EMS service, the greater the level of training required for the EMS providers. Not only is there a large investment of initial training requirements, but ongoing in-service training requirements as well.

Many fire departments use combination staffing to meet the needs of the EMS response dilemma, and fire departments have done it many different ways. Some of the methods include assigning certain duty crews to provide only EMS response services, using duty crews of fire and EMS cross-trained personnel, and using contracted paramedic services (either dedicated EMS or cross-trained personnel).

Changes in Response District Geography

A significant change in response district geography may be another indication that a volunteer department may need to consider a move to a combination system. Although changes are typically somewhat slow in nature, the overall effect of the changes usually means increased demand for fire services. The increased demands can outpace the level of service available from a volunteer staff and thus create a potential need for combination staffing.

Changes to response district geography can come in many forms. One of the most common changes is rapid residential growth, often in the form of single-family, two-family, and multifamily structures. Extremely rapid residential growth, especially in retirement age communities, can increase the demands for fire response, nonfire incidents, and especially EMS response. Rapid residential growth can also increase the need for various fire department support services such as public education services, preincident planning, and mapping of new subdivisions.

Another major geographical change that has affected many fire departments is the development of a major retail center such as a shopping mall, outlet mall, or multiple strip malls. If the initial development is successful, a never-ending stream of additional retail facilities, as well as restaurants, convenience stores, and similar outlets, often continue to expand throughout the area. This type of expansion usually increases the number of automatic fire alarms, motor vehicle accidents, and EMS responses. More time-consuming efforts will be required for inspections and preincident planning operations. Conducting these many services for a large retail complex is a major strain that most volunteer fire departments cannot manage without the help of some type of paid staff.

There are many other geographical changes that could trigger the need for the movement from a volunteer staff to a combination department. This could include a major influx of manufacturing facilities or the development of new transportation facilities such as a new airport, new interstate transportation routes, or new railway facilities. These types of facilities could bring a new series of hazards and new challenges to a fire department. The development of major recreational or entertainment facilities brings a major population influx as well as many of the issues similar to the addition of retail establishments.

Significant geographical changes within a response district can cause different types of impacts on any fire department. Being able to adapt to these changes while still providing quality services is a challenge. In many cases, volunteer departments have found the addition of a combination staff in some form (e.g., duty crews, full-time inspection personnel) have provided the resources to provide the necessary services. The implementation of a combination-style operation allows specific support to be deployed in a specific manner to meet the needs of the changing community.

Increased Service Delivery Needs

The expectation of the fire service to be an all-risk response agency continues to place major demands on all types of fire departments-volunteer, combination, and career alike. The days of just fighting fires are over for almost every fire department. All fire departments are being asked to provide new and improved services, and this increased service demand leads to increased demands on personnel—especially volunteers.

Increased service delivery from a response standpoint includes specialty response covering issues such as hazardous materials and technical rescue incidents. It includes the expectation of not only initial responder-level medical response, but also paramedic-level services. It includes the expectation to provide disaster response, response to carbon monoxide investigations, and response to electrical wires on the ground. It also means the public's continued expectation to respond to a cat-in-the-tree call, a lifting-assistance call, and any other call that no one else will answer.

The old adage, "If you don't know who to call, call the fire department," is true in many different regions across the country. It is an expectation that the fire service provides every conceivable response capability no matter what the size or makeup of the fire department. This kind of mentality has placed a huge demand on the fire service. Not only does the fire service have to respond to every request for service, it also has to prepare for these responses. This includes extensive training as well as maintenance and upkeep of equipment.

Increased service delivery has placed a burden on the fire service in general, but has specifically placed a major workload on volunteer fire department. When the burden overextends the volunteer department's limits, many have looked to the combination solution to provide the extra boost needed to meet expectations. The burden is usually related to a decline in staffing, increased response times, and public outcry.

Changes in Volunteer Demographics

Another major reason why many volunteer departments have opted to move to a combination department is the changes in volunteer demographics. Fifty years ago, many volunteer departments had a large number of volunteers in the 30 to 45 age range and even had a waiting list of younger volunteers. During the 1950s, fire departments provided just basic firefighting operations. Volunteers had more time to meet the fire department requirements and therefore joined and stayed with the department for many years.

Times have changed. Today's volunteers are a cross section of many different groups and have many different reasons for volunteering. The days of local businessmen leaving their shops to respond to incidents is a thing of the past. Today stay-at-home parents, college students, shift workers, and retired professionals have joined the ranks of many volunteer fire departments. Some volunteer fire departments have even enlisted commuters to volunteer in the area where they work during the day, and then volunteer in their hometowns during the evenings and weekend. In many cases, the demands of the volunteer way of life have caused a decrease in the number of active of volunteers, and hence the advent of the combination fire department movement to fill the deficiencies in the staffing.

The increasing physical demands of the different aspects of fire and rescue operations have caused many older volunteers to move away from active volunteer activities at an earlier age. Although there are still a number of senior volunteers, many of these are retiring at an earlier age. Fire and rescue services are physically demanding. Generally speaking, the older members are relegated to driving vehicles and other support activities. There is a need for younger members to do the hard work, but that pool of volunteers has been on a steady decline. The demands of basic firefighting have changed dramatically. Younger volunteers with families find it more difficult to find the time to volunteer because of family commitments.

The days of families having a single breadwinner are a thing of the past. Today both parents work, and thus life at home is more hectic. Demands at work force some wage earners to work long hours. This increased pressure to work long hours to meet basic financial needs leaves little if any time for many breadwinners to volunteer.

Another issue that limits volunteering is liability and the potential of injury. Although fire departments carry insurance, many people feel the risk of firefighting is too great and opt not to participate in such a dangerous volunteer occupation.

A large portion of the younger generation was never exposed to the volunteer process. Although there are commonly several generations of a single family within a fire department, the fire service continues to see an increasingly smaller core group of volunteer recruits. Waiting lists of volunteers for fire departments are the exception rather than the rule. Although there was somewhat of a resurgence in volunteering since September 11, 2001, there are still many fire departments that cannot meet the demands of their communities. In many cases, these fire departments have moved to the combination system.

Need for Additional Support Services and Management

Many members of the general public mistakenly think that fire fighters sit in the fire station all day and wait for the tones to send them out on a call for assistance. But all fire departments—volunteer, combination, or full career—know that this is far from the truth. There are a myriad of routine tasks and operations that must be completed to keep the fire department running. These tasks put a burden on a full-time career department—let alone a volunteer fire department.

Routine tasks such as fire apparatus and equipment maintenance are extremely time consuming yet must be completed to ensure safety and readiness. Some portions of the tasks are regulated. Monthly self-contained breathing apparatus (SCBA) and extinguisher inspections, and annual pump, hose, and ladder testing, are just a few examples of regulated maintenance activities that affect every fire department. These types of activities also require training and extensive documentation. Many growing volunteer departments find that the burden of maintaining apparatus and equipment inventories, as well as facility maintenance, is too great to keep up with. These departments opt to hire personnel to assist with maintenance operations, and in return they obtain additional personnel to respond to incidents. In many cases this is a wise use of resources to assist with lean daytime volunteer staffing.

There are many other routine tasks similar to these that need to be completed by fire service agencies: training program development and documentation, incident documentation, planning, inspection services, and public education services. Although some volunteer departments are lucky enough to have volunteers who can perform these tasks, a number of fire departments have found that the implementation of a combination system is the answer. Many departments in rapidly growing areas have found that combination staffing is the only way they can keep up with the demands of planning and inspection of new construction within the response district.

Yet another area that is stretching the resources of volunteers is the overall management of the fire department. For many fire departments, a paid chief is one of the first positions staffed. In such cases, these formerly all-volunteer fire departments have established that they have a need for a management staff to oversee the fire department. Paid management positions are common for combination departments, especially those with large overall staffs. Human resource management, financial management, and master planning (i.e., developing multiyear plans that will provide the road map for the future direction for the department) are just a few of the many issues that the management staff must deal with. Time-consuming issues such as these are difficult to manage on a part-time basis. A full-time paid staff has more time to deal with these issues.

Need for Personnel Skills Maintenance

Another issue that may tempt an all-volunteer department to move towards a combination system is the ongoing need to maintain the volunteers' skills. Firefighting operations are becoming more technically advanced. To provide quality services, personnel must have ongoing training. Naturally, training costs money, and more importantly, training is time intensive.

Fifty years ago, training for firefighting operations consisted of basic firefighting skills possibly coupled with basic first aid. Today, firefighting operations require initial basic skills training coupled with ongoing refresher training. In many states, this may include required training topics and specific training hours. In many fire departments, the initial training is just the tip of the iceberg. Advanced-level training for specific firefighting operations (e.g., apparatus operations, rapid-intervention training, engine company operations, ladder company operations) is now becoming the norm.

Firefighting training just scratches the surface in many fire departments. The need for EMS, technical rescue, and hazardous materials response training is an additional burden placed on fire departments. For each additional service provided, the amount of required training rises proportionally. As with firefighting training, this includes not only initial training but also ongoing skills maintenance training. Some or all of these skills may be state regulated depending on the location.

Finding the time to attend training is a burden on any fire department, but time and skills are also needed to develop and provide such training. The training officer is a paid position within many combination fire departments. Developing, presenting, and documenting a training program is a burden on a volunteer department that can be eased by shifting to a combination system. In many cases, a training officer enhances training programs, producing better levels of services. Some combination departments have used paid training staff to establish a training academy that trains other fire departments; this can be a profit center.

Size of Response District and Station Locations

The overall size of a response district, including the location of fire stations within the response district, is another reason for a volunteer department to move towards a combination system. Naturally, large response districts, such as those that are seen in rural areas or those that serve large areas such as a township or county, can stretch the limits of any fire department. Extended response times limit the level of services that a fire department can provide. Couple the travel time from the fire station to the fire with the time needed for volunteers to reach the station to start the response, and the outcome is response times that are not effective. The use of the combination system decreases the overall response times.

Another effective use of combination services is having paid personnel in strategically located fire stations within the response district. Paid personnel from a central location within the district can immediately respond to outlying fire stations, which may be staffed by volunteer personnel. In some cases, the combination departments have strategically staffed their fire stations by designating each fire station as the provider of specific services to all stations (volunteer and paid alike). For example, one fire station may be designated as the training academy, another as the SCBA maintenance station, and yet another as the technical rescue station.

Regional Response Opportunities

A newer form of the combination system used by some fire departments is the use of regional response services. This type of operation may provide staffing service ("flying" personnel squads, meaning personnel who work at multiple locations as needed) or specialized response resources (e.g., technical rescue or hazardous materials response) to several volunteer departments in an area. This type of operation provides necessary resources to several fire departments in a cost-effective manner. The ability to rapidly mobilize necessary resources to deal with significant incidents also creates safer operations and better overall service delivery.

Regional response methods may also be adapted to nonresponse operations. Regional service could deliver essential maintenance services, public education services, or a variety of other services. This type of operation saves on expenses during nonresponse periods.

Cost of Providing Services

The overall cost of providing services is one of the most significant reasons why a fire department may

Table 1-1 Various Staffing Models

Agencies may utilize these concepts to design their staffing program:

- Use part-time employees working from a pool of personnel who are paid an hourly rate. Note: The Fair Labor Standards Act (FLSA) precludes same-agency volunteers from working part-time under most conditions. Please refer to the International Association of Fire Chiefs Fair Labor Standards Act document available from the International Association of Fire Chiefs (IAFC).
- Use part-time employees assigned to various hours to address staffing gaps.
- Use full-time employees assigned to various hours to address staffing gaps.
- Use volunteer staffing at various hours, either assigned or voluntary, to address staffing gaps. Note: FLSA rules allow for stipends as long as there is not an hourly component. Refer to the IAFC FLSA document.
- Use full-time or part-time staffing to address specific activities as prescribed by the agency, such as driving vehicles.
- Use joint and regional personnel pools to create a "flying" squad.

adopt a combination system. It costs money to provide volunteer services, and it costs money to provide paid staff services. There are necessary costs to every fire department for apparatus and equipment, insurance, and maintenance and services. A fire department with paid staff has expenses of personnel salaries and other benefits. A combination system is clearly the method of choice when transitioning from pure volunteer staff to full-career staff. Many rapidly growing fire service organizations have found that fully paid staff are needed to meet community expectations. In many cases, the ability to shift from a volunteer department to a fully paid staff is not feasible. A volunteer fire department with a relatively fixed budget and a staff of 35 volunteer fire fighters would find it difficult to change over to a fully paid staff of 35 fire fighters. The cost of salary and benefits alone would likely exceed their total available revenue. Most departments have found that using a combination system to transition to a fully paid staff is effective. This process of transition allows the fire department to incrementally increase staffing levels based on available funding.

Summary

There are many reasons that may lead a volunteer fire department to consider transitioning from a volunteer to a combination department. A number of these reasons have been discussed in this chapter, but this is simply the beginning. A combination department can provide elements of service that a volunteer department (or alternately a fully paid department) may not be able to provide. A combination fire department has a unique blend of characteristics that can meet the needs of many different communities with several unique opportunities. A lack of imagination is the only limiting factor when designing a combination system.

Fire departments considering transitioning to a combination system should thoroughly review the issues leading them to this style of service. They must consider the pros and cons of a combination department and be able to justify their decision to the membership, their customers (i.e., the taxpayers within their service district), and policy makers. Being a combination department increases the overall costs of doing business and increases the complexity of the fire department management structure. On the other hand, combination departments provide a higher level of service and decrease the workloads on volunteers, which may in turn increase volunteering within the department. Do your research. Be open-minded. Listen to the needs of the fire department members. Listen to the needs of the community. Look to other fire departments that have successfully implemented a combination system. The decision to transition to a combination department is not easy, so do your homework, and you can make the grade.

CHAPTER

2 The Search Begins

Fred C. Crosby

Case Study

As the newly appointed fire chief, Ed Johnson wanted his combination fire department to succeed. As he has done throughout his career when faced with a problem, he did thorough research, looking for standards, definitions, and guides involving combination departments. What he found concerned him. There were no national standards for combination departments. The library was lacking useful books on the subject. There was nothing to help guide him to the success that he desired. Chief Johnson was at a loss. How could he move forward with no road map?

Introduction

Chief Johnson faces the same dilemma that every chief of every combination department faces. He is on a quest to find the perfect combination of career and volunteer members to build a fire department that is not only efficient, but more importantly, effective. Yet there are no clear, written standards to follow. We are searching for the best road map.

Finding the Road Map

The first step is to define what a good combination department looks like. However, *good* is subjective, and success is dependent on the needs of each community. A good fire department serves the needs of its community. The first measure of success is how effective your department is in following its mission. The second measure is how well your components work together to enhance and augment one another. A good combination service must find a balance in its use of career and volunteer fire fighters. Both career and volunteer fire fighters bring strengths and weaknesses to the table. A good combination system builds on the strengths of one to augment the other's weakness. It should be one team for one purpose.

In Hanover, Virginia, the search for the right road map began by examining the need for a new approach to leadership. Hanover Fire Emergency Medical Services, or EMS, is a rapidly growing emergency services department in Virginia. The department has cutting edge technology and is appealing to many because of its growth potential.

■ Department History

Hanover County is a rapidly growing county in Virginia. It covers over 512 square miles, supports over 100,000 residents, and has both rural and urban characteristics. There are 12 fire stations, four EMS stations, over 130 career personnel, and almost 600 volunteers.

In 1960, Hanover County hired its first career fire chief. Over the next 4 decades, the volunteer fire chiefs and the Volunteer Fire Association strongly influenced the fire chief on major departmental decisions such as strategic planning, budget, training, personnel management, and equipment.

In 1991, the first EMS coordinator was appointed. His role was to bring together the four EMS rescue squads and to determine the future needs of the EMS system. It was apparent that the EMS call volume was greater than the volunteer resources available, thus paid personnel were added to the EMS department. In 2002, the fire and EMS departments were consolidated to form the public safety department. A new fire/EMS chief was appointed. As the Hanover fire/EMS chief, Fred Crosby has redefined the public safety department's vision and mission. He has issued several special orders to express his expectations to personnel, both career and volunteer. The following particular special order is significant because it illustrates the importance of both the career and volunteer members in the combination system:

> Hanover's system depends on volunteers. This will continue to be the case for the foreseeable future. The loss of volunteers is contrary to the interests of the department and the county. The vision for the department is to create a high performance combination fire and EMS system that serves the current and future needs of the citizens of the county.
>
> All personnel are reminded to consider this fact in their daily actions. Furthermore, all personnel are encouraged and reminded that it is incumbent upon each individual of the department to honestly work and support the vision of a combination system, where all members of the team are treated equally, held to the same standards, and all discourse is handled with dignity and respect.
>
> It is the interest of the department and county to foster a professional and cohesive team of career and volunteer personnel to provide this necessary and critical service to the citizens of Hanover County. All personnel should be aware that this interest is a basic responsibility of each individual and is considered to be a fundamental duty of each position and job description in the department.

Before the merger of fire services and EMS, the administration struggled to hold volunteers accountable to the same training standards and standard operating procedures as the career personnel. Because volunteer officers were selected and treated differently than career officers, a double standard was inadvertently created. This caused tension between volunteer and career personnel. The situation was exacerbated because the career personnel worked during the daytime, and the volunteers worked at night.

The career and volunteer cultures also seemed to conflict. Individuals who volunteer fall into several categories:

1. They are young and just out of school, and they want to start a career in fire services and EMS.
2. They are midcareer and have families.
3. They are retired but do not want to slow down.

Although they take their volunteering seriously, volunteers have full-time jobs and other demands. The constant training and continued education necessary can be difficult and at times demanding for volunteers.

Most volunteer organizations are initially established as fraternal organizations, and decisions are made by the voting members. This concept is in stark contrast with career firefighting where decisions are made using the chain of command.

To ease tension between the volunteer and career personnel, the Hanover fire/EMS chief enforced the same rules and standards for everyone. The career personnel adopted an alternate work schedule, which included several hours of overlap with the volunteers. This allowed them to operate as one single department. They worked together, ate together, and trained together. This greatly boosted morale.

Choosing New Leaders

A leader in a department is the navigator, using the road map to lead the department to success. Officer selection is a key component to developing a premier combination department. Leaders act as mentors and play an important role in developing personnel, setting the department tone, and achieving goals and objectives. In order to ensure that everyone was on the same path, it was necessary to review and update both the career and volunteer officer selection processes for the Hanover Public Safety Department.

In the past, volunteer officers were selected through the electoral process. The electoral process tended to resemble a vote of popularity. It was not based on training standards or qualifications. It did not ensure that the right person would get the job. Leadership is paramount in the fire service. Officer selection plays a significant role in the success of a department. Although the electoral vote may have been useful for volunteers in the past, it has become less applicable today. The fire service has become more technical, and education and experience are essential tools for current leaders. Because the Hanover Public Safety Department has over 700 working members, leaders must have knowledge involving human resources and personnel. The major flaw with the old volunteer selection process is that the most popular candidate may not be the most qualified candidate.

For the department to be effective, all officers, both career and volunteer, must meet minimum training qualifications, have pertinent experience, and be able to perform at high levels. To create consistency in Hanover, the fire/EMS chief formed a steering committee to establish a process to ensure quality and procedural equality in both volunteer and career officers.

The steering committee was a coalition of both volunteer and career officers that reflected an accurate cross-section of the department. The fire/EMS chief's intention was to obtain a broad range of representation. The steering committee developed a plan to alter the selection process of volunteer officers and establish a program consistent with career firefighting that was still acceptable and feasible for volunteers.

With the introduction of career staff and the development of a new public safety department, the fire/EMS chief began to focus more on performance standards. Career personnel are held accountable to high training standards and strict performance measures. The difference between career and volunteer personnel was that career personnel had less influence over the overall direction of the standards and guidelines of the department. Like their leadership selection process, standards and guidelines were not voted on.

However, the volunteer leaders felt that they had lost their place in the decision-making process. This insecurity and the focus on education and accountability led to a feeling of alienation in the volunteers. In essence, the volunteers began to feel that they were being pushed out of the system or not needed.

The steering committee was challenged with developing a promotion process for volunteer officers that was consistent with the career process. This process would not only determine qualified personnel, but there would be a clear understanding of loyalty and dedication to the organization. It was expected this process would be accepted and embraced by the volunteers. By allowing the volunteers to have a seat at the table via the steering committee, the fire/EMS chief earned much-needed respect. Because the volunteers felt they had more control over their destiny, they worked hard on a reasonable leadership solution.

The steering committee developed a career program that examines each individual candidate. The promotion process incorporates a written exam, an assessment center, and performance evaluations. Individuals qualify to begin the process when they have met minimum job requirements and have achieved the desired time in pay grade. The process is an attempt to blend intelligence, experience, education, and performance, in both the professional and practical setting. The assessment center tests the candidate's ability to manage time, resolve conflict, and mitigate an emergency incident. Thus far, the program has been successful, and quality officers have been selected for the department. It has shifted the emphasis from the most senior fire fighter, to the most qualified.

The new officer selection process will be phased in through four steps over 4 years. In the first year, the officers will be selected using the current process. In the second year, the individual fire companies and rescue squads will make recommendations to the fire/EMS chief for the top officers. From the companies' recommendations, the fire/EMS chief will appoint a district chief and two assistant chiefs for each fire station and a rescue squad captain for each squad. In the third year, the junior officers, consisting of the sergeants, lieutenants, and fire captains, will be processed through an assessment center with a pass or fail score. Those who pass will be eligible for an officer position, as selected by the appointed district chief or rescue squad captain. Finally, by the fourth year, all officers must meet or exceed the training standards and qualifications developed by the volunteer training committee.

The development of the new officer selection process was a success. It represented the values of both the career and volunteer members of the department. Both groups consisted of longtime veterans of the department who have held leadership roles. Accordingly, they felt there was a need for change, and the change had to be grounded in training and competency.

Summary

The fire/EMS chief created a steering committee with those who would be affected by changes made. He gave them an opportunity to participate in the decision-making process; thus, they had ownership of the result. He allowed room for negotiation. This flexibility made it easier for the volunteers to adapt to the change, and he earned more trust and respect from the group. Finally, the steering committee had the ability to pitch the new election process to the members because it was their result secured through hard work and determination.

The steering committee created an officer selection process that will favor "combination-oriented" candidates. By respecting the values of both the career and volunteer fire fighters, the department has grown stronger as well as closer to the fire/EMS chief's goal of becoming a model combination department.

In conclusion, the team dynamics and teamwork displayed by the steering committee led to a quality result. The strategy and tactics of the members proved to be successful, and the department has benefited because the proper parties were represented in the decision-making process.

CHAPTER

3 Planning for Success

Fred C. Windisch

Case Study

Chief Joe Huey began his day thinking about how to convince his board of directors to purchase a new fire engine. The usual thoughts surfaced—dependability, drivability, safety, reduced maintenance costs, and motivation for the entire fire department. But there had been board resistance in the past to buying a new engine, primarily because of cost. Joe made the decision to make it happen at that night's meeting regardless of the board's concerns.

Joe made a list of advantages to purchasing a new fire engine. He made a conscious decision to avoid the negatives. In his mind, Joe felt that listing any negatives would show weakness in his argument and would only result in another "no" vote. The chief spent 2 hours preparing a visual presentation he thought would compel the board toward the purchase.

That evening, Joe thought his presentation went off perfectly. There were no questions, so he sat down to reap the rewards. His confidence was swelling. Board member Fred Joyce, a regular supporter of the chief, made a motion to purchase a new fire engine with costs not to exceed $125,000. There was no second to the motion. The chief was stunned. His confidence was reduced to rubble within a few short moments.

Introduction

Although the fire chief in the case study believed that his presentation was going to be a success, his proposal had some fatal flaws. Prior to presenting any major proposal, a leader must address its strengths, weaknesses, threats, and opportunities. Fleshing out all of the anticipated aspects allows the board of directors to make a business decision based on as much information as possible. A strategic plan does just that.

What Is a Strategic Plan?

Strategic plans come in various sizes and shapes. The intent of this chapter is to lay out a model for creating a strategic plan that can be adapted to fit any fire department's needs. Some refer to strategic plans as master plans, business plans, forward plans, or future strategies. Whatever the name, most important is the basic tenant of the strategic plan: include as many stakeholders as is reasonable. This engages all those involved to focus on desired outcomes versus getting bogged down in the details. This type of successful planning minimizes failure. This method is considered backwards compared with the planning methods traditionally used in the fire service. These traditional methods usually involve throwing resources at a problem and then hoping that the resources are sufficient to make the plan succeed.

A strategic plan is a road map to the future. A road map provides various directions and paths for achieving a goal. One management concept is the use of the 80% rule. We sometimes overdesign our plans, and the result is that the plan is never launched or is delayed until it is out of date. The 80% rule allows for consensus, following a road map to the goal.

For example, a team decided that they would drive from Miami Beach to Seattle. All of the team members agreed that the goal was achievable, affordable, and could be done in 6 days. The road map planning concept allowed for the flexibility of taking various routes to Seattle. Some of the team members believed that the shortest route would be the most efficient, while other team members thought that the driving team needed some rest and relaxation along the way. This was the beginning of finding a consensus. None of the team wanted the drivers to be physically exhausted because that would increase the risk of an accident. An open discussion identified the need to stop for regular vehicle safety inspections as well as motel rest periods for the drivers.

The 80% rule was working. The team was satisfied that the goal would be completed, and the drivers would have a safe and healthful trip. The team ensured that the drivers would be allowed to modify their travel route (possibly to include some scenic routes) as long as they arrived within the time frame—with no dented fenders. The team also decided that if anything went wrong, the drivers would contact the team chairman to allow for more flexibility to accommodate issues beyond their control.

This story demonstrates the steps needed to achieve a goal; identify strengths, weaknesses, opportunities, and threats; focus on the mission; and get the job completed safely and efficiently. A strategic plan helps the team focus on the mission without straying, and then designs the road map to achieve the mission.

Fire chiefs are CEOs of their organization. In the past, the fire chief was the best fire fighter, but today there is an increased focus on operational integrity and excellence on the business side. There are far too many "potholes" in our road to success, so the more time we spend defining where these potholes are and how deep they are, the better chance we have at success. Poor planning results in low expectations and failures. Great planning results in great successes.

Focus Groups

Stakeholders need to be defined by the organization's leadership. A strong foundation is an absolute necessity for success. Focus groups should contain both internal and external members, some with institutionalized experience and knowledge. External individuals could include elected leaders, club and organization leadership from the community, a doctor or nurse (especially if the agency provides EMS), an accountant, a mechanic, and others. Care should be taken to define the proposed goals and assure that everyone understands the volunteer nature of the team. The strategic team must be a manageable size, with a suggested maximum of 12–15 people. Make sure to provide incentives to the group such as an occasional dinner meeting, an article in a newsletter, shirts for the team, or other methods to give them notoriety and support.

A team leader should be chosen by the focus group. There should also be someone assigned to record the proceedings. This does not mean that there needs to be detailed meeting minutes. Meeting notes are sufficient—attempt to capture the end results of discussion points, not the discussion process. The written word will help to solidify support when meeting notes are provided to other external community members on a regular basis.

Brainstorming is imperative. It must be agreed on to conduct open and honest discussions—everyone is equal in this team. This team is there to help, and they would not have accepted the challenge if they were not dedicated to the fire department.

Vision Statements

The first task for the focus group is to define the mission of the agency. Mission statements should be short and directly to the point:

> The mission of the Alphaville Fire Department is to provide quality fire, rescue, emergency medical, and special services to the community, and to continually seek improvements in the quality of life of our citizens.

The vision statement builds on the mission statement. Vision statements explain *how* the department is addressing their mission. Vision statements do not have specificity but are guiding principles:

The Alphaville Fire Department's vision is to meet or exceed our customers' expectations. We will utilize our members' expertise in achieving this vision. Our department will deliver the highest-quality service within the limits of the provided resources, and we will protect and improve our members' capabilities on a daily basis.

Another example of two vision statements—one wrong and one correct—are:

- *Wrong:* The Alphaville Fire Department will achieve an International Organization for Standardization (ISO) rating of 3 by 2011.
- *Correct:* The Alphaville Fire Department will work toward improving its ISO rating over a period of 5 years using reasonable resources.

The wrong version is flawed because it has no benchmarks or identified resources. The correct version proposes continual improvement over time utilizing available and planned resources. Next, value statements must be identified, which should be specific to the mission:

The Alphaville Fire Department has these values:

- Provide a safe, healthful, and environmentally friendly emergency response system.
- Support our volunteers and career staff with adequate incentives and awards to achieve top-level performance.
- Meet or exceed local, state, and federal standards for emergency agencies.
- Use a business model that addresses adequate service levels within the limits of available resources.
- Actively recruit the most qualified persons without regard to race, color, or creed.
- Communicate openly and honestly within the organization as well as with the public and related organizations.

A successful fire chief is motivated by common sense. Common sense is reality-based thinking. There are failures caused by a lack of realistic thinking in every different community. A mission, a vision, and value statements, all based in reality, are the building blocks of a department. Overall, the strategic plan must be based in reality.

Cost vs. Benefit

Once again, reality checks are an important part of the strategic planning process. Sometimes the costs of an improvement outweigh the benefits. The strategic planning team can design its plan utilizing the cost-vs-benefit approach.

An example of cost-vs-benefit failure might be, "The Alphaville Fire Department will replace all fire apparatus every 10 years." The vision statement is fundamentally correct in that it uses a time element with the goal of full replacement. The missing link is the cost of full replacement to the specific community's resource base. Buried and unwritten within the vision statement are many benefits of having a reliable and state-of-the-art apparatus fleet.

Everything costs. If the resources are not available, then the benefit of a reliable fleet will not be achievable. Further definition of the costs and benefits is necessary even if the strategic planning team understands and is willing to find the money themselves via a realistic plan.

A capital improvement plan (CIP) is a part of the strategic plan and defines the dollars necessary to achieve the goal. The CIP identifies both. This may include annual funding to a dedicated account; identifying needed apparatus types specific to the community; and a time continuum that establishes specific benchmarks leading up to purchase, delivery, implementation, maintenance, and planning for the next round. Simply put, a detailed plan is needed as part of the strategy.

Another option is to consider a lease purchase/turn-in plan. This method is gaining popularity with the fire service. Research should include a comparison of the existing apparatus replacement plan to the lease purchase plan.

Lease purchasing can also be used to acquire other necessary items—ice machines, copy machines, and a host of other hardware and software. There is also personal protective equipment (PPE) to think about. PPE has a limited life even with excellent care. PPE lease purchases are becoming more popular as costs continue to grow at an unprecedented rate. A PPE master plan can also be a part of the CIP. It is possible to be fairly accurate in predicting PPE longevity and how many replacement sets will be needed per year. A focused PPE plan can identify the necessary resources and associated costs and has the required funding as a part of the plan. Laying out specific plans makes it easier to justify required purchases and lends credibility to the fire chief's capabilities as a great leader.

Other general subjects of a strategic plan could be:

- Information technology
- EMS and dispatching
- Education, training, and training facilities
- Safety and health
- Financial strategy
- Public education and fire prevention
- Human resources
- Organizational structure

Summary

There is an applicable story in one of the fairy tales we learned as kids. Alice is running from the knaves, who want to hurt her. She comes to a fork in the road. The Cheshire Cat grins evilly at her. She asks the cat, "Which way should I go?" The cat asks her where she is going. She says, "I don't know." The cat replies, "Then what does it matter?"

Strategic planning provides a road map. The supporting tactics of each part of the plan become the blueprint for success.

CHAPTER

4 The Importance of Mission

Fred C. Crosby

Case Study

It was a long day for Chief Ed Johnson. He had just finished another grueling meeting with the volunteer leadership, which was filled with conflict and rancor. It seemed to Chief Johnson that he spent all of his time in such meetings, trying to find common ground between everyone. The entire fire department was fractured into small interest groups. Each individual group had their own interests, power structure, and political intrigues. In addition, the volunteers and career fire fighters had their own specific interests. Somehow he was supposed to meld all of these differing interests together in the interest of county administration, and most importantly, in the interest of the citizens that they all served. Yet no matter how hard he tried, he could not find the common ground between the groups. After a day like today, he wondered if he could even get them to agree what day of the week it was.

Introduction

Chief Johnson faces a dilemma that is almost universal in combination systems. By design, the combination system is made up of disparate and often fiercely independent groups. Each of these groups has their own particular interests and agenda. Each group, if allowed, will assert their interests over the interests of the entire fire department. The job of the fire chief is to uphold the interests of the fire department above those of any specific group.

In the past, the chief would accomplish this by ordering the individuals and groups to behave and follow his instructions to the letter. This tactic worked in previous generations, but today, is not an appropriate method for the diverse personalities, generations, and persons that most chiefs lead.

Mission, Mission, Mission

In real estate, it is often said that the key to success is, "location, location, location." In a combination system, the key to success is, "mission, mission, mission." In the fire service, it is often said everyone is here for the same reason, but is that true? If a chief studies the issue, it will be found that fire fighters are not all working for the same reason.

Fire fighters are not there for the same reason because they all interpret that reason differently. They all have the same basic impulses, but they view them differently. Fire fighters all define those basic impulses through the prism of their experiences and through the mentoring that they received when they came into the service. By returning to the most fundamental reasons for becoming fire fighters, it is possible to build a foundation of consensus that will resonate throughout the entire fire department.

■ Mission-Centered Organizations: A Different Paradigm

Everything in the world exists for a specific reason. Everything grows out of a specific need. The fire service is no different. Unfortunately, through the years of evolution, the fire service's purpose of being has not been revisited often enough. By not returning to their origins, fire fighters have forgotten why they truly came to be. This has been exacerbated by different management theories being infused into the fire service over the years. There were thoughts of internal and external customers, as well as what was good for the volunteers or good for the union. The mission became confused by adding in secondary considerations.

The basic and primary mission of the fire service is to serve. Fire fighters are the safety net of the community. The community always calls when they do not know where to turn. It does not matter whether it is a fire or EMS incident, or whether it is a hazardous materials incident or a hurricane, the fire department is always called. Today, the new "Who's going to help us?" in an emergency is the Department of Homeland Security. In the next decade, it might be something else, but something will always come up. Fire departments have an essential and important reason for existing—we go to people's sides on the worst day of their lives and try to make it better. That is the essence of what the fire service does, and everything else, every other consideration that arises, is secondary.

The idea of a mission-centered paradigm for organizational management recognizes this basic fact. This system keeps the mission as the central point, and acts as a reminder that everything else done by the fire service is to support and enable the accomplishment of the central mission (see **FIGURE 4-1**).

In mission-centered organization, everyone understands that their primary interest is to support the main interest or reason for being. That is not to say that secondary issues are ignored. Volunteer issues and personnel issues must still be dealt with. Finance and efficiency are still important. However, these secondary issues are not the most important. When confronting these issues, they must be viewed through the filter of, "How does this help us or hurt us in achieving our mission?"

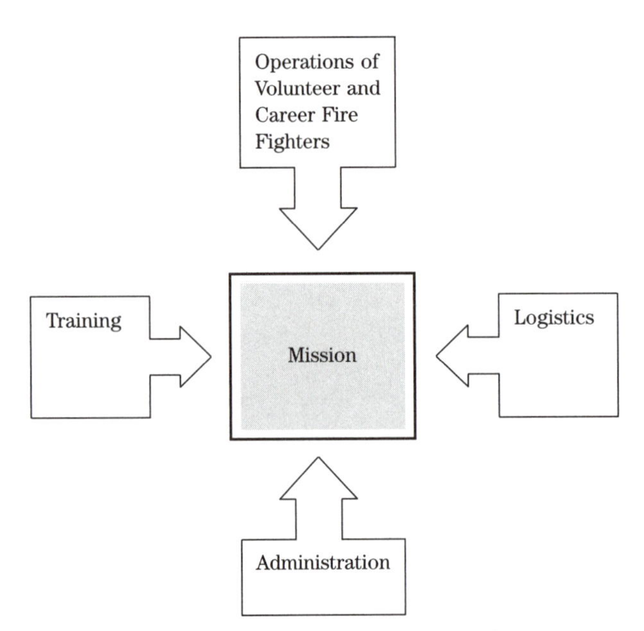

FIGURE 4-1 Mission-Centered Organizational Chart.

It is often heard that fire departments need to take a business approach and learn from the larger world. In today's world, chiefs need to be managers, leaders, administrators, and more. Fire chiefs speak in whatever the "management speak" of the day is. Fire chiefs do need to learn from the business world. The mission-centered ideal of organizational management is right from the playbook of successful business. Successful businesses have a keen awareness of their central reason for being. The best businesses have a laserlike focus on achieving their mission. Successful businesses are driven by what they do and why they do it. This is a lesson that the fire service can and should learn. Fire chiefs should be intensely focused on their central mission. They should infuse that intense focus throughout their organization.

■ Implementing the Mission-Centered Paradigm

Start by asking the question "Why are we here?" As answers appear, the different perspectives will become apparent. Ask this question of everyone and every interest group.

The task then becomes one of building consensus. To do this, the fire chief should bring together representatives of the various interest groups or stakeholders. In a combination department, this will include the volunteers, career personnel, officers, civilian staff, and members of the policy-making management.

Then a fire department must define its reason for existing. This is most often accomplished in the form of a mission statement. For the combination department, the idea of a mission statement should go one step further—it should include performance measures. If it does not, then it is only a slight-of-hand trick of moving the different perceptions from what the mission *is* to what the mission *means*. The fire chief must include some way to define that mission closer to its central theme. The actual performance measure means very little—it could be response times, dollar loss, fire flows, or cardiac arrest survival rates—it just has to be something. There are a couple of other things that should be done to get started—a fire department needs a vision and core values. This combination of a mission statement, which is defined and focused by performance goals and measures, a vision of where the mission is going, and ground rules of behavior in getting there, is so much more useful than a management cliché if used correctly. It brings about the agreement that Chief Johnson is so desperately looking for in the case study at the beginning of this chapter.

Mission Statements

Mission statements are the verbalization of the main purpose of an organization. They should be short, easy to understand, and explicitly clear. The most common mistake is trying to make a mission statement too eloquent. Simplicity is elegance in mission statements—a mission statement that no one remembers or really understands is worthless.

A mission statement should read something like this:

> The mission of the Successville Fire Department is to serve the community and protect lives and property through the provision of professional fire, rescue, and emergency medical services, 24 hours a day.

or

> The mission of the Successville Fire Department is to protect lives and property through proactive prevention, education, and response.

Once a mission statement is agreed upon, that mission should be defined—what are professional fire, rescue, and emergency medical services? What does it mean to serve the community?

Communicating the Mission

For a mission statement to be effective, it must be communicated. It should be the tagline of all department communications. The mission should be something that every member sees on a daily basis. Think of all of the creative ways a mission can be communicated. Maybe it is on the stationery, or perhaps it is on the coffee mugs that members use daily. It could be framed in the station in prominent locations. It can be part of the rookie testing or promotional process. The way a fire chief chooses to communicate the mission is not important. It is crucial that it is communicated often and emphasized as important.

Performance Goals and Standards

Answer the foreseeable questions and the meaning of the mission will be defined. What is a professional fire, rescue, and EMS system? Professional does not mean volunteer versus paid, it means doing the job professionally. It is defined by performance goals and standards. As stated previously, the actual standard is not nearly as important as the simple act of having a standard to achieve. Model standards and other documents to build from can be found at the National Fire Protection Association (NFPA), or they can be built from scratch.

The important thing is that the standard defines what good is for the community. For example, a set of performance standards could be built on response times.

Sample Performance Standards

- Equipped advanced life support (ALS) provider on scene in 8 minutes or less 80% of the time for first priority calls
- Basic life support (BLS) unit on scene in 10 minutes or less 80% of the time for second priority calls
- BLS unit on scene in 15 minutes or less 80% of the time for third priority calls
- Sixteen fire fighters on scene of structural fires within 8 minutes 80% of the time
- Fire flow capability of 3500 gpm established on structural fires within 10 minutes 80% of the time

Vision

As the first President Bush said, "You gotta have that vision thing." Part of defining the mission is knowing where the mission is going. The vision statement is similar to the mission statement in that it is the distillation of all of the lofty goals. Like the mission statement, a vision statement should be clear, simple, and unequivocal in its language. Be careful of poetic terms that leave room for individual interpretation.

A vision statement can read something like, "The vision of the Successville Fire Department is to utilize and improve the dedication and skills of our members and to constantly improve all of our services and operations. We will assist similar organizations upon their request with available resources."

Values

Ideally, an individual's personal values will align with the spoken and unspoken values of the organization. By developing a written statement of the values of the organization, group members have a chance to contribute to the articulation of these values, as well as evaluate how well their personal values and motivation match those of the organization.

Although that statement is true, it may be easier to think of values in this manner: if a mission is why we are here, and vision is where we are going, then values are *how* we act. Values are the ground rules of behavior for a culture. Well-defined and well-enforced values allow us to achieve our mission.

Values are a nonnegotiable item. There is no room for situational ethics or bending of the rules when it comes to values. In a world of gray areas, values are one of, if not the only, black and white area. They should and must be treated strictly.

Think about personal values. What are the values a department must possess to meet its mission? The answer is found in the values statement. Again this does not have to be a long, drawn-out statement. In fact, many of the most profound value statements in our country are the most eloquently simple and stark statements possible. Think about some universally known value statements, such as "Duty, Honor, Country," or the Marine Corps', "Semper Fidelis," meaning always faithful. These short brief statements are so profound that people willingly die to uphold the meaning behind these statements and their organizations. Or maybe it is as simple as the Virginia Military Institute's honor code: ". . . never to lie, cheat, steal, nor tolerate those who do."

Here is an example of a *long* value statement:

Our fire department has these values:

- Provide a safe, healthful, and environmentally friendly emergency response system.
- Support our volunteers and career staff with adequate incentives and awards to achieve superior performance.
- Meet or exceed local, state, and federal standards for emergency agencies.
- Use a business model that addresses adequate service levels within available resources.
- Actively recruit the best-qualified persons without regard to race, color, or creed.
- Communicate openly and honestly within the organization and with the public and related organizations.

Summary

In a combination department, a mission can be the keystone on which a broader consensus is built. For a mission to fulfill this role, the department must make the mission its central theme and most basic consideration. To use "management speak," the mission, with its performance goals, vision, and values, becomes the decision filter for all of the competing interests inherent in a combination department, and thus becomes the intersection of all individual agendas.

A mission, a vision, and values are a powerful tool to achieve one of the most elusive goals for every combination chief—agreement and discussion without conflict and division.

CHAPTER

5 The Role of Standards in a Combination System

Richard A. Marinucci

Case Study

The ABC Fire Department has been in existence for over 100 years. For 97 years, it was an all-volunteer organization, maintaining its traditions while the community around it changed. However, about 5 years ago, the fire department was "forced" to hire its first career fire fighters because of instances of inadequate or lack of daytime response from the volunteer fire fighters.

The four career fire fighters have been good, dedicated employees with a passion for their work. They exhibit competence while responding to emergencies and continue to seek self-improvement through training and education, much on their own time. They are also gaining the respect of many of the volunteers, especially the newer hires, because of their knowledge and expertise. Yet a problem is brewing. The volunteer department has a system of promotion based on the vote of the membership. Because of the department's history, the volunteers maintain control of the overall organization and its chain of command. There are current cases of volunteer fire fighters being promoted to officer positions with only 3 or 4 years on the job. This is partially the result of the turnover within the ranks and a lack of seniority. The volunteer fire fighters highly value the ranking officer positions and continue to promote solely from within their own to fill the vacancies. Furthermore, because of the history of the fire department and its development as a fledgling combination department, the volunteer officers maintain authority over the career members.

Chief Jones, the first career chief of the fire department, has been with the organization for 6 years, and has been watching this situation for a few years. He has noticed the mounting frustration of the career staff, which he views as talented and dedicated. He also recognizes the role of the volunteers and their desire to maintain the status quo. Despite that many of the newer volunteers are beginning to recognize the abilities of the career fire fighters, the more senior volunteer officers are not ready to relinquish any authority to the career personnel. Although the career members have been patient to date, their frustration is becoming more evident, and they are becoming more vocal about it. Chief Jones recognizes that something must be done, but what? Clearly the direction needed by the community is to maintain, as much as possible, the combination system. The career members want an opportunity to utilize their skills, training, and experience. Chief Jones wants to deliver the best possible service to the community.

With the transition to a combination department comes the need to change the way business is done. Everyone recognizes this, but how can it be done to minimize the disruption to the fire department and its daily tasks, gain widespread support of both the volunteer and career staff, and improve the performance of the fire department?

Role of Standards

Standards, policies, and procedures—most fire departments start with a minimal few and develop them when the need arises. The need may be the result of performance issues, external forces (e.g., political pressure, legal issues, risk management), or the desire to standardize the way things are done to be more efficient, effective, and consistent. Standards are used to correct problems and avoid future issues that can affect service delivery. Standards are used to maintain consistency and communicate expectations to personnel with regard to minimum performance. In a homogeneous organization, such as all volunteer or all career, standards are usually easier to draft and implement because the playing field is level for all employees. Because of differences between career personnel and volunteers—mostly in time availability and selection pool—it may be difficult to develop uniform standards that are fair to all. Yet establishing standards for all members is critical to the long-term viability of a combination department.

Standards are important to all fire departments. They establish the norm for operational issues, training requirements, expected conduct, and criteria for promotion. However, it is not easy to draft appropriate standards and have them readily accepted by the entire organization. Usually, most members like things the way they are. A lack of standards allows the freedom to make choices, cuts down on red tape, but can make it difficult to hold people accountable. Making a change to an established standard that currently requires a certain level of performance can be interpreted as a threat to the status quo. This is one reason why standards in a combination department can be a challenge to implement.

The source of standards can be internal or external. For example, the National Fire Protection Association (NFPA) has established standards that are applicable to all types of fire departments. These standards include apparatus, protective clothing, and equipment, which are universal and are usually out of the control of the local department because of laws or other factors (such as manufacturer's potential liability). Other standards from the NFPA may not have the force of law, but do apply pressure to comply (e.g., training standards such as NFPA 1001 and safety standards such as NFPA 1500). Other external standards are published on a federal, state, or local level. There are safety standards published by the Occupational Safety and Health Administration (OSHA), EMS standards published by the National Highway Traffic Safety Administration (NHTSA), and fire fighter certifications asserted by state fire training agencies and local employment policy. It is important to know the standards applicable to one's own fire department.

The type of combination department will affect, to some extent, the establishment of standards in that organization. A combination department can be mostly volunteer, mostly career, half of each, or any mixture thereof. There are also various levels of service that can be expected from both groups. For example, some communities use their career staff solely as drivers and equipment operators. In this case, dual driving standards may not be needed by that combination department. In some cases, the career staff provides paramedic service only. This can have an effect on scene operations, training requirements, and officer criteria.

Operational issues are critical. They are the basis for the delivery of service to the community and establish the quality of service that can be expected. This expectation will affect the budget, preparation of personnel, and potential liability concerns. Operational standards establish which personnel are assigned to respond, the apparatus to take, command and control, and the level of service provided. For the most part, standards regarding emergency response are similar regardless of the type of fire department—all career, all volunteer, or combination. Yet there are some areas that require a variation in the approach to operational issues. These include:

- Minimum response requirements
- Response time or turnout standards
- Response order
- Driving standards
- Authority within the department
- Level of services provided (e.g., EMS, hazardous materials response, special rescue)

Although minimum-response requirements and response-time standards may not differ from all-volunteer departments to combination departments, there are issues that need to be considered. Even if there are career staff on duty, there should be a minimum participation requirement. It is not fair to volunteers or career personnel to allow anyone to "show up" at their discretion. It affects all-volunteer response and can create personnel issues if infrequent responders arrive on the scene with limited prior working relationships with career fire fighters. Volunteer members need to respond and contribute consistently to maintain their skills, work as a team with career fire fighters, and demonstrate their commitment to the goals of the organization. Response by members needs to be tracked and monitored. The core value of the organization is to provide emergency services. This can only be done if everyone participates. A level of participation needs to be established that is fair to the community, the department, and the individuals that are members. The most important thing is to *have* a standard of response, it is not necessarily at what level the response is.

If the fire service is truly interested in service to the community, it must respond in a timely fashion. Again, depending on the makeup of a fire department, there needs to be the expectation of a reasonable response. Career members will have a response time established by staffing levels. If there is to be a combination response to incidents, the companies that are first to arrive need to know when they can expect help. They can then make a judgment that will consider resources needed to successfully handle the incident. Will mutual aid be needed? Should additional stations from the community be requested or should callback of off-duty personnel be initiated? Clearly, a reasonable response should be expected. This is not as easy as it might appear. Occasionally, volunteer personnel will rely on career staff to cover the incident and will not respond as required. Besides the obvious affect on service to the community, it can affect interpersonal relations and the department's ability to function as a team.

In an all-career department, established response is predetermined and understood by everyone. It is important to know what resources are responding and in what order. Career organizations are aware of expected response times and what vehicles are staffed for response. The incident commander (IC) can plan his strategy based on this information. It is equally important for the IC in a combination department to be able to do the same. Therefore, a combination department must have a response plan that includes the vehicles to be utilized and the order of response. It should not be left to the discretion of the responders as to which vehicle to take when more than one choice is available. For example, many fire departments have multiple vehicles within their stations. There can be an engine, ladder truck, rescue, ambulance, or other apparatus. There should be a standardization of response order that would dictate to both career and volunteer fire fighters which vehicle needs to be deployed and when. If left to the individual's discretion, variations will occur based on training and experience of those responding.

Most states have preestablished driving standards that authorize certain fire fighters to drive emergency vehicles. Fire departments should know these rules and take them a step further to ensure that those responding are capable of doing it safely every time. In some cases, insurance companies may have standards that are required prior to obtaining insurance. Each department should check with its insurance provider. Some combination departments use only career fire fighters as their drivers. If this is the case, publish this standard. If volunteers are allowed to drive, set the standard for authorization and which vehicles may be driven by volunteers. There are differences between vehicles that will affect the training needed by the drivers.

Regardless of who drives, the standard of performance must be the same. This would include the ability to operate the pump and elevating platform as part of the driver's responsibilities. It is not just about getting the vehicle to the scene, but being able to operate it effectively on arrival. Each year, 20–25% of deaths on the line of duty are the result of vehicular accidents—responding or operating on the roadway. Make sure that whoever drives the vehicle knows how to operate it safely. It may also be a good idea to limit newer employees from emergency driving. Again, there is a great risk to driving. Make sure the fire fighters who drive, either career or volunteer, are mature enough to do so.

FIRE DEPARTMENT

PERSONNEL PROCEDURE

PRIORITY: 3

RESPONSE PERCENTAGE **NO: 421.0**

EFFECTIVE: __________ **PAGE: 1 OF 1**

FIRE CHIEF APPROVAL:

PURPOSE

To standardize the application of the minimum response requirements for all-volunteer personnel.

POLICY

The current Rules and Regulations indicate that members must respond to 35% (Officers, 45%) of all tone-alerted incidents within their assigned district based upon a 1-year period. This is subject to review by the Supervisor or Fire Chief and consideration may be given to a member's regular work schedule if prior notice has been made to the Supervisor. As part of the review process by the District Chief, members' participation in other Department activities will be considered. Members failing to comply *may* be subject to disciplinary action in accordance with the Rules and Regulations if they do not maintain the minimum response percentage for any continuous 12-month period. Members anticipating a problem because of extenuating circumstances must consult with their Supervisor prior to any problems developing. A member whose response percentage falls below 35% (Officers, 45%) for any given month should consult their Supervisor.

PROCEDURE

I. The administration shall monitor the run percentage of all Volunteer members. Each member is responsible to maintain the minimum response percentage as outlined in the Rules and Regulations. Members are also expected to maintain a consistent response during a 24-hour period (i.e., runs made after midnight require attendance as much as the runs before midnight.) Members anticipating a problem due to work schedules, leaves, or personal matters, shall submit in writing to Headquarters through the Supervisor a memo explaining their circumstances prior to any anticipated problem. If no memo is received or on file, it is assumed that the member shall be able to maintain the required response percentage.

II. Run percentage shall be calculated on a monthly basis, as well as an annual cumulative 12-month basis, always remaining current based upon the previous 12 months. This information will be sent to the Supervisor for distribution to their members. Should a member's monthly response be less than the required percentage and no notification provided, the administration shall contact the Supervisor who will discuss the matter with the member. No disciplinary action will be taken. The member may drop below the required percentage for any given month as long as the rotating 12-month period remains above the minimum requirement.

III. Should a member drop below the required percentage for any given 12-month period, he or she shall receive a letter through the Supervisor indicating the same. This shall be considered a warning. Two warnings in any 12-month period shall result in a member being placed on probation for a period of 3 months. If the member does not improve his or her response percentage as required, he or she shall be subject to further disciplinary action up to and including discharge from the Department. Shall a member be placed on probation three times within a 3-year period, he or she shall be considered a chronic offender and will be subject to further discipline up to and including discharge from the Department. Members with special considerations may be granted a waiver of this requirement if prior notification in writing to Headquarters has been made. Any member anticipating a problem or having any questions should contact their Supervisor before their run percentage becomes a problem.

FIGURE 5-1 Sample personnel procedure—response percentage.

FIRE DEPARTMENT

COMMUNICATION/RESPONSE PROCEDURE

PRIORITY: 1

STATION NOTIFICATION — **NO: 1000.1**

TYPE OF INCIDENT: __________

EFFECTIVE: __________ — **PAGE: 1 OF 2**

FIRE CHIEF APPROVAL:

PURPOSE

To identify the Department's response to an incident based on time of day and type of call.

I. Response

ALL TIMES
A. FULL OUT-STATION + CENTRAL STATION 1. Structure Fires 2. Technical Rescue Incidents **B. TWO FULL OUT-STATIONS** 1. If Station 5 is out of service, activate two full station tones

MONDAY–FRIDAY 0730–1800 (Not including holidays)
A. STAFFED CAREER STATION ONLY[4] 1. Medical emergency/personal injury accident (PIA)[1] 2. Smoke in the area 3. Odor investigations (no evidence of fire) 4. Automatic fire alarms 5. Wires down/arcing wires 6. Hazardous material leaks/spills 7. Nonemergency incident (In service)[2] • Citizen assists • Carbon monoxide alarms • Unauthorized burn/open burning • Open hydrants 8. Fires not involving structures 9. Other incidents not otherwise specified **B. FULL OUT-STATION + NEXT IN-SERVICE STAFFED STATION** 1. Other incidents where the district's staffed station is not available[3]

[1] All expressway incidents will be handled by Station 5 and Station 1 or Station 2 and determined by location.
[2] Next available duty crew.
[3] The Battalion Chief shall be contacted by Dispatch for additional response determination.
[4] Incidents on roadways will require two duty crews to respond.

FIGURE 5-2 Sample communication/response procedure—station notification.

PAGE: 2 OF 2

MONDAY–FR IDAY
1800 - 0730, WEEKENDS, HOLIDAYS*

A. FULL OUT-STATION + CENTRAL STATION

1. Structure fires
2. Medical emergency/Personal Injury Accident (PIA)[1]
3. Smoke in the area
4. Odor investigations (no evidence of fire)
5. Automatic fire alarms
6. Wires down/arcing wires
7. Hazardous material leaks/spills
8. Fires not involving structures
9. Other incidents not otherwise specified

B. CENTRAL STATION ONLY—IN DISTRICT

1. Medical emergency/personal injury accident (PIA)[2]
2. Smoke in the area
3. Odor investigations (no evidence of fire)
4. Automatic fire alarms
5. Wires down/arcing wires
6. Hazardous material leaks/spills
7. Fires not involving structures

C. CENTRAL STATION ONLY—CITY-WIDE

1. Nonemergency incident (In service)[2]
   Citizen assists
   Carbon monoxide alarms
   Unauthorized burn/open burning
   Open hydrants

[1] All expressway incidents will be handled by Station 5 and Station 1 or Station 2 and determined by location.
[2] Next available duty crew.
[3] The Battalion Chief shall be contacted by Dispatch for additional response determination.

*Applicable Holidays: New Year's Day, Martin Luther King Day, Memorial Day, Independence Day, Labor Day, Thanksgiving Day, Day after Thanksgiving Day, Christmas Eve Day, Christmas Day, New Year's Eve Day

FIGURE 5-2 *Continued.*

FIRE DEPARTMENT

COMMUNICATION/RESPONSE PROCEDURE

PRIORITY: 1

EMERGENCY VEHICLE RESPONSE **NO: 1000.2**

EFFECTIVE: ____________ **PAGE: 1 OF 3**

FIRE CHIEF APPROVAL:

PURPOSE

To provide general response guidelines for the response of apparatus, equipment, and privately owned vehicles.

POLICY

It shall be the policy of the Department to respond apparatus and equipment in a manner most appropriate for any given situation. Based on these general guidelines, the senior officer/fire fighter shall be responsible to ensure that the most appropriate piece of apparatus with its specialized equipment respond when available.

PROCEDURE

I. Definitions
   A. The following are recommendations for the response type:
      1. Priority 1—Is an incident which indicates an urgent emergency. The response will include use of lights and siren in accordance with PA 300 of 1949.
      2. Priority 3—Is an incident which is urgent and the vehicles/apparatus responding shall not use lights and/or sirens. The responding unit shall follow all provisions of "normal traffic."

II. Response shall be appropriate to the needs of the incident. The following items should be considered:
   A. The number of apparatus shall only be those necessary to handle the incident.
   B. Additional apparatus and personnel may be requested per incident. This may include:
      1. Apparatus used to protect members on a road or highway
      2. Assist with patient care
      3. Staffing used to handle the incident
      4. Additional equipment is necessary on the incident
   C. The OIC shall also consider "back-to-back" or additional calls when assigning apparatus.

III. Vehicle Response Priority
   A. Fire related incidents:
      1. Priority 1
         a. Any fire or reported fire that involves a structure
         b. Car fires
         c. Large grass fires, or grass fires with exposures
         d. Outside fires in close proximity to a building (Example: Car or dumpster next to a building.)
      2. Priority 3
         a. Dumpster fires without exposures
         b. Small grass fires without exposures
         c. Open burn complaints
         d. All fire alarm calls shall be Priority 3 unless additional information is received that would indicate a fire is present.

FIGURE 5-3 Sample communication/response procedure—emergency vehicle response.

B. EMS response will be based on the priority given by Dispatch. Only initial responding apparatus from the respective station shall respond to Priority 1 unless otherwise directed by the OIC on scene.
C. Smoke/Odor Investigation
   1. Priority 1
      a. Smoke or odor of smoke in a structure
      b. Odor investigation with possible injury/illness
   2. Priority 3
      a. General odors (no smoke) in a building with no signs/symptoms
      b. Smoke/odor outside or not involving a structure
D. Reports of broken natural gas lines where there is a possibility of the gas entering a structure shall be a Priority 1 response. All other broken natural gas lines shall be Priority 3.
E. Carbon Monoxide
   1. Priority 1
      a. Carbon monoxide incidents with medical signs or symptoms of poisoning.
   2. Priority 3
      a. Carbon monoxide incidents without medical signs or symptoms of poisoning.
F. Hazardous Materials Incident
   1. Priority 1
      a. Injury/illness due to an unknown or known substance.
      b. Release of an unknown substance, respond Priority 1 to the appropriate area to evaluate the hazard.
   2. Priority 3
      a. Control of fluids leaking at a motor vehicle response.
G. All other incidents not listed in Sections A–F above shall be responded to at the most appropriate priority and is at the discretion of the officer/senior fire fighter of the responding unit.

IV. Vehicle Response
   A. Based on the type of call, the most appropriate vehicle with its specialized equipment shall respond. If a unit is in service in district and not in quarters, it shall respond to the incident.
   B. On all fire-related calls when the crews are in quarters, the engine shall be the first vehicle to respond.
   C. If an apparatus is out of the station and is not the appropriate apparatus to respond to an incident, the senior member shall request through the Battalion Chief the next due station to respond as well. (Example: If a Rescue is out on inspections and a fire alarm is dispatched, the next in-service engine shall respond also.)

V. General Considerations
   A. Structure Fires—Priority 1 until cancelled or directed otherwise by the OIC. Privately owned vehicles shall respond to the station first until all required apparatus is en route.
   B. Vehicle Fires, Miscellaneous Fires outside a Structure, Wires Down, and Odor Investigations—Priority 1 to the station until the first due district apparatus is en route, then reduce response to Priority 3.
   C. Medical Emergencies and injury accidents dispatched as Priority 1—Privately owned vehicles may respond Priority 1 up to and until the first in district vehicle leaves the station.
   D. Privately owned vehicles shall yield the right of way to all Department vehicles responding Priority 1.

FIGURE 5-3 *Continued.*

PAGE: 3 OF 3

E. Department members are responsible for maintaining their vehicles in a safe operating condition with insurance coverage.
F. At no time shall a POV respond Priority 1 with a non-City employee in the vehicle.

VI. The routine response of POVs directly to the incident scene must be recommended by the Supervisor and approved by the Fire Chief. In those cases where a POV must pass directly by an emergency incident scene, the POV may stop at the incident if such action will not place personnel at additional risk or hinder operations or disrupt the scene.

FIGURE 5-3 *Continued.*

FIRE DEPARTMENT

ADMINISTRATIVE PROCEDURE

PRIORITY: 3

APPARATUS DRIVER CERTIFICATION REQUIREMENTS — **NO: 600.2**

EFFECTIVE: ______________ — **PAGE: 1 OF 3**

FIRE CHIEF APPROVAL:

PURPOSE

To meet the legal requirements of the Vehicle Code as amended which exempts fire fighters from the commercial driver license requirements.

POLICY

All members of the Fire Department must be authorized by the Fire Chief to drive Department apparatus. Members seeking authorization to drive must complete the Fire Department driver training program which consists of six (6) parts as listed in this procedure.

As the Fire Department has no designated apparatus operators, all probationary personnel must be certified to drive the types of apparatus assigned to their particular station (Engine, Squad, Rescue, Medic). Apparatus drivers may be permitted under conditions of emergency response to exceed posted speed limits, disregard traffic control devices, and other regulations governing the operations of a motor vehicle in accordance with the Department's Emergency Vehicle Operation Procedure. Authorization to drive and operate aerial apparatus is not included in this policy.

PROCEDURE

I. Guidelines
   A. The Department reserves the right to review any authorization to drive Department apparatus. Members are entrusted with the responsibility of ensuring their familiarity with apparatus maneuvering and handling features.

II. Driver Training Program Components
   A. There are six (6) components of the Fire Department driver training program. The Training Division will administer the driver training program.
      1. Completion of the hiring and selection process or driver's license record check.

FIGURE 5-4 Sample administrative procedure.

a. It is the purpose of this step to ensure that the Department member maintains a satisfactory driver's record. Human Resources will enroll all Department members in the Secretary of State Driver Record Flag Program.
b. Members without a current or valid driver's license will not be permitted to drive Department apparatus. Authorization to do so shall be revoked immediately upon notification of a member with a suspended or revoked driver's license. Individual members shall immediately inform the Department of any change in the status of their driver's license other than routine renewal.

2. Classroom
   a. The Training Division shall design driver training classroom course work which meets or exceeds minimum requirements and includes all applicable Department practices, policies, and procedures. This will include information regarding mapping, addressing in the City of ________________, apparatus maintenance practices, fuel purchasing procedures, accident procedures, the Department's Emergency Vehicle Operation Procedure, and Emergency Scene Traffic Management.
3. Supervised Nonemergency Vehicle Driving Experience
   a. In order to establish familiarity and confidence, each member shall log sufficient driving mileage so as to become familiar with the operating controls, maneuvering, and handling features of each apparatus type at the assigned station. The student driver must have nonengine nonemergency drive time in the other vehicles also.
      (1) At the direction of the Supervisor students shall drive a sufficient period of time to assure, in the opinion of the supervising officer, that an individual can safely drive each type of apparatus.
      (2) Consideration may be given by the Supervisor with approval of the Fire Chief to members with preservice commercial vehicle driving experience.
      (3) Nonemergency driving experience should include multiple situations the operator may encounter in their response district.
4. Driver Training Low Speed Maneuvering Course
   a. The purpose of the course is to measure the maneuvering skills and performance of the driver candidate in a series of low-speed forward and reverse maneuvering operations.
      (1) The skills course shall be conducted in such a fashion as to allow for standardized evaluation of the student driver's performance. The skills exercises utilized will follow the outline requirements of the Fire Fighters Training Council or the National Fire Protection Association, 1002, the Fire Apparatus Driver Operator Professional Qualifications.
      (2) The driver's skills course should be conducted on clear, dry pavement at a time of year, which will not adversely affect either the road surface or the apparatus, used during the skills course testing.
5. Apparatus Road Test
   a. The purpose of this component shall be to review overall driving performance under normal operating conditions for nonemergency response on the roadways and traffic volumes experienced within the City. The operator shall complete all portions of the driver training program and receive sufficient training in pump operations to meet the minimum performance requirements contained in the certification test.
   b. The driver certification road test shall be completed for each type of apparatus, Squads, Engines and Medics as available to member's station. The road test should consist of a

FIGURE 5-4 *Continued.*

minimum of five (5) miles driving each apparatus in the following traffic conditions listed in above in Section 3, Item a.

(1) The driver certification test shall be administered by the station training officer or a Department officer assigned by the Supervisor. The driver certification test shall consist of (5) sections:
   (a) Vehicle Inspection
   (b) Starting and Safety
   (c) Driving in Forward Gear
   (d) Driving in Reverse Gear
   (e) Pump Operations

c. The driver candidate shall be scored as either satisfactory or needing improvement in each of the categories. The member must receive the satisfactory rating in all sections and for all objectives. After successful completion of the Apparatus Driving Road Test in each of the types of vehicles assigned to the member's station, the candidates then are recommend to the Supervisor for Provisional Emergency Vehicle Driving Authorization.

d. The driver candidate shall complete the practical pumper skills evaluation.

6. Supervised Emergency Driving Experience
   a. Prior to emergency driving, the recruit will complete the supervised driving time.
   b. Each member with Provisional Emergency Vehicle Driving Authorization must obtain a satisfactory emergency response rating in each category. Ratings are to be logged on the Supervised Emergency Vehicle Driving Log form.
   c. The members must complete three (3) Supervised Emergency Response Driving Experiences at their station. (It is intended that three (3) supervised emergency response driving experiences shall be completed in any type of apparatus).
   d. A Satisfactory Supervised Emergency Response Driving Experience rating must be attested to by the Supervisor.
   e. After successful completion of three emergency Response Driving Experiences, the District Chief or other approved Department supervisor shall complete the recommendation for full Emergency Vehicle Driving authorization to the Chief of Department by completing the reverse side of the Supervised Emergency Vehicle Driving Log Form.

FIGURE 5-4 *Continued.*

FIRE DEPARTMENT

SAFETY PROCEDURE

PRIORITY: 1

DRIVER SAFETY STANDARDS **NO: 920.1**

EFFECTIVE: __________ **PAGE: 1 OF 3**

FIRE CHIEF APPROVAL:

PURPOSE

To provide guidelines for the safe and effective operation of emergency vehicles and related equipment.

PROCEDURE

Personnel authorized to respond to emergencies shall be charged with the responsibility of operating personal vehicles (authorized emergency vehicles) and apparatus in a safe manner so as to prevent accidents and injury during the performance of their duties.

I. Introduction
   A. The driver of each Fire Department vehicle has the responsibility to drive safely and prudently at all times. Vehicles shall be operated in compliance with the Motor Vehicle Code. This Code provides specific legal exceptions to regular traffic regulations which apply to fire department vehicles only when responding to an emergency incident or when transporting a patient to a medical facility.
   B. Emergency response (Priority 1) does not absolve the driver of any responsibility to drive with due caution. The driver of the emergency vehicle is responsible for its safe operation at all times. The officer in charge of the vehicle has the responsibility for the safety of all operations.

II. Warning Lights
   A. When responding Priority 1, warning lights must be on and sirens must be sounded to warn drivers of other vehicles, as required by the Motor Vehicle Code. The use of sirens and warning lights does not automatically give the right of way to the emergency vehicle. This requests the right of way from other drivers, based on their awareness of the emergency vehicle. Emergency vehicle drivers must make every possible effort to make their presence and intended actions known to other drivers, and must drive defensively to be prepared for the unexpected inappropriate actions of others.

III. Speed Limit
   A. The Motor Vehicle Code allows emergency vehicles to exceed the posted speed limits. However, Fire Department vehicles are authorized to exceed posted speed limits only when responding Priority 1 under favorable or ideal conditions. This applies only with light traffic, good roads, good visibility, and dry pavement. Under less favorable conditions, the posted speed limit is the absolute maximum permissible. Drivers are responsible for operating in a safe and prudent speed at all times. Special caution should be made when entering and driving through subdivisions.

IV. Intersections
   A. Intersections present the greatest potential danger to any emergency vehicle. When approaching and crossing an intersection with the right of way, drivers shall not exceed the posted speed limit. When approaching an intersection (red light, stop sign), the vehicle shall proceed safely or come to a complete stop. When approaching a yield sign, the driver shall slow to a safe speed and shall proceed through the intersection only when it is determined to be safe. The emergency

FIGURE 5-5 Sample safety procedure.

vehicle may proceed only when the driver can account for all oncoming traffic in all lanes yielding the right of way. If there is any doubt, the emergency vehicle shall wait until the intersection can be crossed safely.

V. School Buses
- A. When responding to alarms, vehicles shall stop for school buses loading and unloading as required by law. If the bus driver gains control of all students, turns off warning signals, and acknowledges the presence of the Fire Department vehicle, the driver may continue to the incident.

VI. General Safety Precautions
- A. Priority 1 response is authorized only in conjunction with emergency incidents. Unnecessary emergency response shall be avoided. In order to avoid this, the following rules shall apply:
  1. The first arriving unit will advise additional units upgrade or downgrade their response whenever possible or appropriate.
  2. When backing up, guides shall be used at all times. Guides shall remain in visual/radio contact with the driver. If no guide is available, the driver shall dismount and walk completely around the apparatus before backing up.
  3. All persons in City vehicles must wear seat belts at all times.
  4. Compartment doors shall not be left open and unattended unless drying in winter months.
  5. Transmission by radio is prohibited while the apparatus is inside the station unless the vehicle is not running and is parked inside of the station.
  6. All personnel shall ride only in regular seats provided with seat belts. Riding on tailboards or other exposed positions is not permitted on any vehicle at any time.
  7. During an emergency response, fire vehicles are not to pass other emergency vehicles. If passing is absolutely necessary, arrangements should be conducted through radio communications. Privately owned vehicles (POVs) shall not pass Fire Department vehicles.

VII. Driving on the Fire Ground
- A. The unique hazards of driving on or adjacent to the fire ground require the driver to use extreme caution and to be alert and prepared to react to the unexpected. When driving apparatus or personal vehicles on the fire ground, drivers must resist the tendency to drive hastily or imprudently due to the urgent nature of the fire ground operations. Drivers must consider the dangers their moving vehicle pose to fire ground personnel and spectators who may be preoccupied with the emergency, and may inadvertently step in front of or behind a moving vehicle. Personnel shall not dismount fire apparatus until they have come to a complete stop and have been instructed to do so by the officer in charge. When stopped at the scene of an incident, vehicles should be placed to protect personnel who may be working in the street, and warning lights shall be used to make approaching traffic aware of the incident. At night, vehicle-mounted flood lights and any other lighting available shall be used to illuminate the scene. Bright lights on oncoming traffic should be avoided whenever possible.

VII. Emergency Response Policy
- A. The Fire Department vehicles shall be operated in a manner that provides for the safety of all persons and property. Safe arrival shall always have priority when driving en route to an emergency incident.
  1. Prompt, safe response shall be attained by:
     - a. Leaving the station in a standard manner
       - (1) Quickly mounting apparatus
       - (2) All personnel on board, seated, and seat belts on

FIGURE 5-5 *Continued.*

    (3) Station doors fully opened
    (4) Compartment doors closed and positively latched
    (5) Use of radios in the station is prohibited
  b. Driving defensively and professionally at safe speeds
  c. Knowing where you are going before leaving the station. Plan your route taking into account traffic, type of road, construction, weather, etc.
  d. Using warning devices to move around traffic and to request the right of way in a safe and predictable manner
  e. Approach the scene with caution
  f. Ensure the vehicle has stopped before dismounting
2. Fast response shall not be attained by:
  a. Leaving quarters before crew has mounted safely and before apparatus doors are fully opened
  b. Driving too fast for conditions
  c. Driving recklessly without regard for safety
  d. Taking unnecessary chances with right-of-way intersections
3. Definitions
  a. Priority 3—Driving without emergency lights or sirens and within normal driving limits.
  b. Priority 1—Driving with emergency lights on and audible warning devices activated.

FIGURE 5-5 *Continued.*

■ Leadership

Who is in charge? This is probably one of the most volatile issues in a combination department. Career members do not like taking orders from part-timers. Volunteer fire fighters view career members as a threat and do not readily wish to turn over the reigns. There may be long-standing members with significant experience and training who have earned the right to be in charge. So what is the answer to the question, "Who is in charge?" The answer should be the most capable person.

That is the easy answer. However, how is it determined who is most qualified? There are many suggestions and different options on how to do this. But it must be done before the incident, before someone questions the authority of an individual, and before a critical decision during an emergency must be made. Clearly defined lines of authority are essential to proper emergency response. Because there are many forms of combination department, there are also many ways to handle the chain-of-command issue. Regardless of how a fire chief chooses to handle it, there will be those that question it, and maybe rebel against it. There is no clear-cut answer to this problem. Who would put a career member with 2 years on the job in charge of an incident when a 20-year volunteer veteran is available? Consider whether to let a 3-year volunteer sergeant be in command when a 15-year career fire fighter is on the scene. These questions are just part of the equation. There is also the consideration of training and education. There are many parts to this matrix. Politics may also come into play. If this is not handled properly, a fire chief risks the ruin of interpersonal relations and, more importantly, the department's service to the community.

So, what can be done? Establish promotional standards based on training and experience. This does not make the problem go away, but gives a foundation to a system that allows the most qualified to be in charge. The prerequisites for promotion need to be a combination of applicable training to the anticipated assignment and meaningful experience. Whatever is decided will need to be flexible in nature. As time passes, the makeup of the fire department will change. It can start as a predominantly volunteer organization and evolve into a career department supported by volunteers. Time and training are the ultimate solutions to this potentially disruptive situation.

■ Training

Operational standards need to be established based on the level of service being provided and also considering special response needs. For example, if the fire department has EMS responsibilities, the stan-

FIRE DEPARTMENT

COMMAND PROCEDURE

PRIORITY: 1

CHAIN OF COMMAND	**NO: 1101.1.1**
EFFECTIVE: __________	**PAGE: 1 OF 1**
FIRE CHIEF APPROVAL:	

PURPOSE

Clearly defined lines of authority are necessary while operating during emergencies. This procedure establishes the organization needed for determination of command.

POLICY

This procedure shall apply to emergency and incident management. It is not intended to represent organizational structure nor is it to replace day-to-day, nonemergency working relationships. For purposes of this procedure, the emergency or incident begins when dispatched and concludes with all equipment back in service and the incident report completed.

PROCEDURE

Members are to utilize the following chain of command while operating during emergencies.

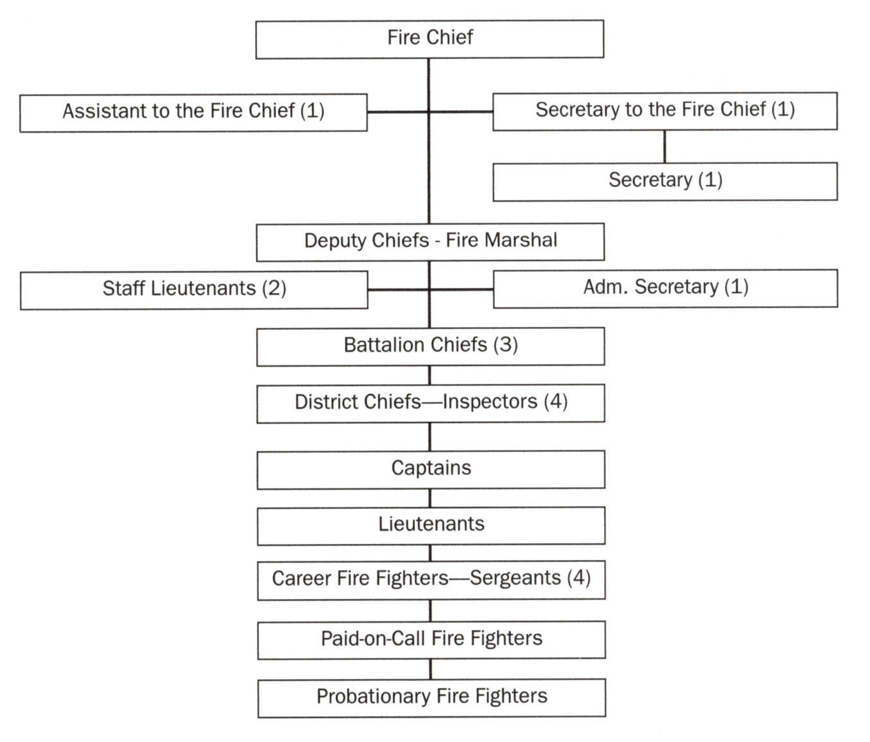

1 - Civilian positions with no suppression or emergency functions

2 - Staff Lieutenants retain fire suppression and emergency response involvement as is the current status, except Staff Lieutenants may assume command of an incident only in the absence of a Chief Officer (Inspector or District Chief and above)

3 - Position is line officer with supervisory, suppression, and emergency responsibilities

4 - Seniority within these classifications will determine rank

FIGURE 5-6 Sample command procedure.

FIRE DEPARTMENT

PERSONNEL PROCEDURE

PRIORITY: 3

PROMOTIONS FOR VOLUNTEER MEMBERS **NO: 420.1**
MEMBERS
(SERGEANT, LIEUTENANT, CAPTAIN)

EFFECTIVE: ____________ **PAGE: 1 OF 2**

FIRE CHIEF APPROVAL:

PURPOSE

To outline the minimum requirements for promotional consideration and for selection in the Volunteer system.

PROCEDURE

I. Process

A. The promotion to the position of Volunteer Officer within the Fire Department will be accomplished through a consistent, organized, and fair process. When a vacancy in the Officers Cadre of one of the fire stations occurs and a replacement is desired, the position will be posted at all stations for a period of 2 weeks. Those members interested shall apply, in writing, to the Supervisor of the station requesting the promotion within the specified time frame.

B. Candidates that meet the minimum qualifications for the position as outlined in these procedures will be evaluated based upon written and oral examinations, seniority, and a review by the Supervisor responsible for the promotion. The written test will consist of questions developed specifically for the vacated position and will address fire suppression knowledge, Department procedures and directives, emergency medical service, hazardous materials, and fire prevention practices.

C. For the oral examination, a panel will be established consisting of Department officers not affiliated with the station requesting the promotions. A single interview board will be used to evaluate candidates seeking a position.

D. The Supervisor will provide input into the selection process by reviewing and evaluating each candidate. The Supervisor evaluation will consider hands-on operation, physical capabilities, response record arriving on the scene, performance on emergency scenes, support of Department, Supervisor and officers cadre, attendance at special events, and commitment to the Department.

E. The final score will be calculated with the following percentages:

1. Written Examination: 35%
2. Oral Examination: 35%
3. Supervisor Evaluation: 30%
4. Seniority: 1% per year, up to 5 years

(Seniority points shall accrue from the point a member is eligible for a promotion.)

F. The candidate with the highest score will then be recommended to the Fire Chief for promotion. In the event that there is only a single candidate for a vacancy, the candidate shall be evaluated against previous department candidates' accumulative scores for the same position and must have a passing score of 75% on all portions of the selection process.

G. Individuals promoted to the rank of Sergeant, Lieutenant, or Captain shall serve a 1-year probationary period in rank from the date of appointment.

FIGURE 5-7 Sample personnel procedure.

II. Prerequisites for Officer Positions
  A. In order to be considered for promotion to the position of Sergeant, Lieutenant, or Captain, Department members must meet the following requirements. Equivalency will be evaluated by the Fire Chief.
    1. Sergeant
      a. Three years on the Department
      b. Fire Fighter II
      c. Fire Officer I
      d. State licensed EMT
      e. Good history of community service
      f. Satisfactory evaluations and service record
      g. Run percentage greater than 45% during the 12 months prior to the posting of the position.
    2. Lieutenant
      a. Sergeant's qualifications
      b. One year as a Sergeant
      c. Fire Officer II
      d. Satisfactory Sergeant evaluation and service record
      e. Run percentage greater than 45% during the 12 months prior to the posting of the position.
    3. Captain
      a. Lieutenant's qualifications
      b. Lieutenant for 2 years
      c. Fire Officer III
      d. Satisfactory Lieutenant evaluation and service record
      e. Run percentage greater than 50% during the 12 months prior to the posting of the position.
  B. Officers promoted in an acting capacity must achieve rank requirements within 12 months of promotion.

III. Response Expectations
  A. Officers are expected to maintain a run response percentage of 45%
  B. Officers will further maintain a run response percentage of 45% between the hours of 1800–0730.

FIGURE 5-7 *Continued.*

dards are different for the first responder, EMT, and paramedic and may vary depending on whether the fire department transports patients. If all members of the fire department are expected to participate, they need to meet the same level of licensure and certification. If not everyone is expected to be a part of all levels of service, then their standards for training and licensure will be different. Make sure that everyone is aware of the expectations of the fire department. This discussion is also applicable to the other services that may be provided, including hazardous materials response and special rescue. It is not necessary for both career and volunteer fire fighters to participate in all responses, but if they do, then the training and response standards must be the same for all and should be based on the expected performance.

Training requirements are both easy to establish and hard to implement in a combination fire department. This is not a result of the willingness of the participants, but of the time available. Career members can receive training while on the job and have overtime opportunities. Volunteers have their full-time paying jobs to consider as well as commitments in their personal lives.

There may also be state standards to consider. In some states, there is a higher level of training

FIRE DEPARTMENT

TRAINING PROCEDURE

PRIORITY: 3

MINIMUM DEPARTMENT TRAINING REQUIREMENTS | **NO: 600.1**

EFFECTIVE: __________ | **PAGE: 1 OF 1**

FIRE CHIEF APPROVAL:

PURPOSE

To list the minimum levels of training, license, or certification to be obtained during the member's probationary period and maintained throughout the member's employment with the Fire Department.

PROCEDURE

I. General Training Requirements
   A. All members of the Fire Department are required to obtain the following minimum training requirements during their probationary period.
      1. First Responder license
         a. Professional Rescuer Basic Life Support Certification
         b. Automatic External Defibrillator (AED) Certification
      2. Fire Fighter I and II Certification
      3. Hazardous Materials First Responder: Operation Level Training
      4. Fire Department: Apparatus Driver Certification
      5. Fire Department: Apparatus and Pump Operator I & II
   B. Emergency Medical Technician Basic license must be obtained within 30 months of the date of employment.
      1. Failure to obtain an EMT license may result in disciplinary action.
   C. All Fire Department members are required to maintain the minimum training requirements of the Fire Department except as approved by the Fire Chief.
   D. Newly hired members may request to be granted equivalency under this procedure. When presented with proper credentials, the Training Division will make recommendation to the Fire Chief, who may grant equivalency or require that the newly hired members complete elements of the Fire Department training program. The Training Division shall respond to all equivalency requests in writing.

II. Continuing Education
   A. All fire fighters shall be responsible for attending 80% of all training hours. This will be based on a 12-month rotating year.
   B. Employees shall be responsible for keeping specific records for their EMT-B, Specialist, or Paramedic Continuing Education. If an employee finds a problem with their CE's, the employee shall notify the Training Division.
   C. Failure to maintain the minimum training or continuing education requirements may result in disciplinary action.

FIGURE 5-8 Sample training procedure—minimum department training requirements.

FIRE DEPARTMENT

TRAINING PROCEDURE

PRIORITY: 3

ESSENTIAL KNOWLEDGE AND SKILLS EVALUATION — **NO: 605.0**

EFFECTIVE: ____________ — **PAGE: 1 OF 1**

FIRE CHIEF APPROVAL:

PURPOSE

To provide a program to ensure that members have maintained the essential knowledge and skills required of the position.

PROCEDURE

I. In order to verify member's proficiency on various types of skills, the Training Division will develop various scenarios to evaluate an individual or company's skills. These skills will simulate skills as performed in the field.

II. Each member will be required to perform the functions associated with each essential skill. The procedures may be outlined in the following:
   - A. Essential Skills of Fire Fighting (Including Student Applications)
   - B. Manufacturers Operating Manuals
   - C. Department Policies and Procedures
   - D. Medical First Responder, Emergency Medical Technician, and Paramedic course objectives/task sheets

III. The member will also complete a written test, which may be provided at training sessions. Members will be required to answer 90% of all questions correctly.

IV. Remediation may include retraining and subsequent retesting on a specific area in which the member had difficulty within 30 days.

V. The Training Division may provide evaluation criteria and/or study aids prior to evaluations.

FIGURE 5-9 Sample training procedure—essential knowledge and skills evaluation.

required for a career fire fighter than a volunteer. This is more based on the available training time than the hazards to be faced. Because of this circumstance, it may appear to some that the career fire fighter is superior. This may or may not be the case, but this assumption is a potential problem and must be addressed. The simple solution is to have the same requirements for all personnel, regardless of status as career or volunteer. This can be done if the time commitment needed is considered. It is possible for volunteers to obtain training; it simply takes longer to do so because of scheduling. Basic training must be equal for all. Specialized training can vary. However, all members must receive the training needed to perform the duties expected of them. These acceptable differences are mostly isolated to fire fighter training. In EMS, state licensure dictates the level of training. There is no discrimination of training for a paramedic in regard to their status as career or volunteer. They must do what is necessary to obtain the license as well as attend continuing education to relicense. Most states have no equivalent licensing for fire fighters. However, there are similar requirements for special response services such as hazardous materials. Check the state's requirements to make sure that the department complies. When there is a choice, training standards for career and volunteer must be the same. This would include fire fighting, officer requirements, and incident command.

FIRE DEPARTMENT

PERSONNEL PROCEDURE

PRIORITY: 3

CAREER INSPECTOR/ FIRE FIGHTER ASSIGNMENT — **NO: 420.2**

EFFECTIVE: __________ — **PAGE: 1 OF 1**

FIRE CHIEF APPROVAL:

PURPOSE

To outline the qualifications and process for promotion in the Department.

PROCEDURE

I. Qualifications

A. It shall be the policy of this Department to promote career fire fighters who meet the minimum qualifications and maintain such qualifications to the level of Inspector/Fire Fighter and shall be recognized therein as officers and shall be placed within the officer cadre and assume a rank within the chain of command.

B. A career fire fighter shall be promoted to the rank of Inspector/Fire Fighter when he or she meets the following qualifications:

1. Four years of career service with the Fire Department
2. Fire Fighter II
3. A.E.M.T. license with ACLS (Advanced Cardiac Life Support) Certification
4. Formal Fire Inspector Training (Community College, State, National Fire Academy, or other approved course)
5. Fire Officer III
6. Hazardous Materials Level II - Operations
7. Fifty college credit hours from an accredited institution

II. Exemptions

A. Current Inspector/Fire Fighters shall be exempt from the above minimum qualifications assuming they meet previous requirements. Career Fire Fighters who have not previously attained the rank of Inspector/Fire Fighter must meet all qualifications prior to promotion. An Inspector/Fire Fighter may be subject to demotion based on an unsatisfactory performance review. Any challenge to a demotion shall be subject to the grievance procedure as outlined by the current bargaining agreement between the City and the union.

III. Step Down in Rank

A. An Inspector/Fire Fighter reserves the right to step down to the rank of Fire Fighter at any time and shall submit to the Fire Chief his or her intent of such action in writing.

IV. Waiver

A. The Fire Chief reserves the right to waive any qualification in order to promote a member when it is in the best interest of the Department. The position of Inspector/Fire Fighter is a rank designation that should not affect the rate of pay as determined by the bargaining agreement between the City and the union.

FIGURE 5-10 Sample personnel procedure—career inspector/fire fighter assignment.

On a side note, training standards can create a healthy competition, pushing both volunteers and career personnel to a higher level of performance. Make certain that this remains positive. If members do not approach this with the proper attitude, it can create a "better than you" situation based on perceived performance. Strong leadership is needed to keep all members focused on the mission of the fire department.

■ Who Is in Charge?

The question of leadership remains one of the most difficult in a combination department. So much depends on the makeup of the organization and the personalities of those involved. Ultimately, there needs to be a system, and there are several viable options. Career members often become sensitive around this issue. Although they accept the position knowing the rules, they may change their perspective as they gain experience, and the volunteer membership experiences attrition. The belief of many career personnel is that they are in their chosen profession and should have control of their destiny. Many volunteers would argue that they were aware of the circumstances when they were hired. This is a situation where emotions can run high and rational decisions become difficult. The fire chief wants two things: the right person in charge, and minimal personnel issues. There is no easy answer, but there are a few helpful suggestions. Keep the issue flexible to change as the fire department changes. Utilize personnel from both the career and volunteer ranks to help develop a fair and equitable system. Neither group will be satisfied if a directive is given without their input. Finally, base the promotional system on solid standards and qualifications.

Summary

Standards are necessary to maintain order, discipline, and good conduct. These expectations are applicable to all employees regardless of status. To minimize problems, these standards should be consistent to the greatest possible extent. Although equal standards are continually promoted whenever possible, there are circumstances that make career personnel and volunteers different. One of the major issues is time available, which must be kept in mind as standards are created. Don't make it impossible for a volunteer to comply. Conversely, hold steady on important issues that affect service levels. Though not always relevant, a fire chief should be aware of various motivations for performance. For example, career members may attend training for overtime pay. Volunteers may be collecting certificates of training. It is hoped that performance enhancement, meaning gaining the skills that will assist in completing the job better and more safely, is a good motivator. Regardless, the objective is to get members trained at the highest level possible.

CHAPTER 6

Training in a Combination System

Fred C. Crosby and Wes A. Cole

Case Study

Mike has been the training lieutenant at the Chatham Volunteer Fire Department for about 2 years. He has done a great job scheduling monthly drills, and volunteer attendance has been improving. This year, the department hired eight career fire fighters to cover the daytime shift while the volunteers are at their paying jobs. Because the career fire fighters were previously volunteers, Mike saw no reason to adjust how he conducted training.

The career fire fighters have been working for a while now, and they are starting to complain because the training is at night when they are off duty. To stay sharp, they have started to do their own training. This has sparked some disagreements between the career and volunteer fire fighters. The career fire fighters have even threatened to notify the local Occupational Safety and Health Administration (OSHA) office.

Mike thinks there is nothing he can do—he is only one person. There is no way he can train the career fire fighters during the day, the volunteer fire fighters at night, and do his regular job. He has threatened to resign as the training officer because of the controversy and extra work. The chief is left with some fairly tough decisions.

Introduction

Team sports are a large part of today's society. Across the country, people watch and play football, baseball, basketball, soccer, and hockey. It appears that the teams who do the best each season are those that are the best trained, the best disciplined, and the best conditioned. The head coach is responsible to ensure that each player on the team is properly prepared for the game. In firefighting, the chief is the head coach and is responsible for the training, discipline, and conditioning of the department's fire fighters.

There are coaches who punish players for failing to attend practice with running laps or doing push-ups. Consider the message sent if any fire fighter who did not attend and participate in training on a regular basis was left at the station on a call. Fire chiefs can send that message, but must to be ready and able to make training available to both volunteers and career personnel on an equal basis.

Training in a combination department can be similar to juggling. It is common for the training officer to become caught in the middle and thus frustrated with trying to meet increased demands with minimal resources. It is critical for the fire chief to understand these changing dynamics and provide the consideration and resources necessary to properly address training.

Training sets the baseline for department performance and serves as the cornerstone for nearly every function. One of the philosophies essential to a successful training program and fire department is that *everyone* in the department must be properly trained to carry out their duties. It does not matter if it is a small, single-engine volunteer company or a large combination system, everyone must be properly trained.

To help facilitate this level of training, minimum training standards should be established to ensure consistency. In some regions, it is simple to establish such standards because they are mandated by local authorities. However, in many places, there are no training standards in place. Whatever the case, it is important to review training standards and ensure that they are reflective of all the positions in the fire department (**TABLE 6-1**).

In some departments, achieving such training standards can seem nearly impossible. It is common to find small, rural departments where the training officer says, "How can we possibly do this training? Our members have jobs during the day and families on the weekends." This is a reasonable concern, but experience has shown that the standards can be achieved by any type of department. It is important to never underestimate the commitment and abilities of volunteer fire fighters. They are highly capable when both the department and training systems are supportive.

The training system also gives a first impression of the fire department to new fire fighters. This is the department's chance to make an initial imprint of its values. Training also serves as the catalyst for long-term organizational change. When a department wants to adjust its values, the process starts in training. It may take a generation for significant changes in fire department culture to take hold, but it starts on the first day of training for the brand new fire fighter.

Stakeholders in the Training Process

One of the most significant changes required for training in a combination department is the addition of new stakeholders. The first stakeholder is the public (**TABLE 6-2**). It is critical that the abilities of the department's fire fighters reflect the needs of the community. If there are no high-rise buildings in the area, then high training should take a backseat to other specialties, such as rural water supply. This is not to say that high-rise tactics are not important, but they are not the *most* important in a rural area.

As a department transitions from volunteer to combination, one of the most vocal stakeholders, the volunteer fire fighters, will remain a vital component. It is important for the combination department's training system to sustain the training programs for volunteers as well as those for career personnel—one should not be favored over the other.

One of the most challenging issues to consider when addressing volunteer training needs is scheduling. Although volunteers are highly capable, they will face challenges in attending some training, and care

Table 6-1 Job Duties and Training Standards

Job Duty	Training Standards
New volunteer	• Department orientation and safety • Infectious disease control • Equipment locations on apparatus • Basic nonentry job tasks • Hazardous materials, awareness level
Fire fighter	• NFPA 1001 equivalent (Fire fighter I and II) • Hazardous materials, operations level • Fire fighter survival
Fire officer	• NFPA 1041 for equivalent position
	• Lieutenant: Fire officer I • Captain: Fire officer II • Chief officer: Fire officer III

Table 6-2 Stakeholders in the Training Process
Stakeholders in the Training Process
• Citizens in the community • Volunteer personnel • Career personnel • Volunteers seeking future employment

must be taken to ensure that they have reasonable opportunities to participate. For volunteer fire fighter training, nights and weekends are the optimal choice, but there are other options available.

Training officers should consider distance-education methods. There are a variety of options such as print-based, self-study programs, CD-ROM, and Internet-based options. The training officer should research the most cost-effective and convenient options for volunteer training. The key word here is options—creativity in thinking can develop into a comprehensive and customized system for the community.

Per their job description, career fire fighters are a captive audience when it comes to training. To receive a paycheck, career fire fighters must show up where and when they are directed. However, operational coverage quickly becomes an issue when assigning career fire fighters their training. If you leave career fire fighters in service during training, they may not accomplish their training goals because they are constantly running back and forth on response calls. If you take them out of service, then you run the risk of being short staffed and possibly missing a call altogether.

The best option is to budget the necessary overtime to ensure that there is no short staffing as a result of training fire fighters. The Campbell County Fire Department in Wyoming took an aggressive approach to meeting this need through budgeting. For every four fire fighters hired, the chief allots the equivalent of one full-time salary to cover overtime costs related to training. Other chiefs have chosen to use staffing formulas to address this issue. For example, for each full-time position hired, they hire an additional 0.25 positions. So if four career fire fighters are hired, one more fire fighter would need to be hired to cover time away for training and other leave time. This essentially means that one fire fighter will be off duty every shift day for one reason or another. Whichever option is used, it is critical that while career fire fighters are sent out for required training, proper coverage is maintained at the fire station.

In addition, bringing on career fire fighters may affect how a department's current training program is run. There may be state, local, and federal regulations and standards to adjust to when adding career fire fighters. In many states, the career fire fighter system is a well-defined process that addresses the opportunity for position advancement within the organization.

The Volunteer Training Needed to Transition to a Career Fire Fighter

The final stakeholder to consider is the volunteer who aspires to become a career fire fighter. Volunteers often want to make a profession of firefighting. It must be established whether the training obtained independently by volunteers will count in the career environment. This is done differently around the country, and each department must decide where it stands on this issue. Initially, it is appealing to accept such training because it saves time for the training officer. However, another point to consider is how to ensure the quality of training. Unless volunteers are trained within the department, there is no quality assurance. These fire fighters may have certificates, but this does not necessarily confirm the quality of their training.

Hanover Fire and EMS in Virginia has taken the stance that they will accept training from select training academies in the area. However, if a volunteer has a certificate from an unapproved academy, then that volunteer must go through the department's own basic academy. Local volunteers converting to career personnel are permitted to bypass the basic academy, because the volunteer and career academies in Hanover are essentially the same. Each agency needs to evaluate how to handle this issue. One solution is to allow the prospective volunteer to test on various disciplines, compared with the all-or-nothing approach, which may be inefficient considering that the volunteer has already invested a lot of time and effort in whatever training program he or she attended.

Tactical Equality

Tactical equality is a concept in which positions are defined by their function and necessary knowledge, skills, and ability—not by career or volunteer status. In this approach, responsibility and authority is consistent with the traditional chain of command, not dictated by career or volunteer status. This concept seems to be a common thread among highly successful combination departments. The concept helps to demonstrate that both career and volunteer fire fighters are of equal importance to the department.

Historically, many of the tensions between career and volunteer fire fighters have been generated because one group was receiving something that the other was not. If all personnel are trained equally, this substantially reduces the impetus of any potential disagreements. The concept of tactical equality is a tall order in most departments, but when achieved, it greatly enhances interdepartment relationships.

Instituting tactical equality requires an audit of the entire department to identify where responsibilities need to be equalized between career and volunteer personnel. In its truest sense, tactical equality means that both volunteer and career personnel have an equal role in the department. It is important to note that declaring tactical equality should be a conscious decision among both volunteer and career personnel. Though the responsibilities may seem attractive at first, after closer scrutiny, some volunteers may prefer that the duties not be exactly equal.

In some cases, it may be beneficial to develop tactical equality at the operational fire fighter level first, as this is fairly easy to facilitate with a comprehensive training program. When management and leadership responsibilities are examined, it may be revealed that the career personnel shoulder more of the load. This says nothing about the abilities of the volunteers, it only points to the large amount of time required to obtain the necessary training and carry out management duties.

Embracing Regulations and Standards

It is not uncommon to hear training officers complain about the variety of regulations and standards that guide fire service training. As departments transition from all volunteer to combination, they may discover new rules and guidelines that apply. It is critical to research which requirements may impact the department, as ignorance is not a reasonable excuse to avoid these training requirements.

Rather than dread such regulations, be aware that there are numerous benefits to having guidance. Standards establish consistency across fire service and help to establish the industry as a profession, rather than labor. This may not seem to be a significant issue at first, but it can seriously affect negotiations for salary and benefits. Human resources personnel often view jobs in one of three categories: labor, paraprofessional, and professional.

Labor is just that—task-oriented, skilled labors that provide a service. This could apply to anyone from trash collectors to auto mechanics. Paraprofessionals typically require some type of professional certification, along with an associate's degree or equivalent. Professional positions typically require a bachelor's degree or higher. Salaries and benefits correlate to the amount of education and certification required. It helps the fire service as a profession to be placed in the paraprofessional category at minimum. There are still fire chiefs who spend hours arguing that fire fighters perform at a higher skill level than the average laborer. Although regulations and certifications can be bothersome, they do serve a larger purpose that helps every fire fighter in the long run.

The first step in identifying regulations is to contact the local Occupational Safety and Health Administration office (OSHA); not every state is an OSHA state, but there may be other local agencies that regulate career firefighters. A state may have its own OSHA designation as well. Check with the local department of labor. Ask each agency for a letter describing their requirements for career fire fighters. This will provide a firm set of training requirements. OSHA regulations carry the weight of law and must be followed.

Professional standards, although not law, do establish the best practices within the industry. There will not be any criminal liability for not following these standards, but a good attorney could hold a department liable in a civil case if something went wrong. The National Fire Protection Association (NFPA) has established a broad scope of standards for the fire service (**TABLE 6-3**). These standards are developed by representatives from the various stakeholders in the fire service, and each standard goes through a rigorous validation process.

The NFPA 1710 and NFPA 1720 have historically created some grey areas for combination departments. The language in the standard indicates that a department should follow 1710 if it is mostly career personnel and 1720 if it is mostly volunteers. Either way, there must be a clear decision on which standard each combination department follows, as it will sig-

Table 6-3 Common OSHA and NFPA Regulation and Standards

Common Laws and Standards	Description
OSHA 1410.910	Respiratory Protection Standard • Describes requirements for respiratory training • Establishes requirement for "2 in–2 out"
OSHA	Workplace Safety: Community Right to Know
NFPA 1001	Professional Qualifications for Fire Fighter
NFPA 1041	Professional Qualifications for Fire Officer
NFPA 1403	Requirements for Live Fire Training
NFPA 1404	Respiratory Protection
NFPA 1410	Training Standard on Training Evolutions
NFPA 1710	Standard on Deployment in Career Departments
NFPA 1720	Standard on Deployment in Volunteer Departments

nificantly affect the way the department trains and responds.

New-Member Training

A course should be designed that provides new fire fighters with detailed safety instructions and gives insight on how to be a productive member of the department. This training program should be based on how each department plans to use the new fire fighter until the new member becomes fully operational. Some departments may allow new volunteers to participate in a limited capacity on the fireground while waiting to complete basic training, while others will not. Whichever the case, training should reflect the future duties of the new volunteers.

Such programs range from 8-hour programs conducted on a Saturday to elaborate 24-hour programs. Some departments have adopted an online training component to indoctrinate new personnel into the department. The training officer must ensure that the training program makes a positive impression on new volunteers as they join the department.

Mentoring is a newer concept in new-member education. Assigning an existing and dedicated member to a new fire fighter as a mentor is a great way to assure that they understand the methodology of the organization and the reasons behind it. Mentoring does not have to be a complicated, policy-driven activity. Establishing a program can be as simple as encouraging the membership to become involved with the newer members, be friendly, and guide them for a defined time period.

Basic Training

Basic training comes in many forms, depending on the region of the country. Some departments handle all of the basic training, while others send new fire fighters away to a state academy or other organization to complete NFPA 1001 training. However it is done, basic training sets the stage for every future event in the fire fighter's career, whether it be volunteer or paid. Some departments have adopted formal academies to provide basic training. Large combination departments may run several volunteer academies of up to 300 hours in addition to several career academies during a calendar year.

It is important to establish the proper terminology for each situation. When a department is using full tactical equality, the appropriate terms would be *daytime school* (typically career) and *nighttime school* (typically volunteer). It is possible that career fire fighters may be sent to a nighttime school and vice versa for the volunteers. Current staffing and departmental needs dictate how resources are utilized.

When new fire fighters are sent away to receive their basic training, it is imperative that the training officer examine the curriculum used. Just because other agencies are given the right to train outside fire fighters, the fire chief still has the ultimate responsibility to ensure that his or her fire fighters are trained correctly. Take an active interest in the basic training process.

As mentioned previously, fire fighters may come to the department with the necessary certificates required for the job, but this does not necessarily mean they have been properly trained. The training officer must actively research the training backgrounds of new fire fighters and determine if they have the proper knowledge, skills, and abilities to be successful in the department.

Fire Officer Training

Fire officer training can be far more complex. It is important to know what is desired before creating the training program. The training officer must work with the fire chief and other leaders to define what a fire officer means to their department. Then, the training officer can work to create that training program. The training officer may need to seek assistance outside of the fire service to supplement the fire-officer training program.

The Importance of Regionalization

It is critical that the training officer seek assistance at the regional level to ensure that the training program is successful. In many cases, particularly for newly forming combination systems, departments will hire only a handful of career fire fighters. It may be difficult to train a small group, and the process can be a waste of resources. By communicating with neighboring departments, opportunities to consolidate training can be identified, thus getting more out of the training budget. Departments can share instructors, facilities, and administrative costs for both volunteers and career personnel.

Summary

Training is a critical component in the success of any fire department. The fire chief is responsible for the training, discipline, and conditioning of the department's fire fighters. There are many training options; take the time to research and discover what is best for each department.

CHAPTER

7 Public Administration and Budgeting

David B. Fulmer and Bruce Mygatt

Case Study

Chief Ed Johnson had learned some hard lessons, yet he still did not understand the makeup of his municipality. Why was the government administration formed the way it was, and what were all of the bureaucratic rules? Why is it government cannot function like business? How could he explain the way government worked to his deputies and battalion chiefs? They just wanted results, not layers and layers of bureaucracy.

Chief Johnson thought he was prepared for being chief. He had experience at all levels except chief. He had attended and graduated from a fire science program, but no one had explained the way government actually worked.

Introduction

American government is a result of evolution. America is often referred to as the great experiment because it is a system that began on principal and evolved to meet the needs of the current society. The system is alive and still evolving and changing.

The basic premise of government in the United States may be best explained by the Tenth Amendment to the Constitution:

> The powers not delegated to the United States by the Constitution, nor prohibited by it to the states, are reserved to the states respectively, or to the people.

This premise relies on the theory that the best government is local and that the federal government should be limited in scope. The framers of our system were revolutionaries. They held a healthy cynicism toward a strong central government and created a federal government that was based on many state or colonial governments. The federal system is actually an amalgamation of many entities at the state and local level.

When the fathers of the country formed the federal government, they ensured that state and local governments remained in place. The highest level was the federal level, but the other layers of government were kept in place so that all citizens could be involved in government. Tip O'Neil, congressman from Massachusetts and former speaker of the house, stated that, "All politics are local." Indeed, all government and all public administration are local.

State Governments

State governments began as colonial governments and were managed as separate entities by the British Crown. After the revolution, these entities came together and formed a central authority with the US Constitution.

In general, matters within state borders are the concern of state government. This includes criminal laws; property laws; and the management of business and industry, property, and public utilities. State governments must be democratic and cannot adopt laws that conflict with federal law or the US Constitution. State governments are broken into the same system of checks and balances as the federal government. There are legislative, executive, and judicial branches of all state governments.

For the most part, the three branches of state government operate the same as the federal branches. The chief executive is the governor. Most are elected for a 4-year term on a popular vote, although some states elect their governors for 2-year terms. In addition, in most states the chief executive has actual power because the legislative branch is usually limited to one session per year. In all states except Nebraska, the legislative body is broken into two groups, referred to as bicameral. There is a house of delegates, or assembly, and a senate. Like the federal House and Senate, the terms of state house members are shorter than those of state senate members.

A unique byproduct of the evolution is that there are both states and commonwealths within the United States. Massachusetts, Kentucky, Pennsylvania, and Virginia are all commonwealths. Although there are some theoretical differences between states and commonwealths, there is no real difference because all must conform to the same general systems of governance.

In most states, there are now agencies that govern or regulate fire and emergency medical services (EMS). The level of involvement varies widely from tight regulation to loose control.

■ The Dillon Rule

In 1868, Judge John F. Dillon wrote what became a doctrine of the relationship between state and local governments. In this ruling, he wrote that the power of municipalities was limited to those powers, "expressly granted, necessarily or fairly implied, or absolutely indispensable (sic)." to local governments. This doctrine was accepted into the general rule of law through case law over the years. The rule limits the scope of local governments and grants more power to the state.

Today 39 states employ the Dillon Rule to limit the scope of local government. The remaining states operate under the Home Rule doctrine, which grew as a result or backlash to the Dillon Rule. In home rule states, the state's constitution has been amended to limit the state's ability to enact local rule of law, and the authority has been granted to local entities. This difference in doctrine has a large effect on local government from state to state.

Local Governments

Local government is a product of evolution and comes in many different shapes and sizes. There are two basic kinds of municipal government that the fire service must be familiar with: city and county. Most local governments only have legislative and executive branches of government. In most cases, the judicial branch is solely a function of state and federal government.

Cities can be categorized by whether they are incorporated or unincorporated as well as by size: city, town, or township. Depending on the state constitution, the differences in authority and responsibility on the local level can be drastic. In general, there are three basic forms of local government. There are many derivatives, but all fall into one of the following broad forms:

- Council–Manager
- Mayor–Council
- Commission

■ Council–Manager

In the council–manager form of municipal government, there is an elected council. The council then hires a chief executive (i.e., city manager) to carry out the policies of the council. In this form, the council serves much as a board of directors would in a corporation, while the city manager operates as the CEO. In the council–manager form of government, the city manager operates with a great deal of autonomy in managing the daily functions of government. In this form, the council is usually composed of five to nine members and chooses a mayor or president to be the head. This form of government is popular throughout the country and is often used in cities with populations of 10,000 or more.

■ Mayor–Council

The mayor–council form of city government is often called the strong mayor form. In this form, there is a council, but the chief executive is the mayor. In this form, the mayor is elected by the entire city. Theoretically, the mayor represents all of the city, rather than any particular section or interest. The actual power of the mayor as the chief executive varies widely from city to city.

In the mayor–council form of government, the mayor has the authority to veto legislation passed by the council, to hire and fire department heads, and to present a budget for consideration, as well as conduct other executive-branch duties. Unlike the council–manager form of local government, the mayor (i.e., chief executive) does not serve the will of the council.

■ Commission

Around 1900, the third form of government was invented in Galveston, Texas. In the commission form of local government, commissioners are elected but do not represent a specific area such as a borough, ward, or district. They are all elected at large. Each commissioner is then responsible for a specific function of the government such as finance or safety. One of the commissioners is selected by the commission to be mayor or chairperson. The commission form of government does not have separate legislative and executive branches. The commission blends the two basic branches into one.

County Government

Forty-eight of the 50 states have functioning counties within the state government. In Louisiana, counties are called parishes, and in Alaska they are called boroughs. The two states that do not have functioning counties, Connecticut and Rhode Island, have counties in name, but do not have county governments. There are three basic forms of county government:

- Commission
- Commission–Administrator
- Council–Executive

■ Commission

Like the city government commission, the county commission is a group of elected officials who are responsible for both the legislative and the executive functions of the entire county.

■ Commission–Administrator

This form of local county government is similar to the council–manager form of city government. The administrator works at the pleasure of the commission or board of supervisors. The authority of the administrator varies widely and is granted by the commission or board of supervisors.

■ Council–Executive

Similar to the council–mayor form of city government, in the council–executive form of county government, the chief executive is elected by the voters and has broad authority. This may include the ability to veto council action.

Budgeting in Public Administration

One of the most complex issues facing fire service leaders is budgeting. It is an extremely time-consuming and difficult process that drives everything in the department. Not withstanding the rules, regulations, limitations, and laws that guide and sometimes restrict the budgetary processes across the country, the basic premises of budgeting remain somewhat consistent.

The financial stability of any department often serves as a measuring tool for determining the resources available for completing their mission. Every department is fueled by finances and therefore necessitates that a budget, even a rudimentary one, be developed and utilized.

Although a simplistic comparison at face value, a typical household is an excellent example. A household typically has a fixed income derived from a biweekly income source. The income source and the security of the income determine the quality of life of the household. Decisions are made to ensure that sufficient funds are available to purchase food, pay utilities, provide health care, buy clothing, etc. Without a firm grasp of income versus expense, the household would live paycheck to paycheck. A budget assists in properly managing available financial resources and assists in establishing priorities within the household. The fire service is no different.

■ Public-Sector Budgeting

In an attempt to understand budgeting, it is helpful to review the three most common forms of budgets utilized:

- Line-item
- Program
- Performance

Line-Item Budget

Line-item budgets are most commonly used as the operating budget of a department. These budgets consist of logical divisions within the overall budget and are represented by lines in a budget document. Within a line-item budget are many components that are utilized as planning tools for future budget-planning processes.

It is not uncommon for a line-item budget document to contain several years of budget information. By creating a historical snapshot of expenditures, a department can track and forecast budgetary needs based on an evolving organization. It also provides a percentage variance of actual expenditures versus the approved budget. This shows the department's experience in the previous fiscal year and often serves as an indicator for requesting additional funds if needed, or decreasing funds if a line-item has surplus funds.

Program Budget

A program budget is utilized to fund a function rather than to delineate funds to a particular line item. Oftentimes it may involve broad categories such as public safety, parks and recreation, or perhaps fire suppression. A program budget is useful in demonstrating the cost for providing programs within a specific division of government. When used in conjunction with a budget narrative, program benchmarks, and other performance criteria, it provides a simplistic overview of the financial needs for program delivery. The example listed below (**TABLE 7-1**) reflects what a typical program budget might include for a normal fire and EMS department.

Performance Budget

A performance budget is based on dedicating funds to predicted benchmarks or performance measures. For example, a department may fund a fire prevention and public education program with $5000 with the intention of educating 1500 children at a cost of $2.50 per child and installing 125 smoke detectors at a cost of $10 each. By utilizing this type of budget, a department can closely monitor the success of a program because of the performance measures and dedicated funds earmarked for specific purposes. **TABLE 7-2** utilizes the example of a child safety-seat inspection program.

It is helpful to list the assumptions that were utilized to determine the calculations in the performance budget. In this case the assumptions are listed below:

1. The personnel costs are for the lieutenant of fire prevention to dedicate 10% of his regularly scheduled hours (208) to child safety-seat inspections and installations. The hourly rate for the lieutenant is $26.31 for 2005. FY 2006 and FY 2007 reflect a 0.0325% cost of living increase (COLA) in the hourly rate.
2. The division conducted 300 inspections and/or installations in FY 2005, 350 in FY 2006, and it is projected that they will do 400 in FY 2007.

Personnel Services

When it comes to public and private sector budgeting for emergency service departments, personnel services have proven to be one of the most cumbersome areas with regard to financial resources. Even volunteer departments may incur personnel ex-

Table 7-1 Example of a Program Budget

Fire Department	FY 2005 Actual ($)	FY 2006 Budget ($)	FY 2006 Actual ($)	FY 2007 Budget ($)
Administration	56,000	57,000	56,500	57,500
Fire suppression	230,000	240,000	245,000	250,000
EMS	175,000	185,000	184,300	190,000
Fire prevention	5500	5800	6000	6300
Public education	4546	6000	7700	8000
Support services	151,105	146,500	147,000	178,224
Communications	294,059	218,643	200,643	226,121
Human resources	47,000	54,000	34,000	54,000
Total	**$963,210**	**$912,943**	**$881,143**	**$970,145**

Table 7-2 Example of a Performance Budget

Cost	FY 2005 Actual (S)	FY 2006 Budget (S)	FY 2007 Budget (S)
Fixed personnel costs	5472.48	5650	5834
Personnel hours	208	208	208
Child safety-seat installations	300	350	400
Unit cost (per installation)	18.24	16.14	14.58

penses. Costs associated with personnel include, but are not limited to:

- Wages, salaries, and stipends
- Workers' compensation
- Health care
- Life and/or health insurance
- Pension or retirement system
- Length of service awards and longevity programs
- Medicare
- Medical surveillance

Budgeting for personnel services will vary depending on the composition of each department as well as local, state, and federal regulations affecting personnel, which apply regardless of whether personnel are volunteer, paid-on-call, part-time, or career.

Wages, Salaries, and Stipends

The composition of an organization and its staffing plan will guide the need for financial resources dedicated to personnel services with regard to wages, salaries, and stipends. Chapter 5, Human Resource Management, discusses issues such as wage scales and benefits packages, which play an instrumental role in the budget development for personnel services.

It is also important for the department to fully understand other aspects involving personnel derived from collective bargaining agreements, and local, state, and federal laws. These are covered in detail in Chapter 5. Issues such as the Fair Labor Standards Act affect compensation issues, such as overtime, which will ultimately have to be accounted for in personnel services. In addition, many forms of local and state governments have stringent compensation and benefit requirements for volunteer, paid-on-call, part-time, and career personnel. Your legal counsel can be extremely helpful in identifying compensatory requirements involving personnel services.

Workers' Compensation

The fire service is among one of the most dangerous professions and with that comes inherent risk to personnel. As a result of work-related injuries, workers' compensation claims, and insurance premiums, departments are forced to dedicate monies to fund statutory requirements for workers' compensation premiums. Typically volunteer personnel premiums are higher than those of part-time and career employees, based on historical claim data. Departments should be aware of workers' compensation laws in their respective states and any applicable laws that involve reporting of injuries, compensation of injured workers, or premiums for personnel.

In addition, departments should work hand in hand with their state bureau of workers' compensation, insurance carriers, and risk managers to ensure that preventative measures are in place to limit their liability and manage their risks. Many states have programs in place that roll back premiums for safe workplaces or assist organizations in managing their risks. These organizations can also assist in budget preparation because they are most familiar with the calculations needed to determine contributions based on personnel composition and the type of compensation mechanisms.

Health Care

Healthcare premiums have become a major concern for most employers as a result of the volatile healthcare system. However, when establishing staffing and compensation schedules, caution should be taken to ensure that local and state regulations and guidelines are followed. It is not uncommon for states to establish work-hour thresholds that if exceeded require employers to offer healthcare benefits to part-time personnel.

Medical Surveillance

A rising personnel cost for the fire service involves medical surveillance. Cardiovascular disease is one of the largest factors in line-of-duty deaths to all members of the fire service. In addition, many states have enacted cancer presumption laws because of the high rate of cancer among fire fighters from encountering carcinogens in their regular work environment. With the rising awareness of preparedness as it relates to weapons of mass destruction, terrorism, and public health emergencies, organizations have expanded inoculation programs for fire fighters, many now requiring ongoing updates and medical surveillance. Additional resources that may assist with budgetary planning are:

- National Fire Protection Association (NFPA)
- NFPA 1500—Standard on Fire Department Occupational Safety & Health Program
- NFPA 1582—Standard on Comprehensive Occupational Medical Program for Fire Department
- NFPA 1583—Standard on Health-Related Fitness Programs for Fire Fighters
- International Association of Fire Chiefs (IAFC)
- Fire Service Joint Labor Management Wellness and Fitness Initiative
- Guide for Implementing the Fire Service Joint Labor Management Wellness and Fitness Initiative

Operational Expenses

While developing the operational budget for an organization it is important to have a firm grasp on the day-to-day operation of the organization. In Appendix E, it is evident that the sample organization utilizes a line-item budget that consists of 26 accounts that are further divided into subaccounts. An example of an operational expense is Account #50700 Operating Supplies, which is further divided into seven accounts ranging from #50701 *Office Supplies and Postage* to #50707 *Special Ops. Group Operations*. This particular group of line items breaks down the organization into functional groups, which have been identified as major divisions, and provides dedicated funds for the operational events that are specific to each division.

It is important to realize that operational expenses will change from year to year based on organizational priorities and new or aging infrastructure. For example, Account 50904 *Fuel* indicates that the organization was 14% over budget in 2004, which was a result of unanticipated spikes in the petroleum market. Costs such as utilities for the facility could also be affected. New or aging apparatus can affect costs of maintenance, parts, and supplies. Although no one can predict volatile petroleum markets or unanticipated engine failures or accidents, proper planning can help ensure that adequate funds are available for day-to-day operations. Operational budgets should not account for accidents and unanticipated events. To do so would arbitrarily inflate a budget and would result in large annual surpluses, which, in theory, tie up otherwise available funds.

Capital Planning and Expenses

What constitutes a true capital expense is defined by each respective organization or those within that governing body. The definition is almost always tied to a dollar threshold and does not usually include reoccurring expenses but is more for equipment purchases. Common capital expenses for fire and EMS agencies are:

- Fire and medical apparatus
- Communications equipment
- Self-contained breathing apparatus
- Personal protective equipment
- Medical equipment
- Facilities

An essential function of capital planning and budgeting is to develop life cycles for equipment and facilities. Capital items are usually large dollar items that require organizations to reserve funds for several fiscal years to make these purchases. It is not uncommon for capital purchases to be in the planning cycle for 3–5 years prior to being purchased to ensure that adequate funds are available. In determining the life cycle of equipment, apparatus, and facilities, it is important to evaluate the organizations strategic plan. The strategic planning process was discussed in Chapter 3, including important aspects such as customer expectations, essential services, and cost versus benefit. These aspects demonstrate the level of service that is expected, from which capital investments and priorities are determined.

A useful resource in establishing life cycles and estimating purchase prices for equipment, apparatus, and facilities are the vendors that an organization may work with. Vendors are accustomed to working with emergency responders to develop proposals and costs and also have a firm grasp on their services and the typical inflation of goods and services, which is a vital factor to an accurate budget proposal.

Preparing and Defending Budget Proposals

It is essential that a budget proposal is built on accurate information and is closely tied to organizational priorities that were identified in the planning process. These priorities should be linked to items such as a strategic plan, the mission, and values. Items contained in the budget proposal should not come as a surprise to those who ultimately approve the proposal. Within the overall proposal, key components will be highlighted.

Budget Narrative

A budget narrative typically serves as the executive summary, which describes the budget package. This narrative provides an opportunity to explain the accounts within the budget. A typical narrative highlights the increases or decreases in line items and provides justification for each. The narrative also establishes priorities for the organization and should help justify the need for financial support for the different operational aspects of an organization. Without a clear and concise budget narrative, the reader is left with numbers that may or may not make sense. Through a clear and concise narrative the reader develops a picture of the organizational priorities and the planning that went into the documents.

Revenue Estimates

The revenue stream for an organization will vary depending on whether the organization is a nonprofit, private, or government agency. It is important to understand the legal means of collecting and spending revenues for your organization and the methods of accounting for those revenues.

The most common revenue stream for many public sector emergency service organizations is property tax, which is derived from the total assessed valuation of the taxing district. It is important to work with the county auditor, assessor, and finance department to determine the revenue projections from year to year. Property taxes are often affected by the collection rate, which may be limited by local or state legislation. Clearly, this ultimately caps the revenue available. Another form of a property tax is for the use of a levy, which is common for many fire and EMS districts, townships, and other forms of government and requires that the taxing authority get voter approval to renew, replace, or raise the mileage of a levy.

Another revenue source that is gaining popularity is billing for services such as EMS transport, response to motor vehicle accidents, hazardous materials response, and malicious calls. These nontraditional revenue streams, although growing in popularity, require agencies to properly educate their elected officials and customers. It is not uncommon for taxpayers to view these alternative revenue sources as unnecessary and, in essence, a double tax.

Many organizations utilize fund-raising mechanisms to augment the revenues received by their tax base. Ice cream socials, pig roasts, bingo, car washes, open houses, car shows, and spaghetti dinners are all examples of fund-raising mechanisms utilized by organizations.

Another limited revenue source, depending on your local and state legislation, is investments and interest income. Each jurisdiction should be aware of the legal requirements of investing public and private funds and take caution while researching options. It is always a sound practice to consult the guidelines of your organization as well as other agencies such as the county auditor, state auditor, or the Internal Revenue Service.

It is easy to see that the revenue stream can be quite diverse as can the expected income derived from those revenue sources. Let us examine how an organization can estimate their revenue. Tax-based revenue sources from property taxes can be estimated most effectively by relying on county auditors, local assessors, and finance directors who typically prepare and certify these revenue sources. In conjunction with this information, demographic data from the census bureau and economic development date from the local economic development coordinator or chamber of commerce can assist in predicting the decline or growth of both residential and commercial growth. This serves a dual purpose. It serves as a benchmark for both anticipated service delivery and anticipated revenue from property and sales taxes. One of the best indicators for predicting revenue is historic data, especially when viewed against any anomalies such as economic recession or growth. For those agencies that utilize nontraditional revenue sources such as EMS transport, MVA responses, and so on, the historical data regarding calls for service and collections rates are the most accurate indicator for future revenue. The same holds true for those agencies that rely heavily on fund-raising events.

■ Budget Management

One of the best tools for managing the budget is policies and procedures that provide guidance for personnel regarding purchasing procedures. In addition to providing guidance, it also serves to delineate the roles and responsibilities that employees have for purchasing and budget accountability. Another helpful tool for budget management is the use of periodic budget reviews. Organizations routinely use quarterly reviews to ensure compliance with approved budgets and to pinpoint surpluses or deficits in particular portions of the overall budget. Random audits that examine budget management, program management, and accounting practices can help identify issues involving expenditures and assist in ensuring that funds are spent from the appropriate accounts.

Summary

This chapter has provided a basic overview of the varying forms of government that emergency service organizations operate in. Regardless of the type of organization (volunteer or combination) successful emergency service administrators are intimately familiar with their local, regional, state, and federal forms of government. The ability to identify the political structure, recognize the mutually dependant relationship that exists between politicians and emergency service administrators, and skillfully navigate and manipulate the political system for the benefit of service delivery are all necessary skills for emergency service leaders. A failure to develop these skills will undoubtedly hinder a leader's ability to operate within the political environment where decisions regarding laws, standards, guidelines, funding, and policy are routinely made.

This chapter has also serves to provide a basic overview of budgeting, which is, or should be, present in every emergency service organization. The American fire service is represented by a gambit of organizational structures ranging from private incorporated fire departments to federal fire departments. There are organizations that survive from fund-raiser to fund-raiser as well as those that are extremely well funded. Regardless of the funding source or the amount of financial support an organization may have, there is a level of fiduciary responsibility and accountability to members of the organization, shareholders, taxpayers, etc. The rules that govern the fiduciary responsibility of the administrator will vary depending on the organizations charter, as well as local, state, and federal laws and regulations. Ultimately the emergency service administrator is the steward of those funds and is responsible for ensuring that those funds are utilized in an effective and efficient manner in accordance with the legal requirements and policies of his or her organization.

CHAPTER

8 Human-Resource Management in Combination Systems

Dan Eggleston

Case Study

Chief Ed Johnson is at his wit's end. There seems to be tension between the volunteer members and Chris Smith, the newly hired career fire fighter who works the day shift. The East Overshoe Volunteer Fire Department has served the town of East Overshoe since 1950. The department has about 50 volunteer members and five daytime career staff. The department operates an engine, a ladder, and an ambulance and last year, the department responded to 1100 calls for service.

East Overshoe has had a full-time chief since 1995. Because of the lack of daytime volunteers, the department hired its first career fire fighter in 1998 to fill the need for a daytime driver. Over time, the department has hired additional daytime fire fighters to supplement the volunteer staff. During the previous year's budget determinations, Chief Johnson worked with the town manager and town council to approve one additional career daytime staff position. This additional position would allow the department to fully staff both an ambulance and an engine during the daytime hours.

The department went through a rather rushed hiring process because everyone knew that Chris Smith was the person for the job. The level of excitement was high soon after Chris was hired. Chris had been one of East Overshoe's best young volunteer members. Chris won the Volunteer Fire Fighter of the Year award 2 years in a row and was one of Chief Johnson's best volunteers. Everyone was pleased with the decision to hire Chris as Overshoe's new daytime career fire fighter.

Four to five months after hiring Chris, Chief Johnson noticed that the number of the daytime volunteers spending time around the station was dwindling. A few of the volunteer members had mentioned in passing that Chris seemed distant and crass. Finally, one morning Chief Johnson received a call from Pat Jones, one of East Overshoe's long-standing daytime volunteer members. Chief Johnson and Pat are longtime friends, both joined the fire service in 1975, and have enjoyed working together ever since. When Chief Johnson heard Pat's tone of voice on the other end of the phone, he knew something was wrong.

Pat began the conversation direct and to the point. "Ed, you've got a cowboy on your hands. That Chris Smith left the station en route to a call this morning and left Mike Beal and I standing in the bay getting dressed. We confronted him when we got back, and Chris said, 'You volunteers need to be quicker next time,' and laughed as he walked away." There was a long pause on the phone and then Pat spoke up, "Ed, other volunteers have complained to me about the same thing. That's why the daytime volunteers have not been showing up. You need to fix this or Chris will drive everyone away." Chief Johnson decided that he had reached a breaking point, and that he was going to talk with Chris.

Chief Johnson usually arrives at the station early in the morning, sometimes around 6:30 a.m. This gives him a chance to have coffee with the volunteers who are ending their shift and to talk with the daytime career shift as they arrive at 7:00 a.m. There is the usual ribbing of one another and telling of tall tales. Everyone enjoys the fellowship around the kitchen table. Ed kept an eye on his watch and at 7:00 a.m sharp, Chris is nowhere to be found. Around 7:10, Chris comes in the kitchen from the apparatus-bay floor. Chris passes the crowd and does not say a word. Chief Johnson can feel the tension in the room. Chief Johnson speaks up, "Chris, I need to see you in my office before you get started this morning." Chris and Chief Johnson make their way to the corner office.

After Chief Johnson closes the door and sits down, he runs down the list of complaints that he has received about Chris and gives him a chance to respond. Chris immediately complains about a number of things he does not like about East Overshoe Volunteer Fire Department. Chris continues that he is so frustrated with the lack of professionalism that he has applied with Metro Overshoe, an all-career department 30 miles north. In fact, Chris has successfully completed the prescreening and is scheduled for a final interview next week.

Chief Johnson is shocked. He thinks to himself, "What went wrong? Chris was a model volunteer that everyone loved. How could I have avoided such a failure?"

Introduction

Departmental leadership in today's environment requires a highly skilled and talented management force combined with a solid human-resource (HR) management system and process. Leadership of fire fighters in a combination system of volunteers and career personnel is even more challenging because there are greater areas for potential conflict. This chapter focuses on the most important areas involving leadership and management in a combination department. Specifically, the chapter covers the following areas:

- Partnerships
- Recruiting career employees
- Compensation and benefits
- Performance management
- Promotions

Partnerships

The greatest asset a fire department has is its personnel, both career and volunteer. Most successful departments usually have a talented, motivated, and satisfied staff. Fire service leaders are typically at their best while leading a group of fire fighters on an emergency call. However, leading a team during nonemergency work can be more challenging than a third-alarm fire. Responding to personnel needs, expectations, and legal rights has become more demanding and complex over time. Fire service leaders need to partner with their HR experts to help navigate through and around the myriad of obstacles.

Human Resource Manager

The first step in gaining a better knowledge of human resource (HR) management is to establish a working relationship with the director or person in charge of the local HR department. This HR manager is educated and experienced in labor laws, work analysis, staffing, training and development, appraisal, compensation, and labor relations. Although HR staff have experience with personnel matters, many may not be familiar with the structure and processes associated with fire service, especially a combination department. Therefore, it is advised that the department leadership schedule time to provide an overview of the department to the HR manager before diving headfirst into a particular issue.

The best place to start is to review the Fair Labor Standards Act (FLSA) with the HR manager. It is most important to review Section 7k of the FLSA as it relates to the department's current work schedule. A significant portion of newly formed combination departments are not structured to meet the requirements set forth in Section 7k of the FLSA. In the 2004 fiscal year, more than 265,000 employees received back wages as a result of wage and working-hour investigations in Fair Labor Standards Act (FLSA) cases.[1] Understanding and applying wage and working-hour polices can be difficult. Seek professional assistance from the local HR department and attorney's office before developing a policy.

It is also advisable to review current department policies and guidelines as they relate to personnel matters. It is not uncommon to have internal department policies that conflict with local personnel policies. Department and local personnel policies should be well integrated to avoid gaps, conflicts, and overlaps in policy. Whenever possible, personnel policies should be maintained and updated centrally by the personnel department.

Lastly, the fire service leader and the HR manager should review the training and educational requirements of career and volunteer staff. Helping company officers develop skills and abilities through leadership training and education is a cornerstone to any successful combination department. Although most skill-based training and education is conducted within the fire service, the locality will sometimes offer employees soft skill training and education.

Other Fire Service Leaders

During the development of a combination system, fire service leaders sometimes fail to recognize that they are not blazing new paths. Oftentimes, other fire service leaders have struggled with the same issues and have successfully developed useful strategies.

Joining and participating in professional fire service associations is the best way to meet and converse with fellow fire service leaders. Most states have fire-chief associations and conduct annual conferences that are designed for networking. The International Association of Fire Chiefs (IAFC), especially the Volunteer and Combination Officers Section (VCOS), is a valuable resource for information about combination departments. Most professional organizations maintain a Web site with links to past articles and publications about managing and leading combination departments. Just recently, the IAFC published the *Guide to Managing Volunteers for FLSA Compliance*. This document is available for purchase from the IAFC.

The National Volunteer Fire Council (NVFC) is also a good resource for policy-related issues. The NVFC is a nonprofit association that represents the interests of the volunteer fire, emergency medical, and rescue services and serves as the information source regarding legislation, standards, and regulatory issues.

Educational programs and events are also some of the best venues in which to network and learn from others. The National Fire Academy has a wide range of educational opportunities meant for fire service leaders. Most courses incorporate group participation. Many students leave the academy with newfound knowledge and a long list of professional contacts.

Additionally, experienced and established fire service leaders will often volunteer their services as a coach or mentor. A mentor is an experienced fire service leader who is willing to help and guide those who are less experienced; so take advantage of their knowledge. It is important to find a professional who shares values, work style, and sense of humor with the person they are mentoring, but at the same time, he or she should be someone who will provide a challenge. The best mentors are people who are excited about the fire service, who are willing to spend the time to help leaders and their respective departments develop over time, and who will achieve a sense of personal satisfaction from seeing others succeed.

Recruiting Career Employees

Successful fire departments are made up of a group of motivated individuals striving towards a common vision. Successful departments are able to fulfill the needs of the individual while simultaneously achieving the mission of the department. This give-and-take approach can be accomplished in an environment that is supportive, educational, and enjoyable. Challenging a department leader can be intimidating, but it is important for the leader to genuinely allow others to identify different and improved ways of functioning.

A fire service leader has the responsibility to carefully select the best individuals to complement the

[1]Department of Labor. Wage and hour maintains high enforcement levels in fiscal year 2004. http://www.dol.gov/esa/whd/statistics/200411.htm.

team. Recruiting employees should not be rushed or carried out haphazardly. Taking the time to follow an objective methodology will yield better results. Cutting corners or relying solely on emotions can allow a misguided hiring that will often result in a difficult parting.

There are a number of federal laws pertaining to employment. Therefore, it is advisable to seek professional assistance when navigating through the hiring process. Some of the applicable employment laws are as follows:

- Title VII of the Civil Rights Act (Title VII)
- The Americans with Disabilities Act (ADA)
- The Fair Labor Standards Act (FLSA)
- The Family and Medical Leave Act (FMLA)

Where to Start?

Recruiting employees is similar to selling a product. Like a consumer, a prospective employee will evaluate the product (i.e., the fire department) for value (i.e., benefits that he or she will receive). Why would someone want to work for one department versus another? What does the competition offer?

Although compensation and benefits are often considered the number one value among job seekers, personal development opportunities and the quality of the working environment are equally important. It is important to identify the benefits of working for a department to develop an effective advertising campaign for it.

Before starting, it is important to develop a well-planned schedule that covers the entire hiring process. As with any hiring, it is advisable to consult a HR specialist to verify that the procedure and tools used throughout the process meet all legal requirements.

For most small combination systems, hiring a career employee is a significant event. Existing career and volunteer members add value to the process because they can greatly assess the fit of an individual to the organization. Involving a cross-sectional group of career and volunteer members throughout the process will help to gain support from the remaining members and generate a sense of ownership in the final selection. Members involved in the selection process are more likely to help ensure that the new employee is successful in the department.

Advertising

Many departments embark on an extensive and expensive advertising campaign only to find that the best fire fighter candidate is within their own volunteer force. It is likely that a qualified and eager volunteer is waiting for his or her opportunity to apply. Serving as a volunteer gives the member the opportunity to gain valuable training and experience in fire service. The department they volunteer with may eventually employ them. The training and experience a volunteer receives will enhance their ability to compete during the selection process.

Although a department may have well-qualified volunteers who could fill the role of a full-time career employee, it is important not to bypass the screening process. Volunteer members should compete for open positions in a fair and equitable manner. The objective of advertising is to generate a qualified pool of applicants that can be further evaluated to determine the best fit for the department.

There are several venues in which to advertise a job opening. Advertising in newspapers and journals can be expensive. The cost for advertising depends on the length of the ad and the time for which the ad runs. However, an advertisement can be carefully summarized to reduce the amount of words and save significant dollars. As mentioned earlier, seek assistance from an experienced HR manager on the best way to word a written job advertisement.

Many online services will also advertise job openings. Charges for online advertising range from no fee to fees comparable to that of printed advertisements. The goal of recruiting is to create a large pool of candidates for further screening. An online advertisement has the potential of reaching a wider audience than printed media and should be utilized to broaden the advertisement's reach.

Screening and Interviewing

A successful advertising campaign can generate a number of qualified candidates. Before the first candidate is considered, it is best to develop a *screening matrix*. A screening matrix helps to create an objective process and is designed to determine which applicants meet the minimum qualifications and who has the most relevant job experience. The matrix categories should be designed around the job description duties as well as the knowledge, skills, and abilities needed to successfully perform the job. The matrix should be used throughout the screening process.

Once the initial screening is complete and it has been determined which candidates meet the minimum qualifications, some departments choose to conduct a physical ability test. The physical ability test is a practical exam used to test a candidate's physical ability to perform a job task related to firefighting.

It is important to involve a HR specialist and obtain legal advice before developing and administering a physical ability test. There have been several lawsuits challenging the fairness of physical ability tests. The International Association of Fire Fighters (IAFF)/IAFC Joint Labor Management Wellness and Fitness Task Force developed the Candidate Physical Ability Test (CPAT) as a fair and valid evaluation tool. The CPAT is designed to assist in the selection of fire fighters as well as ensure that all fire fighter candidates possess the physical ability to complete critical tasks effectively and safely. See the IAFF or IAFC Web site for more information.

In addition to the physical ability test, some departments also choose to administer a general ability test. General ability tests typically measure one or more broad mental abilities, such as verbal, mathematical, and reasoning skills.

The next step in the screening process is to generate a manageable number of candidates to interview. The interview is probably the most commonly used assessment tool. Structured interviews often utilize standardized questions, trained interviewers, specific question order, a controlled length of time, and a standardized response evaluation format. The screening matrix can be used during the interview process to help gauge candidate responses. It is important to ask follow-up questions that are relevant to the job to clarify and explore issues brought up during the interview. It is unlawful to ask questions about medical conditions and disability prior to a conditional job offer. Even if the candidate volunteers such information, the interviewers are not permitted to ask about the nature of the medical condition or disability. If a candidate volunteers such information, refocus the topic so that emphasis is on the ability of the applicant to perform the job, not on the disability.

The majority of departments form an interview panel to screen potential employees. Before involving inexperienced department members in the hiring process, it is advised to clarify the expectations and educate the group on the process as well as proper screening and interview techniques. It is also important to inform the candidate that they will be interviewed by a panel.

Depending on the number of applicants and the number of positions open, the committee may choose to conduct one or two interviews. The committee will need to determine how many candidates will be contacted for the first interview. There is no exact number; however, the objective is to interview a pool large enough that the final interview round will include two to three people for one position.

Once the initial interviews are conducted, the remaining candidates should be further screened by checking references. Reference checks are used to verify education, employment, and achievement records. This particular step could be performed earlier in the process depending on the number of applicants. Most applications require that a candidate provide at least three references. Typically, references supplied by the candidate will result in positive reports. However, it is advisable to go beyond the list of references supplied by the candidate and interview previous employers and coworkers. Keep in mind that questions asked during the reference checks must relate to the candidate's ability to perform the job.

At this point, a small pool of candidates should be ready for the final screening. The committee may choose to conduct a final interview to follow-up on topics discovered during the first interview and reference checks. As with the initial interview, it is recommended to use a structured process with job-specific questions.

■ Job Offer

Many departments require a physical exam, drug screening, and criminal background check before employment. Some choose to conduct a physiological exam as well. It is important to note that a contingent job offer must be provided before a candidate undergoes this level of screening. The *NFPA Standard 1582: Medical Requirements for Firefighters* should be used as a guide to evaluate a candidate's medical condition. It is advised to consult an occupational health professional for assistance.

A conditional job offer should be spelled out in a letter to the candidate. The job offer should include contingencies (e.g., physical, background check), start date and location to report, starting salary, and benefits. The exact start date should be worked out with the candidate and include adequate time for notice to his or her current employer.

■ Orientation

Employees develop long-lasting impressions of their department during their first several weeks on the job. During this time, new employees learn how a department values its employees. Because many combination departments are not structured to offer an extensive new-member program, it is advisable to give an orientation program to familiarize the candidate with the people, equipment, and environment. After all, much time and energy has been invested on behalf of both the candidate and the department at this point. A well-planned and well-executed orientation program can help clearly define expectations and better ensures that problems do not emerge early in the process.

Some departments choose to partner a new employee a more experienced employee, called a work associate. An associate is someone who partners with a new employee during the first 3 to 6 months of employment. The associate offers advice and guidance to help foster and promote the professional development of a new employee. The associate is an experienced employee who is well trained and educated and is familiar with department policies and guidelines. The associate is also someone who the new employee can rely on for advice and encouragement and can help the employee fit in as a contributing member of the team.

Compensation and Benefits

Compensation is payment to an employee for performing his or her duties and is usually provided as base pay. In some cases, a *pay-for-value* system is established that provides additional pay above the base pay for certain forms of education or certification. A person's pay should be based on his or her job within the department and a market survey of wages for comparable jobs. The complexity of a job is determined through analysis using the job description as a guide. A job analysis gathers information about the nature and level of the work performed and the specifications required for a person to perform the job competently. Once a base pay is determined, an appropriate pay range is selected. Pay ranges include the minimum and the maximum amount of money that can be earned per year for a particular position.

Most fire-fighter positions are considered nonexempt employees and so receive overtime pay for hours worked beyond normal working hours as defined by the department's policy and the Fair Labor Standards Act (FLSA), Section 7k. Note that nongovernmental agencies such as nonprofit agencies do not fall under the Section 7k exemption; therefore the 40-hour week is the required system. Calculating annual pay and breaking down the hours worked into pay periods and time periods can be performed to address this issue. There are three major factors that are considered in determining whether or not an individual holds an exempt position: discretionary authority for independent action; percentage of time spent performing routine, manual or clerical work; and earnings level.

Departments must be able to pay their employees a fair wage to remain competitive. Determining the appropriate pay level is often established by matching similar jobs offered by other like-sized departments. It is vital to match jobs properly and according to the job analysis. A mismatch can skew the survey results. For this reason, it is often advisable to seek the assistance of a knowledgeable compensation and benefits specialist.

Some departments develop a pay-for-value system that compensates employees for certain educational or certification achievements. A pay-for-value system is often used to offer lateral career development opportunities for fire fighters without relegating each person to the traditional supervisory path. Small combination departments that offer limited opportunities for advancement through the normal supervisor ranks can offer a pay-for-value system as an added benefit. Some departments find that it is more cost-effective to offer pay-for-value compensation for such training levels as paramedic and hazardous material certifications than it is to hire minimally qualified fire fighters and pay to send them to school.

In addition to competitive salaries, a department should offer a variety of benefits as a part of its total compensation package. Choosing the right mix of benefits as well as highlighting the value of the benefits can be challenging. The following employee benefits are usually offered as a minimum:

- Medical insurance and dental insurance
- Retirement and life insurance
- Paid leave (e.g., sick, annual, sick bank, holidays)
- Employee-assistance program (a confidential, short-term counseling service for employees with personal problems that affect their work performance)

In addition to the basics, some employers offer the following additional benefits:

- Flexible spending account
- Tax-deferred compensation and annuities

- Free or discount admission to parks and school sporting events
- Special rates and reduced fees at local banks and credit unions
- Membership discount at gyms

Research suggests that employees do not understand the value of a benefits package even though, on average, organizations spend about 41 cents for benefits for every dollar of payroll. Employees undervalue their benefits for many reasons. Employers sometime poorly communicate the value, employees usually have little or no choice in what kind of benefits packages they receive, and employees misunderstand the market value of benefits. Begin communicating with employees about benefits. The communication process should be ongoing, not simply one event.

Because benefits are a large portion of the total compensation package, departments should strive to offer benefits that are valuable to the employee and the department. Because the personal needs of employees can range significantly, some departments provide a *menu approach* of employee-chosen benefits rather than only a one-size-fits-all offering.

Performance Management

Effective performance management is critical for creating and maintaining a high-performing department. Employees need to understand their role and the expectations their leaders have for their performance. Departments should support the career development of its employees and motivate them to take control of the direction of their careers.

Supervisors should assist employees in assessing their knowledge, skills, interests, and performance, as well as their personal and career goals. Feedback on an employee's performance (formal and informal) is most valuable when it is given regularly throughout the year. There should not be any surprises to the employee at the formal annual evaluation.

The primary purpose of a performance review is to inform employees about their abilities, contributions, and level of performance, and to offer constructive advice as to how they might improve. An effective performance review system is designed to:

- Maintain or improve each employee's job satisfaction by showing an interest in his or her development.
- Serve as a systematic guide to planning further improvement in job performance.
- Provide a written, objective opinion of an employee's performance.
- Assist in determining and recording special talents, skills, or deficiencies.
- Provide an opportunity for each employee to discuss concerns about his or her job.
- Assemble data to use as a guide for such purposes as wage adjustments, promotions, training opportunities, disciplinary action, reassignment, and dismissal.

At a minimum, a formal employee evaluation should be given annually. Twice a year is ideal. It is important that the evaluation process be designed to measure an employee's performance in the main duties for the position. Employees should fully understand the evaluation process well before the review.

Promotions

To be promoted from a fire fighter to an officer requires a wide array of talent. As the fire service expands its range of services, company officers are going to be required to possess more knowledge, more skills, and more abilities.

Most combination departments provide the same range of services as those offered by larger departments that are staffed with more personnel and have more resources. Company officers may be required to know and understand in detail certain services such as advanced life support, hazardous materials, technical rescue, and investigation. High-quality company officers will successfully achieve the mission of the fire service, and the community will in turn support the fire service.

Evaluating and selecting company officers can involve a detailed process that requires several days of evaluating knowledge, skills, and abilities. *NFPA 1021: Standard for Fire Officer Professional Qualifications* outlines the recommended knowledge, skills, and abilities expected of every company officer in seven categories: general knowledge, human-resource management, community and government relations, administration, inspection and investigation, emergency service delivery, and safety.

Some departments utilize an assessment center method to evaluate officer candidates. Assessment centers are one of the most effective means to predict whether a candidate will succeed as a company officer. A job analysis and NFPA 1021 are useful guides to determine which elements of the candidate's critical knowledge, skills, and abilities will be measured throughout the process. As an example, the following critical measures may be evaluated:

- Emergency scene management
- Personnel management
- Oral and written communications
- In-basket exercise
- Customer relations
- Preincident planning
- Emergency service delivery
- Ethical issue resolution in the fire station
- Individual and group interviews

Once the critical measures are identified, exercises are developed to evaluate the candidate's ability. Each critical measure is assigned a weighted value. An overall scoring sheet is used to capture each candidate's score. It is important to define the desired responses for each critical measure to help objectively measure each candidate's responses.

Usually, an outside group of trained and experienced assessors are used to evaluate the candidates. The assessors observe the candidates working through a series of exercises. Candidates are assessed and assigned a score for each area. Once the process is completed and the score sheet is tallied, the list is usually forwarded to the chief for a final recommendation. The chief's decision will be much easier to make if he or she is confident that the promotional process has been well planned and executed and that the results have been determined without bias.

Choosing the next company officer for a department is one of the most important jobs of a fire service leader. A solid investment in the process will result in a wise hiring choice that will lead to a more productive and effective department.

Summary

Is it possible that Chief Johnson could have avoided the negative situation with Chris Smith if he had been better prepared? Is it possible that Chris Smith would have not been hired if a more thorough hiring process had been followed? Is it possible that Chief Johnson could have exercised better leadership abilities by helping Chris Smith develop into a productive East Overshoe fire fighter? They are all possible. However, Chief Johnson discovered that learning by trial and error is not the best strategy, especially when it relates to human resources.

It is the fire service leader's responsibility to guide department members toward the accomplishment of common goals. This process is dynamic, situational, and often involves a merging of values and perceptions. To meet this challenge, fire service leaders must commit to a lifelong learning process and stay abreast of the latest trends in human-resource management. Fire service leaders must strive to be the best in issues related to hiring, firing, training, and other personnel issues.

CHAPTER

Strategies and Tactics for Success

Fred C. Crosby and John R. Leahy, Jr.

Case Study

Chief Ed Johnson has been the new chief for 6 months now, and the honeymoon period is over. Being the chief is not at all what Ed had envisioned. He thought that he knew what it would be like to be the chief. After all, he had been a deputy chief for the past 5 years. Surely that experience had prepared him. But it had not—it seemed like everywhere Ed turned, he was in the middle of a fight or under attack. He had to fight everyone in administration, from the finance director to the personnel manager, to the purchasing officer, to the board members to get the necessary tools for his department.

Then he returned to his department to find that his effort was not appreciated. His members fought everything he was trying to accomplish. More than that, for the first time in his professional life, Ed understood what it was like to be truly alone. Chief Johnson is tired and depressed. He is questioning his own ability and his survivability in the job. If only there was a guide, a set of survival rules for the chief, something he could turn to.

Introduction

Simple rules are often the best, even though they can be the most difficult to follow. As with the original Ten Commandments, these rules are unequivocal in their language and intent. The idea of a code is that it is not situational. Ethics and rules should be static and secure; they are the foundation on which you build your life and actions. A leader who is not consistent will never be successful. Thus every fire chief should live by these ten commandments for a fire chief.

The First Commandment: Thou Shalt Build a Strong Relationship with Managers and Elected Officials

A sergeant major of the Marine Corps was recently quoted saying that good leadership involves good "followership." The first commandment touches on the value of duty. As leaders, we have a duty to follow those who are charged with leading *us*. A fire chief who cannot respect his manager or elected officials cannot gain the respect of those he seeks to lead. In other words, good officers must first be good soldiers.

This principle does not advocate blind obedience. Along with the duty to ultimately follow and respect the decisions of managers and elected officials, comes the obligation to assure that they receive good counsel from you. They ultimately make the decisions that affect the good of your fire department. It is up to you to make sure that those decisions are informed decisions. By accepting this simple philosophy, you can build a relationship with your superiors, which allows you to guide and impact the decisions that they make and fulfill your responsibility of followership.

Tactics

Have Effective Communication with Each Individual

In building relationships with your manager and your elected officials, you must build a method of effective communication. Communication is most effective when it is ongoing and from both parties. If you start from a point of understanding the position of the person with whom you are communicating, then you can make your argument much more effectively. For those who have been instructors, remember the first rule you were taught. Start at the level of the student. To communicate your vision, your point, and your agenda, you must know where to start. You must listen more than you talk to communicate effectively.

Never fall into the trap of contacting your elected officials only when you want something. That tactic is a surefire way to fail. Elected officials fully understand that they need you and your support, but they also do not desire a whiner who calls only to complain and ask for things.

Conduct Individual Tours

Emergency services are difficult for outsiders to understand. We have our own culture, our own language, and our own eccentricities. Providing a service 24/7 is foreign to many people. Add to that the high level of expectation, and you can see how difficult it is for others to understand how we perform our vital services. It is difficult for most of us to remember what life was like before we caught the fire fighting bug. To understand an outsider's view of emergency services, try to remember your first day. It was scary. *You* were an outsider.

In order for managers and elected officials to support our needs, they must have at least a rudimentary understanding of what we do and how we do it. The best way to provide this understanding is to show them. It is important to do this individually. By avoiding the group type of tour that every fire officer has conducted, you will have the undivided attention of the person you are attempting to reach. You will find that the person is more relaxed, more open, and more likely to ask questions. You can then key in on those questions. You will also find ample opportunity to understand the person's individual interests. It may be taxes, or it may be the questions that the elderly face. Remember that each of these officials is first and foremost a person. By understanding the official's particular motivation and interest, you can communicate about your common motivations and interests.

Know Your Officials

The idea of this tactic is to build a relationship with your local officials. By knowing their families, their special occasions, and important dates, you can relate to them on a human level. You know the interest you have in your own children or grandchildren—this is an area of humanity that you can share. It is a point on which you can build a relationship.

It is difficult to fight or ignore those you know on a human level. It will be more difficult for officials to fight with you or ignore your point of view on a particular issue if you have an interpersonal relationship. It is often said that it is not *what* you know but *who* you know. That is true to a large extent. By knowing your officials, you have an automatic advantage.

Make Yourself Available

If you are remote or difficult to find, your superiors will look to others to respond to their needs. Make sure that your secretary knows the voice of the manager and officials and set protocol that tracks you down quickly and easily when you are away from the office. Share your home and cell phone numbers, and never make the mistake of sounding too busy or unenthusiastic about receiving a call from one of your bosses.

Take the time to listen, and make sure that you understand the request and the reasons behind it. This will make it easier to respond correctly the first time. It is important that you build your credibility by answering requests quickly and correctly.

Your superiors and officials have to know that they can count on you. Being the go-to person for managers and elected officials is a true advantage in the world of local government.

Politicians and county or city managers like to win. They also desire that the people they associate themselves with be fixers. Setting yourself up as a fixer makes you indispensable. Being indispensable is good for job security. Remember that much of what you are doing to survive on a daily basis is actually setting the stage so that you can survive, and even win, on your worst days. Credibility building is done over time and with success.

Keep Them Informed

You want your superiors and officials to realize that they are an important part of your team. They should know that their support has real-world results. Let them know of the victories that they help foster, both small and large. For example, one fire department has a tradition of picking up the district politician and having him or her ride the last block on all new fire-vehicle deliveries. By doing this, the official sees firsthand how his or her support for your budget made a tangible improvement. It drives home the point that his or her decisions have real-world implications, using a direct and fun method.

Submit Routine Reports on Activities

Routine reporting brings accountability into the equation. By doing so, you are acknowledging that your fire department is accountable to the larger system. It also builds the knowledge base of those you are trying to convince to support you and your fire department. If they understand your daily functions, then they can make the jump to understanding your needs. It is important to remember to tailor these reports to your audience. The idea is to make the presentation more appetizing.

Remember that local officials are inundated with material to read and understand. The shear volume of material that is put before them is overwhelming. It is important to convert the information into a sound bite. Use visual presentations as much as possible. You want your reports to be skimmed and understood. The executive summary is the most important.

Attend Functions

Whether you like it or not, your position is high profile. It is important to be seen and to represent the fire department. Whether it is a ground breaking, a ribbon cutting, or a retirement party, it is good practice to attend these functions. It is also an opportunity to impact the image your manager and your officials hold of you and your department.

Most people do not realize how much business is actually carried on at these types of events. By simply being present, you may be able to answer a key question that could swing a vote in your favor. The world we live in is one where information and action are in perpetual motion. Managers and officials are forced to be minute managers. You may need their ear when they are at the opening of the new business or school. If you do not attend these events, then you miss your opportunity.

Make Introductions

We all know that whether we hide it or not there is a place deep within every fire chief that warms to the sound of the word "Chief." It is a measure of status and a mark of achievement that is hard won. The title is a culmination of effort and is important. Why would you think that it is any different for a city manager or for an elected official? These individuals have worked long and hard and sacrificed much to achieve their position. It is not only a simple matter of respect to introduce them by using their formal titles. By taking the time to acknowledge their status, you reinforce yours. You can only receive respect if you freely give it.

Utilize Their Skills

No matter what you may personally think of politicians on a whole, it is pure folly to underestimate the ability of the individuals who have been elected to public office. You will find that each individual has skills, and a smart fire chief recognizes that those skills can be of great personal benefit.

One skill that almost all elected officials possess is the ability to find consensus. Although most fire chiefs will tout their own abilities in this regard, the savvy ones will recognize that not all chiefs are good at this task. Chiefs are fairly good at ordering consensus, but not always so in actually building consensus.

One of the first things a chief of a combination system has to do is bring a vision to the fire department. In crafting this vision, you have to have a consensus agreed on by all of the players if you have any hope of actually seeing your vision brought to fruition. A political official may be the perfect answer to this dilemma. It is their stock and trade to build consensus from disparate groups and agendas. Their skill in this arena may be the ticket that you are desperately seeking. If you get to know your officials and their skills, then you will be able to utilize their abilities. The most surprising thing may be that you will find them not only able and willing partners if you enlist their help, but also to be enthusiastic. Soon they will be staunch allies.

Make Staff Available

In Fairfax, Virginia, they recognized the unique status of the neighborhood firehouse and began attaching district offices for the local Board of Supervisors' offices. This close relationship of community service is proving beneficial not only to the politicians, but also to the fire service.

Other fire departments lend their staff for all sorts of duties. There is a fire department that utilizes their battalion chiefs to deliver the agenda packets twice a month to officials' homes. Whatever the task, when allowable, it is mutually beneficial to let your staff assist the manager and the officials. The only real quagmire to avoid is to remain apolitical.

One of the key advantages to the fire service is our unique ability to be a generalist. Some say that if a spaceship landed on main street, that the public would call the fire department because they would not know what else to do. If you are useful in many different ways, then you are useful to many different people and circumstances. In today's world, the fire service has to compete with schools, law enforcement, and even parks for their local slice of funding. If you want a good chance at getting a large slice, then you need to be useful and even crucial to the success of many different things. Capitalize on this advantage to the fullest extent possible. You should always be the first call they make for help.

Have a Relationship with the Financier

Any fire chief who does not know by now that budget drives policy is living in a fairy-tale world. Even though it goes against the grain of our instincts, we must realize that money drives many, if not all, of the decisions in the end.

This places the traditional fire service in a difficult position. With fires and fire deaths on the decline for the last couple of decades, we face the realistic possibility of being considered a footnote by the financial types who actually run public policy through the budget. Our only hope is to find ways to make what we do relevant from a financial perspective.

If you are going to be successful, then you must understand the perspective of your financiers. You must learn to communicate with them in their language. This requires awareness and a relationship that is foreign to most chiefs. Clearly, most fire chiefs do not get into this business saying, "Oh boy, one day I'll get to distill life and death decisions to a spreadsheet." However, that is what you are now required to do. You must be able to put good business practices in place to gain the respect of the finance wizards who hold the key to the coffers of government.

Have a Relationship with the Department Attorney

Like finance, the world of legalities is real and vitally important. This is another area where most chiefs feel foreign. It is dangerous in our occupation not to give proper consideration to the legal impact of our actions or nonactions.

One local fire chief wanted to make the jump from chief to assistant administrator. The administrator looking to hire him called his previous manager. That manager paid the candidate a compliment of the highest order. The manager said that the chief was a great fire chief who had learned to always talk to the city attorney before taking certain actions.

This illustrates how chiefs are viewed by their bosses. We are people of action who sometimes lack the ability to think through a situation in a legal manner. The department attorney can give you the counsel that helps you make a sound decision. These legal experts can also keep you out of trouble.

Create Honorary Members of the Department

Making managers and officials honorary members of the fire department drives home the point that they are part of your team. It is called buy-in. They will feel an ownership that they would not experience from sitting on the sidelines. Remember that the whole point of the first fire chief commandment is to build a beneficial relationship with policy makers. By tying their success to your success, you will ensure that they act as an advocate for you and your department.

Be Open-Minded

If you execute commandment one well, you will be a victim of your success. You will have engaged and built a strong relationship with your manager and officials. They are now interested, supportive, and involved. Expect that with involvement will come suggestions and even criticism. It is important to be open to these suggestions. You may even find that some suggestions and comments are beneficial. It is a fatal mistake to ignore these suggestions, expecting that these people will be happy to blindly follow all of your ideas. All of the work you put into bringing them onto your team will be squandered if you do not recognize them as true team members. You will also quickly learn that the worst enemy is a disenfranchised friend.

You must also accept deserved criticism. No one is perfect, and even with the weight and wisdom of five bugles, you will make mistakes. Be prepared for the criticism that follows such missteps. Readily accept your foibles. Taking responsibility for them only builds your credibility as a leader and earns the respect of those around you. You should also be prepared to accept and defend against unfair criticism. Simply because someone else would have made a different decision, that does not mean that your decision is necessarily wrong.

Learn People Skills

Without people skills, you will not last long as a leader—no matter how technically skilled you may be. Being chief has more to do with people than with technical ability. A chief's job is similar to herding cats; if you lack the skill, then the cats will run all over you. Take the time to study and improve your people skills.

Body language and appearance are often underestimated. If you think this is not important, then think of the people who turn you off. They are usually dressed inappropriately, or they have body language that conflicts with their message. A person can say all of the right things and then ruin their message through bad body language or appearance.

The fire service is a unique and often misunderstood culture that has its own eccentricities. The problem comes when we try to communicate with "normal" people. The next time that you are at a gathering of fire fighters or EMS staff, take a moment to observe. Look at the t-shirts. What county manager or elected official would get the inside joke of a skull in a Maltese cross or a big hose shirt? Your dress conveys an image. Think about the image that you want to convey. Do you want an image of respect and credibility, or do you want the image of a backwards baseball cap?

The Second Commandment: Thou Shalt Provide Sound Personnel Management Practices

None of these commandments is more important than the others. These commandments should all be followed. If you have strong outside relationships, but do not provide sound personnel management systems, then you will fail.

A sound personnel management system is a generic term that covers a great deal of territory. It can mean everything from good documentation to fair labor and promotional practices. There is no way to cover all of the necessary information to define sound personnel management in total. Here are some tips that may help.

■ Tactics

Know Your Staff

This sounds so simple, and yet it is one of the more complex steps in providing sound personnel management. To really know your staff and support them requires a great deal of time and effort. It also requires a real desire, which cannot be faked. If you try, you will be spotted immediately.

A seasoned fire chief realizes that each of his staff have their own unique desires, goals, and objectives, or put another way, their own agendas. You must find ways to make *their* agendas fit *your* agenda. Look for how their success can aid your success. You should also find ways to settle the conflicting agendas between staff members. Always look for common ground.

Know Your Employees

This sounds easy at first, yet it may prove almost impossible if you have a large department. Although you may be able to understand the desires, goals, and objectives of a small department of 50–60 members or employees, you cannot have that same level of understanding with a staff of several hundred. In the case of large numbers of employees, your only real hope is to understand and feel the generalities. After all there are some goals, desires, and objectives that are almost universal. Who for instance does not want to be appreciated? Who does not want their work to make a difference? Who does not seek to be of value?

The best way to understand these generalities is not to forget where you came from. Almost every chief started as a fire fighter. If you take the time to sit down with a representative group of your employees, you will be reminded that the desires they hold, their goals, and their objectives are not all that different than yours were when you were in their position.

Know Your Officers

There is absolutely no way you can guide your officers if you do not have knowledge of their duties and responsibilities. The theory that says if you can manage anything you can manage everything has absolutely no application in the fire service. You do not have to be able to do their duties as well as they can. It means that you have to understand what they are attempting to do if you hope to lead them. If there are areas that you do not understand, learn them.

Mentor

Consider each individual in your organization as an educational project. One of your main jobs is to educate your members. Education is an extremely effective tool. Through education you can change attitudes, implement change, and affect behavior.

Most fire departments have recruit academies. Think about the real purpose of these academies. They all teach basic firefighting skills and EMS, but they also achieve other goals. The academy, if used well, imprints the recruit with the values of the organization. It instills the discipline system that fire fighters will follow for their entire career. The academy starts the new fire fighter on the right path.

If your recruit class system does not have a mentoring component, then you must commit to finding another way that the agency's mission, vision, and values can be institutionalized from the beginning.

Continuing education should be designed to hone basic skills and to keep fire fighters on the right track. As they progress, education should be designed to make them find their own way, challenge the accepted practices, and be innovative.

Besides education, there is the issue of patterns. You want your members to build patterns using the values and objectives of your organization. Patterns are not something that can be achieved by education alone. Proper patterns can only be achieved through the process of mentoring. Mentoring in today's fire service does not get the attention that it deserves. The fire service of the past did a much better job of mentoring than we do today.

To mentor properly, your department should understand that almost every member is a mentor to his peers, and every officer is a mentor to his members. It is like the old navy saying, "Not on my ship, not on my watch." If someone wrapped a hydrant wrong in the old days, then everyone worked on that skill once they had returned to the station. It is important to foster that same mentoring attitude in your department.

Personal Accountability

Although it is not always obvious, the value of personal accountability is not an extinct principal. A system that builds on personal accountability will automatically have accountability throughout all the strata of the group. Accountability goes far in creating a department that gets the job right the first time. If one of your values is quality, then the first step is to institute a system of personal accountability. It is an extension of the "not on my ship" philosophy.

One of the best ways you can institute this idea is to provide the necessary support to see it through. Support an environment that rewards and applauds personal responsibility rather than punishes it. If you punish every small mistake, you foster an environment of fear and intimidation. You teach your members that it is better to hide the bad from you. If you create an environment where mistakes are tolerated and used for betterment and learning, then you create a positive environment.

Teach your members that it is acceptable to make mistakes as long as there is accountability taken for their actions, and they take the necessary steps to correct the situation and guard against recurrence. To fully drive the idea of personal accountability home, you should institute an honor code. The honor code defines the punishable mistakes. All other sins are forgivable with the proper actions. The honor code of West Point is simple, "I will not lie, cheat, or steal, nor tolerate those who do." The simple elegance of this code, with its definition of the unacceptable breeches, is unmistakable.

Keep Employees Informed

An informed workforce is an engaged workforce. Make every effort to get the word out on what is going on in your department. Remember this is the fire service where there is ample downtime in the stations. In each of your organizations, there is a formal communications system and an informal communications system. If the formal communications system fails to keep pace with the informal, then rumors will run rampant.

Also remember that fire fighters are perhaps the quintessential team players. They travel, eat, sleep, work, and play in packs. This is a good trait because it enables them to complete their job. But like all things good, it does have its dark side. The dark side of team culture is that with it comes a herd mentality, and once the herd starts a stampede it is difficult and dangerous to stop. The clever chief will spend time on communication to keep the department informed.

Recognition Programs

Positive reinforcement is a proven technique that is used in many different ways in the business world and in schools. In the fire service, we are often slow to recognize the daily acts that lead to our success. We reserve our honors for extreme acts and often only recognize sacrifices posthumously. The daily acts of heroism are viewed as part of the job.

In a system of personal accountability, it is important to recognize the successes. It is helpful to reward members with positive reinforcement. If you are going to hold people responsible for their actions, then you must be prepared to hold them responsible for the good as well the bad. You should treat each appropriately. Try to catch your members doing good. When you do, recognize and reward it. If you reinforce good behavior, they will want to repeat their actions. A reward can be as simple as a big thank you, or as specific as a monthly recognition discussion at the staff or officers' meeting.

Use Outside Agencies

If you reach out, you will find an ample amount of community organizations that are not only willing, but actively looking, to support public servants. Lions Club, Rotary, Veterans of Foreign Wars, and even community associations are receptive to recognizing fire service. In addition, there are state and federal award programs in place that are free or cheap to institute in your department.

By looking to outside agencies for help, you can stretch your budget to include these types of programs. You may be able to create a budget that you have never had in the past through donations. Through this you can also create partnerships with your community. You may find that having friends in these kinds of organizations is invaluable the next time you need political support to achieve one of your goals.

Experience will show you that within local government, it is a small number of people who are actually impacting the decisions on a daily basis. Most citizens are relatively uninvolved and uninterested in the daily grind of their government. It is the activist who you want to befriend. Members of community clubs and associations are the activists of today's local government culture. Having a partnership with them may provide you the edge you need at a crucial point in the future.

Physicals

A healthy workforce is happier and more productive than an unhealthy workforce. There have been numerous studies that prove this. Healthy people tend to work more effectively and have better attitudes towards work and their organization. Obviously this is desirable for any organization. If you do not believe this, then look at the statistics on line-of-duty deaths. How many fire fighters die on the fire ground because of heart attacks? What we do is extremely stressful and dangerous. We have long periods of inactivity punctuated by moments of high intensity.

This makes the idea of wellness even more important. There is a concept that most of us learn eventually—you cannot save anyone else until you save yourself. Build a wellness program into your daily existence. Physical training, routine physicals, and programs that teach and reinforce lifestyle choices are immensely important. It is also an easy sell to your superiors because it saves a great deal of money in the long term. A good wellness program can increase productivity, decrease absenteeism, cut workers' compensation costs, and decrease health insurance costs. It is so simple and logical that it is astounding that wellness is not universal in the fire service.

Alcohol, Drug, and Tobacco Programs

Part of any good wellness program is a component that recognizes the value of attending to mental health. Depression, alcoholism, and drug abuse are much more prevalent in emergency services than we care to admit. Emergency service personnel live at the intersection of the underbelly of our larger culture and normal life. We see things and handle situations on a daily basis that people in nonemergency professions may only have to encounter once or twice in their lives—if at all.

Our failure to admit this fact creates a situation in which most of us internalize emotions. We all deal with this internalization in different ways. Some drink too much. Others may be in an almost constant state of mild depression, and others may turn to illicit drugs. We may also exhibit these tensions by choosing to engage in risky sports or other activities that remind us that we are alive and well, at least for the moment.

We are all adrenaline junkies. The real question is this: is that a trait we brought with us, or is it something we have learned as a coping mechanism? Recognize that there is a mental-wellness need and address it before it causes problems. Establish programs that will aid your members through rough patches.

Do not be afraid to take this step. Your members may be resistant at first, and concerned about the personal and professional relationship. However, after the first instance of helping someone, they will recognize this assistance as strong leadership.

Mission and Vision Statements

Mission statements should be the most simple and straightforward distillation of the purpose of your organization. If it is simple and straightforward, then it will be an ideal that all members can remember and repeat. A good way to test this is to place a question in every promotional exam that requires the individual to recite the mission statement and explain its meaning.

Standard In-House Rules

People want and need direction. For some, this is a difficult concept to accept. Many of today's management concepts advocate an open architecture in process management. The theory behind these concepts is that the more open you leave your processes and procedures, the more you foster innovative thinking.

Unfortunately this approach is not always successful for the fire service. You are trying to build a cohesive team that performs fire ground tasks in a predictable manner. Opportunity for innovation can, in some ways, be a downfall when it is created through freelancing. For every issue you face, you must first ask the question of whether you want movement or motivation.

The difference between movement and motivation is the difference between a linebacker and a lineman on a football team. You want the linebacker to be all over the field, creating chaos, and jumping on every ball. That is motivation. It is creative, unpredictable, and chaotic. On the other hand, a lineman should block designated players to open predictable lanes and to move when and where he is told. That is movement.

For issues in which you want movement, you want a well-designed, easily understandable playbook. In the fire service, this playbook is made up of your rules, regulations, and standard operating procedures (SOPs). The challenge is to define movement without killing motivation. It is a balance that can only be achieved through experimentation. Look at your rules and regulations as something fairly static that will not change often, if at all. Your SOPs and Standard Operating Guidelines (SOGs) are fluid and can and should be changed easily as lessons are learned. Remember, the more you communicate the more they will understand.

In-House Committees

This is an easy and straightforward tactic to grasp. It is so self-explanatory that to explain it in any detail is almost insulting to the reader's intelligence. Then the question arises, if the need for this tactic is so evident, why have so many departments failed to implement these in-house committees?

Most departments have an accident review committee to review driving infractions. Expand the role of the accident review committee. Make this committee not only responsible for review and remedial action on vehicle accidents, but fire ground and in-station injuries and accidents as well. The next step of including prevention and health activities is so logical that the committee will probably recommend it to *you*.

Safety awareness is an attitude that must be cared for, fed, and fostered. It requires a proactive approach rather than a reactive approach to achieve any real success. It is also an issue in which a peer committee can be more effective than a top-down approach.

The Labor vs. Management Relationship

To chiefs who are required to have this relationship because of collective bargaining laws, this may seem to be a no-brainer. Think of it rather as a tactic in which you should have a *good* labor versus management relationship. There are also many chiefs in right-to-work states, some with strict laws written against organized labor within public safety, so that the chief is prohibited from formally recognizing the labor organization. To these chiefs, the need for a good relationship may not be as self-evident.

In either case, a good mutually beneficial relationship between labor and management is crucial. Do not fall into the trap of only thinking of this relationship once a year when contract negotiations must be done.

Labor, more directly organized labor, must be a consideration in your daily deliberations. By having a good relationship in place, you can have necessary open and honest dialogues before finalizing a decision. In making a decision or implementing a change, it is much better if any fighting is done early in the process. This creates a team approach, rather than an adversarial environment.

Adversarial environments are almost always bad things for you, for labor organizations, and for the department—no one wins. Any situation that requires a winner for resolution also requires a loser. Losers seldom forget the loss and seldom accept it. The loss is most often considered the loss of a battle, which cements the need to win the war. Professional "wars" for fire chiefs are a destructive and distractive phenomenon. Good ideas and chiefs are destroyed by wars, so avoid them whenever possible.

Union Contracts

The requirement to follow union contracts is not only a legal issue; it is an issue that will define the relationship you have with the labor organization. You need to follow the letter of the contract, as well as the spirit of the contract. Too many chiefs make the mistake of trying to live only by the letter of the agreement and not the intent. The chief who takes this route is setting up future trouble. This kind of behavior only sets the stage for more acrimonious negotiations.

The labor organization is not unaware. If you try to outsmart the organization by living only to the letter of the agreement, you will create an environment of distrust where each negotiation turns into an adversarial meeting. Adversarial labor and management relationships are bad for everyone involved. It hurts the labor organization, it weakens the chief, and worst of all, it is detrimental to the mission. Labor organizations are like any other group of people. The vast majority simply wants to do the right thing and get a fair deal. They want to be treated with dignity and respect. There may be a small minority with the sole purpose of creating trouble. These are the anarchists, and they only want to fight. They are destructive, and they are also easy to spot.

The real problem is when one of the anarchists gets elected to office. When this occurs, the chief, the union, and the department are in for a roller-coaster ride. There is little that you can do in this situation other than maintain the high road. Although the temptation is to react, you must stay above the fray. The easiest way to do this is to remember that it is a temporary problem. The reasonable people in the union will tire of the constant bickering and non-productive leadership and replace the ones causing trouble with people of goodwill. Your task is to remain calm, do the right thing, and wait the process out.

There is another type of union member often mistaken for an anarchist—the revolutionary. The revolutionary is willing to slash and burn, but these people have more on their mind than simply creating trouble. They want change. They can also be a good influence for the chief, the union, and the department, if, and only if, you can recognize them.

The other large mistake you can make when dealing with labor organizations is to run scared. By doing so, you are abdicating the rights and responsibility of management. There is a simple equation that must be remembered in labor–management relations. Labor organizations have rights but so does management. If either side abdicates their rights, then the equilibrium required for a good relationship is lost.

Management's job is to require certain performance standards and to deliver through the work of others. This has to be done in a manner that is both cost effective and efficient. The job of a labor organization is to keep management in check. Together, the two sides, which do not always agree, share one important job in the fire service: to look out for all fire fighters.

When anyone in this complex relationship forgets their job and allows it to become personal, the entire process falls flat. The smart chief shows respect to both labor organizations and management. Both roles are honorable endeavors and should be treated as such.

Personnel Skills

Good fire fighters are generalists. Fire fighters of today are renaissance men. Most fire fighters possess many skills that have nothing to do firefighting. They may be craftsman such as carpenters, plumbers, or electricians. They may be more cerebral, expressing it as sculptors, poets, writers, singers, and musicians. The one thing all fire fighter's share is an incredible can-do mentality.

Most fire fighters love nothing more than the challenge of being told they cannot do something. This is a certain invitation to take the challenge head-on and succeed. This trait gives an unbelievable advantage to the creative chief who can tap into these skills and attitudes. However, most ignore this ability and crush motivation and freedom.

Think about this from a business perspective. What executive would not want a workforce that is willing to do anything to achieve its goal? What executive would not want a company where the employees feel such a bond toward one another that they care for each other when they are sick, down on their luck, or broke? What executive would not want a group of people with diverse skills that can succeed in almost any task?

The fire chief is handed this workforce on a silver platter. In a combination department, this gift is amplified by the unbelievable diversity of members within the volunteer ranks. The chief will find computer experts, bankers, nurses, and every other occupation there, waiting to be put to good use.

If you use the tactic of being familiar with your employees, their hopes, and their dreams, you will also recognize their skills. Utilize those skills and the diversity of individual backgrounds to create the strongest organization possible.

Enhance the Department

You can use volunteers in your fire department in many ways. Nonemergency volunteers can add a dimension to your department that may literally mean the difference between success and failure. There are obvious routes such as community emergency response teams (CERT) and fire corps. But have you thought of partnering with the American Association of Retired Persons (AARP) or other groups? If you use these resources, you will find a wealth of experience and skill. You will find volunteers skilled in administration and business. You will also gain wisdom through experience. Firefighting is an extremely physical skill making it a young person's game. Because we are a youth-oriented profession, we tend to be caught in a never-ending cycle of reinventing the wheel.

To use the military corollary; recognize the value of seasoned, grizzled veterans and the benefit they provide to the youth. We could use the wisdom that can only be understood by actually living it. Age and experience does not guarantee wisdom. However, wisdom cannot be achieved without age and experience. You would do well to remember this simple fact and seek out the wisdom that all organizations need.

Mix All Members for Training, Association, Respect, and Teamwork

Though combination departments are common, how many are really combination? On close inspection, you will often find that what is called a combination system is in reality a dual system, or worse, a "duel" system. This is when a paid department and a volunteer department coexist within the same space but are not interwoven.

A true combination department melds these two systems into one cohesive system. You cannot achieve this by only instituting a system of separate but equal. If you want a combination department, you have to build the appropriate culture.

This is a difficult and long-term process. It may require a generation to fully implement. The question for you is this: are you ready and committed to take your organization on this course? If you are, then you must begin somewhere.

A journey of 1000 miles begins with the first step. A good first step is to mix all of your members. They should know each other and build relationships with one another. By taking this first step, you will create a generation that knows nothing but a combination approach. That generation will then acclimate the next batch of recruits. In time, your entire organization will understand that their culture is a blend of both volunteer and career members.

Utilize Employees as Mentors, Proctors, Big Brothers/Big Sisters, and Marketing

This tactic goes hand in hand with utilizing all of the talents that your employees bring. One of the main talents of your members is the ability to pass on knowledge, skills, and wisdom that they have gained. Yet some chiefs overlook this tool and rely only on officers to mentor and train. These chiefs miss out on the opportunity to build a department that values learning and growth for every member of the organization.

Any one of your employees may serve as a mentor. Any fire fighter can also serve as a big brother or sister. When institutionalized, this process can become a natural part of your organization's daily life. This tactic is a powerful tool in building relationships between your career and volunteer members. A career member whose big brother was a volunteer will undoubtedly view volunteerism in a positive way. Conversely, the volunteer who was mentored by a career member will better understand and respect career members. This tactic is a surprisingly potent way to ensure that the next batch of members, both career and volunteer, do not fall into separate cultures.

The implementation of this idea needs to be structured for success. If the values and lessons are not properly defined, then there is a risk that the wrong lessons and values will be taught.

Visit Stations

The fire service is built on a paramilitary model. This model provides the structure and strength needed for the fire ground. It is the best system model for the fire service. However, it is rigid and structured when compared to business models in use today.

As with any system, the paramilitary model has its weak points. One of the major weaknesses that you must overcome is that a structured rank system tends to insulate the chief. There is a common complaint that the chief is out of touch and does not know what is going at the member level.

The sad fact is that if you are not vigilant, then the complaint will come true. You must realize that everyone in the fire department is a "fixer." We pride ourselves on handling every situation that comes our way. If you only rely on your formal communication chain, you will only hear of problems when your subordinates have found them to be unfixable. Until then, you are insulated from the daily trials of your members, because, it is being handled. You need to realize that if a problem is unfixable for your subordinates, then chances are by the time it crosses your desk, it is impossible to fix. Thus, you need to be visible and out in the field on a regular basis. You must fight the irresistible pull of the black hole that is the office. You must visit your stations, engage your members, and see firsthand what is going on. General Patton had a rule that is applicable to today's fire service: all commanders must visit the front daily. You also must visit your front. This is crucial if you want any chance of having an affect before problems become unfixable.

There are several pitfalls that you should avoid when being active in the field. First, you have to guard against the natural temptation to micromanage. Remember you are a fixer too, that is your background, and your nature is to jump in when you see a problem. Temper that urge.

Also, you must be ready to hear and see the good, the bad, and the ugly. If you seek the truth you shall find it, and you should be prepared. It can ruin an otherwise perfect fantasy of how well your department is doing and how all of your ideas are universally loved and respected.

Third, watch out for the red rag syndrome. Every fire fighter knows this syndrome. It is based on the idea that if you stay below the officials' radar, then they will go back to their offices and leave you alone. So what do fire fighters do? They carry a red rag in their back pocket, and every time they see an official they stop and start polishing whatever is closest. When asked how everything is going, the pat answer is, "Things are great, Chief." The chief can then retire to the cloistered office reassured that all is well in the world. When you fall for the red rag syndrome, you are deluding yourself and setting the organization up to fail. You are also wasting your own and your employees' time in the facade of "all is well." You need to see everything, and you must insist on living in the real world when you visit the firehouse. You must guard against the syndrome by asking tough questions, engaging your members, and spending the time and energy to understand their difficulties.

The Third Commandment: Thou Shalt Provide Good Organizational Management

Think of your organization as a living, breathing entity. You need to nurture it. Your organization is a collective of individuals. At times, it may seem as if it has its own independent personality. You need to recognize this phenomenon and be prepared to attend to it as needed. As is the undercurrent of all of the commandments, the successful chief is proactive and stays ahead of the organization's needs. You must always place the department ahead of your ego.

■ Tactics

Have Clear Lines of Authority and Responsibility

An organizational chart needs to be more than boxes and lines on paper. It should be the blueprint for how your organization makes decisions and takes action. You should ensure that there is a clear understanding of authority and responsibility throughout the organizational hierarchy.

Clouded or unclear organization in a highly structured environment like the fire service will deliver confused and uncoordinated actions. Like an orchestra, the different areas within your organization need to be coordinated to produce the perfect piece of music. You are the conductor of this orchestra. If the horns are playing one song and the woodwinds are playing another, the piece will sound horrible. In a combination system this can be doubly important because you are blending two different styles of music into one piece. It is like trying to combine a jazz band and a classical quartet. If not blended well, it is noise.

Volunteer Fire Departments

Volunteer fire departments are an interesting mix of the paramilitary model and the fraternal social model. Although these fire departments have fire officers who must be able to give orders, they also have infused democracy into the organization. This mixed model approach is much like the all-volunteer army at the start of the Civil War when battalions and companies elected their officers. The military abandoned this model and moved to the model of officers being promoted based on skill and ability rather than by a political process. That abandonment led to even greater success for the US Army.

You may find yourself with an organizational design that is a compromise between the volunteer model and the career model. You may have an officer corps that you have no control over because the officers answer to a political process rather than a promotional process. You may also find yourself mired in committees, advisory boards, and other groups. This is a scenario that is fraught with danger for you and is a recipe for failure for the entire fire department.

Although difficult, the best cure is to cut this process from your organizational design. Even though it may sound horribly egotistical, an organization can have only one chief. Any other arrangement leads to confusion and fighting between factions. Many combination departments have not corrected this design flaw. It is precisely this reason that so many combination systems are in constant conflict.

Although it is always good to have participatory management, where all of the members have input and are engaged and motivated, it is imperative that there be only one chief. This is a fine line, but it is a line that must be traversed.

Monthly Reports for Accountability

To manage a mission-centered system, a fire department must not only know its mission, but also how well it is meeting its objectives. Monthly reports can be used as a guide to recognize how well the department is performing and where it needs to improve. For this system to work, the organization must have clear goals. The reports need to measure specific points of accountability by company, by shift, and by individual.

Conduct Quarterly or Biannual Reviews

Most fire departments have a formal periodic review of their service delivery systems. However, many only use this tool every five years. In today's environment, waiting that long between reviews creates a situation where the fire department can only react rather than act. It is always better to act than to react. You will find that this statement is applicable to every aspect of emergency services from the fire ground to the council chambers. In any setting, you are better served by being a proactive force.

The only way you can be proactive is to constantly evaluate (a process sometimes referred to as metrics). Your system for delivering services is one of the key components that enables you to perform your mission. Contrary to popular belief, your system is not static. It should be a fluid and changing system that evolves on a daily basis to better itself. It should also be a system that not only allows, but celebrates, experimentation.

Nothing is constant except change. It is folly to think that the way we do our business is the one thing that does not change. The fire service needs to recognize this basic fact and create a system that is fluid to change.

The best way to accomplish this is to formalize a review of your system on a regular basis. Try new ideas, discard the failures, keep the success, and then start over again. You need to be an active participant in this process and not relegate something as important as this to a committee. Committees are fine tools for gaining advice and input but it is foolish to abdicate the responsibility of a final decision to a committee. You are much more than an administrator who delegates throughout the management structure. Ultimately, the responsibility for the decision is yours, and the decision must remain yours.

However, it is a foolish chief who does not listen carefully and consider the counsel of committees and officers. The wise chief will more often than not take the advice given. The wise chief will also remain firmly in his leader position. In the end, the final decision is the chief's alone.

Utilize Consumer Surveys

The fire service is a closed society that is difficult for the public to understand. We have our own language, our own code, and our own brotherhood. None of these translate well to the life of an insurance agent or a day trader. Yet the insurance agent, realtor, and day trader are the people we serve and also the people who we depend on for our support. The fire service needs to be a part of, rather than apart from, the community. It is important to understand the pulse of the community and how you may learn from the community's attitude to guide your department. A sound way to accomplish this is through surveys. You should survey the people who have actually used your fire service. What were their impressions of your members and their actions? Was the response timely? Did the response meet the needs of the user? You should also routinely survey your business community.

Have Ties to Outside Organizations to Maintain Support

The importance of membership in community organizations is often underplayed (see Chapter 13). By being actively involved in associations and advocacy groups, you can gain so much. These organizations are incubators of thought. By having your members involved in these groups, you guarantee an influx of ideas and enthusiasm on a regular basis.

It also allows you to maintain your proactive stance. If you wish to have any control of your environment, you must be involved and help move that environment in the direction you wish. The time and the effort it takes for your members to be involved in these organizations is a miniscule price for the rewards.

Another less recognized advantage is the power that knowing people brings. By being involved, you will build the relationships discussed in the first fire commandment. You do not want to be meeting your state and federal officials for the first time when you need to ask for a favor. By being members of these community associations, you can often meet these people in a friendlier environment.

Host In-House Events

You may have a great idea that is not well received in your community. This same idea might be presented by an outside expert and embraced as a wonderful idea. Bringing in outside expertise can be extremely beneficial to you. It can allow you to infuse new ideas and change and can build your credibility as a visionary.

This is an especially powerful tool when you need to build political support to implement a necessary change. By allowing politicians to hear your message from recognized experts, it bolsters the argument and can be the difference between success and failure. This approach is especially important in the inbred department. An inbred department not only resists outside ideas, but it rejects them at face value. This is a dangerous kind of department, and the outside influence may be the ticket to spur thinking and innovation. Innovation is the lifeblood of an organization.

Seek Alternative Funding

While discussing the importance of innovation, you must also consider how to fund this new thinking. You may find that local government structures are reluctant to spend money on experimentation. Such initiations may be seen as dubious at best. However, the resourceful chief will find ample opportunity to fund this activity through grants. By using alternative means to fund these initiatives, you will be able to spur the activity you want in spite of the constraints of local funding.

You will also find that by doing this once or twice and having high-profile successes, the local opposition to you taking risks will be greatly reduced or may melt away altogether. You will find that your manager and officials will look favorably on your initiatives and even fund them when times are good and there is room in local revenue.

The Fourth Commandment: Thou Shalt Provide Sound Fiscal Management

A lack of sound financial management and control ruins more chiefs than almost any other mistake. No mistake is less forgivable or easier to spot by local administration and the public at large. Money makes the government function, and if you do not tend to the fire department's budget, you will find that your department quickly deteriorates.

Finance can also be seductive and dangerous in its charms. The larger the fire department, the more this is true. Sometimes it seems that budgets in the millions of dollars are only numbers on paper. There is a tendency to forget that these numbers are actually real money and not one cent of it is yours. Add to that the reluctance of your staff to disagree with you, and you have set yourself up for a major fall. The only way for you to protect yourself is to put in place sound fiscal policies that apply to everyone in the fire department, including yourself.

Sadly the fire service is not immune from the ills of the outside world. People may steal and abuse funds, and you must put in place processes that reduce the chances of this happening as well as deal with it when it does in a quick, severe, and efficient manner. The simple requirement of a receipt for everything enforces a policy of, "spend it like it is your own."

■ Tactics

Utilize In-House Personnel

There are mixed feelings about this idea. Some people will extol the value of using your own people to hold costs down. Others strongly oppose this message, and in some areas of the country, labor agreements may make the idea impossible. The best advice is to use personnel in roles other than their primary job judiciously and when necessary for success. However, doing this as a general practice may be detrimental. The bottom line of this tactic is that some things can only be accomplished cheaply, and some things should never be done cheaply. You should use this tool sparingly and cautiously, without ignoring that it is available to you.

Establish Sound Practices

Regarding finances, fire departments can learn much from the business world. Purchases and payments should require approval of more than one person in your organization. The same is true with the acceptance of any money. Authority should be limited and spread among people only as necessary to assure that business can be conducted.

Inventory systems should be in place and equipment should be tracked from order to disposal. Inventory should be tracked by assignment, and individuals should be responsible and held accountable for specific pieces of equipment. Auditing must be a regular and ongoing process. Purchases, expenditures, and revenue should be reviewed regularly by personnel not involved directly in those processes. External audits should be conducted annually to ensure sound practices.

Exercise Fiscal Restraint

Many finance personnel have a legitimate issue with the fire service because we generally do not care what something costs. We are totally results oriented, and we believe that no one does anything as important as we do day in and day out. So who cares what it cost? However, the rest of the world does care. The finance personnel are the ones who must explain our excesses to the rest of the world, and they are naturally resistant to the position the fire department may place them in. The well-served chief will not only remember this perspective, but respect it.

There are times when we can spend budget money without regard to cost. However, those times are few and far between and generally are only during a major emergency. All other times you need to remind your purchasing personnel to not spend excessively.

A recurring issue in volunteer systems is a weakness in the paper trail. It is imperative that every purchase has a receipt. In today's business world, this error has hurt many a fire department.

The Fifth Commandment: Thou Shalt Not Ignore the Need to Manage the Infrastructure

While remembering to keep your member, communication, and financial matters in order, do not forget to keep your physical structure in order as well. Fire and EMS services are somewhat unique in that our workplace is also a place where we live.

■ Tactics

Maintain and Improve Your Space

The people who are stationed in your buildings will care for them if you simply appoint them the tasks. More importantly *enable* them; give them the tools and the ability without too many hurdles, and they will accomplish great things. The secret is basically a matter of human nature—after all who does not want their house to be the best on the block?

Provide Positive Reinforcement

Take the time to recognize the achievers. Give awards or special recognition to those stations that are properly maintained. Fire fighters live in a social, competitive culture that sometimes resembles a sibling rivalry and relationship. Use this dynamic to achieve results. By recognizing the achievers, you can motivate the underachievers to be better through positive means rather than by only punishing the offenders.

The Sixth Commandment: Thou Shalt Integrate Your Department into the Community

The idea that a fire department should be integrated into its community should not seem like a new vision to anyone. The problem for most chiefs is that although it is not a new theory, they have not put it into practice. As a whole, we often talk without action. Once the volunteer fire departments were the hub of many communities, but today we have allowed our departments to become isolated from their environment. A recurring theme of this chapter is that we must relearn some of our core values if we are to be as successful as we were in the past. Being a part of your community rather than apart from your community is one of those lessons.

■ Tactics

Be Actively Involved with Community Leaders and Activities

There are many community groups, leaders, and activities waiting for you to take advantage of the opportunities for inclusion. It may be the Lions Club or the Ruitans or other civic groups. It may be as simple as community associations. All of these groups have leaders who influence your community and who can be a major source of support for your organization. These groups also have normal activities that can be a source of exposure for your department. Get your members out of the station and involved in the community.

Market Your Department

Clearly, this is simple statement, but most departments fail to do it. Any other business understands that a key to its success is marketing themselves or their product. They also understand the importance of branding. Branding is the process of creating a carefully designed reaction to a brand or identity symbol. Everyone recognizes certain brands with a glance. Why? They have seen the logo everywhere with a consistent message accompanying the logo. Has your department branded itself?

Think about all of the possibilities: coffee cups, hats, shirts, key chains, trading cards, you name it and it is out there. Develop a simple idea that you want your community to associate with your fire department. It may be dependability or quality or any other concept that you believe is applicable. Take that simple ideal and apply it to all your actions and methods of communication.

Being a recognized brand in your community will create large dividends. You will see better support in budgets, equipment, and volunteer recruitment, all as benefits to branding.

Make Your Buildings Community Centers

Open up your buildings to the community for events that bring the public into your home. It could be hosting birthday parties while conducting fire safety demonstrations to opening up your meeting rooms for groups, meetings, and activities. Think of the ways you can offer up your services to draw the community's focus to your fire department.

The Seventh Commandment: Thou Shalt Plan

There are a million clichés about the need to plan. You have probably heard most of them, for example, did we plan to fail or fail to plan? However, clichés diminish the importance of planning by trivializing the process. Planning is essential. Planning brings people together and creates important consensus on the major issues facing your department. Planning builds support, and creates teams out of disparate groups. Planning is the foundation for progress.

Keep in mind, the plan is not the end product. It is simply one of the many products of the process of planning. The process itself is the most important, and the savvy chief will recognize this and use the process for all its benefits.

■ Tactics

Develop Immediate, Intermediate, and Long-Term Plans

Develop plans with reasonable timelines that are achievable. Do not get carried away with long-term planning; the truth is that no one's foresight is good enough to see much more than 3–5 years. Today's world moves so fast and the environment changes so quickly that even the best visionary would be incorrect in trying to predict trends too far in advance.

There is book after book about planning processes and methodology. However, few recognize what good plans can really accomplish for a fire chief. Good plans do several things; first, they lay the framework or the blueprint for where you are going or what you are building. In a combination system, this equates consensus, which is hugely important to the department's success. Secondly, it can be a powerful tool for building support. If you treat your finance personnel, administration, and elected officials as stakeholders and have them help you develop the plan, then you automatically gain their support for the means necessary to implement the plan.

However, be careful not to make the beginner's mistake. Be totally transparent while you are planning. If an idea is going to be expensive, tell the stakeholders immediately. If something is controversial, then tell them who and what interest groups will bring opposition. Nothing could be worse than for the supporters to feel like they were lead astray or tricked. You will find that most politicians, administrators, and accountants want to do the right thing and are willing to take some risk as long as they understand what they are getting into.

Align Your Budget to Your Plan

An important part of planning is the consideration of resources. Money is a necessity for almost everything you will want to achieve as chief. Without it, the task will be almost impossible. Build all of your budget requests around your plan and tie them directly back to your plan. This simple act will solidify the support of your budget because the decision makers will be able to see that you planned ahead.

Build Your Immediate Plans with Quick Wins in Sight

Nothing breeds success more than success. If you develop your immediate plans with a few simple, quick wins in sight, then you have set the stage for more success in the intermediate and long-range sections of your plan. Your support will also be concrete when it is built on a foundation of success.

Government accountants want to know that the money is used as intended and for good. Most also have a keen sense of frugality—they dislike what they would view as waste. Administrators want consistency and predictability—they dislike the unknown and unpredictable. Politicians like to win—they dislike the idea of losing. When you build your plans with your budget aligned and offer predictable results and successes, then you have provided the key elements required for your plan to succeed with the government. When one plan works, it is easy to then propose new plans and directions.

The Eighth Commandment: Thou Shalt Aggressively Pursue Public Relations and Communication

Good public relations are a recurring theme in this chapter for one simple reason; public relations are extremely important. Although most people do not like to admit it, perception shapes reality so much that it almost equals reality. You may have the best, most efficient department, but if no one knows that, you have nothing. Without public relations, you will have no support, success, money, or any of the other tools necessary for your fire department to thrive. You will also be constantly overshadowed by the government agencies that do understand this important concept.

■ Tactics

Brand Your Department

As said earlier, branding is an extremely important basis for marketing. Branding creates a recognizable trademark that speaks to the essence of your organization. Branding is like the brands on cattle—it identifies you. Take the time to create a brand that speaks to the message you want the community to know about your department. Remember to bring what you do down to its essence.

Once you have created your brand, make sure the trademark or symbol is on everything from the trucks, to the building, to the letterhead. The brand logo should be consistently applied as the center of everything that identifies your fire department. It should be on all of your free promotional items, on all of your press releases, and on all your prevention materials.

Submit Regular Articles to Local Media

The local media will often be glad to offer you a public outlet for your fire department. This could be in the form of weekly, monthly, or quarterly articles; it could be "Fire Fighter of the Month" snapshots; or it could be routine radio or television spots. The only thing limiting you is your imagination. Whatever venue you use, the concept is to keep the brand continually in front of the public. They should see you and know who you are.

Appoint a Public Information Officer

Most fire departments understand the concept of the public information officer (PIO) and utilize it well in emergency situations. What is missed is that the job can and should be used for much more than emergencies. A good PIO is the image consultant for the fire department. The PIO continually keeps the brand in the public eye, keeps the department on message, and shapes the community's perception of your department.

A good PIO will build relationships with the media that will offer returns on a daily basis and in emergency situations. The establishment of a good public relations program will create the community support you wish for. It will also get your message out and understood during critical times. Consider using an outside firm, or utilize a fire corps concept to assure that a PIO is a fixture in your system.

The Ninth Commandment: Thou Shalt Develop Strong Intergovernmental Relations

Although we easily recognize the importance of community partnerships, we often overlook the in-house partnerships that are also vital for achievement. We've spoken about some of these in previous commandments, but it is important to consider intergovernmental relations alone.

■ Tactics

Think of the Commonly Overlooked Partners

We recognize the need to have strong relationships with financial and the administrative officials, as well as the politicians who reside at the top of the governmental food chain. However, there are other partners within the government who can be extremely useful. Perhaps an alliance with your purchasing department will be the key to your success. Have you thought about the solid waste department? When you have a major storm or a tornado and debris removal becomes a huge part of your job, you will realize that everyone is important to you. No fire department has the resources to handle everything that may come its way today. By having preexisting relationships and strategic partnerships with some less commonly thought of governmental agencies, you can draw on necessary resources when you need them most.

The Power of Regional and National Groups and Associations

Today, a fire chief is so busy that you have to limit your involvement in some activities. Speak to any chief, and you will find the common theme is that there are not enough hours in the day. Most chiefs seem harried and overworked. So why would you take the time to be a part of regional and national associations? However, it *is* valuable to you. First, it allows you to help craft the future rather than react to it. Second, networking is powerful and critical. In the currency of today, knowledge is power. The knowledge you gain from these groups and networks is extensive. The ability to make a phone call and get an answer is valuable.

Remember, administrators like consistency and predictability. Most chiefs have pitched an idea or proposal and been greeted with, "What are other departments in the region doing?" If you want to know what other departments are doing, be active in your regional groups.

Also, if you ever need a powerful ally in difficult times, bring in an expert from another part of the country who has confronted the same issue you have and who successfully met the challenge with the tactics that you are proposing. This creates instant credibility for you and your proposal. These allies can also often say what you need to say but cannot.

The Tenth Commandment: Thou Shalt Not Ignore Personal Professional Development

While you are busy tending to the needs of the department, the community, and your members, you might overlook the needs of your own professional development. Evolution is a simple concept—you either grow and evolve or die and become extinct. As much as we would like to believe that we have reached the epitome of personal development when we earn the vaulted five bugles, the truth is we have not. The choice is yours to make, to continue to professionally grow and develop, or wither and grow stagnate. One way leads to success while the other does not.

■ Tactics

If It Is Good Enough for Them, It Is Good Enough for You

What is your fire department's emphasis? Safety? Perhaps you require every fire fighter to take "mayday fire fighter down" classes and rapid intervention team (RIT) training. If so, you should also take these courses. Maybe you are training your department on leadership. Take the same training you require your members to take.

Reach Out for New Opportunities

Look for new areas of interest. Take courses that will offer you insight into other areas of government. The options are truly limitless. The important thing is to continue to learn and expand your thought. You must be many things. You must also have a flair for the intellectual. The thought process is as important as knowledge. Become a learning person, attend the National Fire Academy, join and be active in your local, regional, state, and national fire chief associations. Attend conferences, but do not be a "tire kicker"—become engaged and attend the learning experiences that every conference offers.

Summary

These are simple rules that you can follow. These simple rules are a map to success that most people only gain through years of experience. As with all rules, these rules are meant to be living and growing guides. Do not simply follow the rules blindly; use them to guide and help you.

The ten commandments of success for the fire chief are the following:

1. Thou shalt build a strong relationship with managers and elected officials.
2. Thou shalt provide sound personnel management practices.
3. Thou shalt provide good organizational management.
4. Thou shalt provide sound fiscal management.
5. Thou shalt not ignore the need to manage the infrastructure.
6. Thou shalt integrate your department into the community.
7. Thou shalt plan.
8. Thou shalt aggressively pursue public relations and communication.
9. Thou shalt develop strong intergovernmental relations.
10. Thou shalt not ignore personal professional development.

CHAPTER

10 Strategic Management of Change

David B. Fulmer

Case Study

Three communities, Alphaville, Betaville, and Zetaville, run a joint emergency-medical district, which operates as a third service for the delivery of paramedic treatment and transport services. This joint emergency-medical district has been in existence for over 2 decades and was formed at a time when all three communities were more rural than they are now. Over the last decade, shifts in demographics have increased the populations in all three communities, but mostly in Alphaville. The joint emergency-medical district has an oversight board composed of elected officials and citizen appointees from each community. The allocation of expenses for the joint emergency-medical district is determined using a percentage formula based on community size and use of services. Alphaville, the largest community, accounts for 65% of operations and therefore is responsible 65% of the costs.

In 1984, the Alphaville fire chief conducted a feasibility study on the consolidation of the joint emergency-medical district with the Alphaville Fire Department. This study revealed a potential for expense savings by reducing the duplication of administrative services. Also medical personnel could be placed in the highest demand areas. Although the fire chief of Alphaville had enough political traction to initially research the issues and draft recommendations that would alter the current system, which would result in increased efficiencies and financial savings to Alphaville, there lacked enough political will and capital to alter the current system. This lack of political will stem from the overwhelming willingness of Alphaville to subsidize the operational costs and service delivery of Betaville and Zetaville. When the decision to alter the existing model needed to be made, the key decision makers found it more beneficial to continue with the regional effort at the cost of Alphaville. The recommendations provided by the Alphaville fire chief were never implemented.

In 1986, the Alphaville fire chief was asked to provide input regarding the consolidation of fire and emergency medical services (EMS). Again, recommendations were made that would result in cost savings by consolidating the two organizations and training personnel in both fire and EMS. The recommendation suggested that Betaville and Zetaville contract for emergency medical service with Alphaville Fire Department. Similar studies were conducted in 1992 and 1997, after which similar recommendations were made, but never implemented. Once again, when the decision to alter the existing model was considered, the key decision makers felt it more beneficial to continue with the regional effort at the cost of Alphaville.

Although the faces of key decision makers changed, the political climate and the willingness to subsidize Betaville and Zetaville remained constant. The political decision makers of all three communities had sweat equity and political capital in the existing model along with a sense of being vested. The overall political assumption was that there was no reason to tweak a system that benefited all three communities, even though it was not equitable to all.

In 1999, the new Alphaville fire chief was directed to study his department's participation in the joint emergency-medical district. Research revealed that there was a growing ability within the joint emergency-medical district to staff a second ambulance. Also, there was an increase in the requests for mutual-aid medic units to handle calls within Alphaville, which resulted in increased response times and loss of run revenue. The new Alphaville fire chief recommended that Alphaville withdraw from the joint emergency-medical district and that the EMS functions of the joint district be brought under the scope of the Alphaville fire department. The recommendation outlined the cost savings to Alphaville as well as the increased ability to respond to emergency incidents with appropriate staffing and decreased response times. These recommendations suffered a similar fate to the previous studies and recommendations. Although the fire chief of Alphaville had been directed to initiate the study by the mayor, there was a lack of understanding or appreciation for the history of this issue. The fire chief had researched the issue, was aware that similar attempts had been unsuccessful, and yet was obligated to objectively research the issue and render recommendations to the mayor. The fire chief was explicit in the report about the history of the topic, the commons themes that were identified through numerous studies, and the political and organizational ramifications of subsequent studies and failing to implement recommendations. Another obstacle was the EMS chief, who had no ambitions of moving the EMS district underneath one of the two fire departments because it was viewed as a loss of authority and autonomy for their organization. In addition, the joint EMS district's oversight board would have ceased to exist, which was also viewed negatively.

Introduction

Unfortunately for many fire departments, the old cliché, "180 years of tradition unimpeded by progress," still rings true. This has been the root cause of many issues faced by the fire service in America, specifically in volunteer and combination fire departments. There are plenty of examples in corporate America of businesses that failed to heed the signs of impending changes to the market only to be forced out of business or left scrambling to regain market share. For example, Goodyear and Firestone were once corporate giants in the tire industry and were forced to retool their entire corporations to compete with changing technology and foreign competition.

A common definition of insanity is said to be, "doing the same thing over and over again and expecting different results each time." In some respects, the fire service is the epitome of that definition. Examine the changes that have affected this profession, such as protective clothing, enhancements to apparatus and equipment, and other health and safety measures. It is clear that the majority of those changes have been mandated or come at the cost of the lives of fire service members. With that said, within fire service there are still those who fail to embrace changes for safety and welfare and who long for the days of leather lunging, open cabs, long coats, three quarter boots, and beer in the fire station. It is incumbent upon the leadership of every agency to assure that he or she is making every reasonable attempt to assure that the business and operational model is working toward improvement versus status quo.

There are many issues that are catalysts for change in fire departments, including changing trends in the field of emergency medicine, new roles in homeland security, cultural diversity in the workforce, and changing demographics. The failure of fire departments to proactively identify a changing environment, implement internal change, and institutionalize change, results in reactive approaches to changes, which are typically detrimental.

Change Management Model

Fire departments are accustomed to preparing for emergency-response incidents. Fire departments engage and complete many steps including preincident response plans, plans for buildings, mutual-aid agreements, mutual-aid box alarms, initial and ongoing evaluation, after-action critiques, and quality-improvement and quality-assurance activities. Although proficient at implementing these steps on the fireground, many fire departments fail to recognize the importance of these steps in nonemergency activities. The steps that departments are accustomed to implementing at emergency incidents are the same steps that departments can and should implement to address nonemergency activities.

The change management model is a four-step process taught in the National Fire Academy's Strategic Management of Change Course and examines the following steps:

- Analysis
- Planning
- Implementation
- Evaluation and institutionalization

■ Analysis

The first phase in the change management model is analysis. This involves an introspective look at the fire department. To review how this is done, see Chapter 3. Often times, the analysis phase serves as a reality check for the fire department to reexamine the current operations and weigh them against performance measures such as mission and vision statements, internal and external expectations, and other benchmarks (e.g., response times, staffing levels, budgets).

By examining the case study, we can walk through the analysis phase. The case study states that similar studies were conducted over several years. The common theme among the studies was their failure to effect change, which could likely be related to the analysis step. Each year, the catalyst for change was the consolidation of the third EMS service delivery agency into the Alphaville department. One of the obstacles in the case study was a strong political will among the elected officials of all three communities to maintain the joint EMS district as a symbol of regional cooperation. Another obstacle was the EMS chief who had no ambitions to move the EMS district underneath one of the two fire departments because there would be a loss of organizational autonomy that had existed for many years.

■ Planning

The second phase of the change management model is planning, in which fire departments evaluate the steps necessary to move the department from its current state to the anticipated state. This process includes several steps, which the success of the entire project hinges on.

One of the steps is the evaluation of forces that positively or negatively affect the change process. For sake of discussion, we will classify the forces as *enabling* (i.e., for the change) and *disabling* (i.e., against the change), both playing equal roles in the success or failure of the change movement. These forces can be both internal and external forces. Enabling forces are those that are in support of change, and they often serve as catalysts in the change process. Disabling forces are those that serve as barriers, roadblocks, or hidden pitfalls to change and prevent or impede the change process.

Vision Statement

A vision statement serves as the goal for the fire department. It states the direction for the process, which ultimately results in the desired change. Thus, it is important that the vision statement be clear, accurate, and achievable. As a road map, the vision statement needs to be flexible enough to allow for alterations along the way that are necessary because of unforeseen roadblocks. The vision statement also needs to be clear and consistent to facilitate the navigation of the fire department from the known to the unknown.

Determining a reachable goal through the use of a vision statement is important because it serves as the "light at the end of the tunnel" for the team and the fire department. The vision statement is the yard stick by which progress is measured. The vision statement should be developed by the person in the department who is avidly working for change, which is most likely the fire chief or his or her designee. After all, the vision statement serves to recruit team members, invigorate them, and sustain their passion throughout the change process. There is no prohibition against the change management team making adjustments to the vision statement.

Assembling the Team

Like any team, the key components are the players and the coach. There are several approaches to developing a change management team, so it is for each department and its members to embrace the manner they feel is best suited for them. Fire departments can choose a team that is composed of:

1. Chief officers
2. Chief officers and company officers
3. Representatives from the entire organization (e.g., chief officers, company officers, line personnel, and support personnel)
4. Optional: Others who have a stake in the department such as dispatchers, medical directors, contract mechanics—specific representatives who may be affected by the change

A team composed entirely of chief officers is less likely to receive support from midlevel managers and line personnel, as they often perceive senior managers as being out of touch with the day-to-day operations of the fire department. The same is true of teams composed only of members from the officer corps. A team composed of members that are representative of the entire fire department is best suited to gain support from all members. This method ensures that everyone has a seat at the table and that all interests will be heard.

■ Implementation

The most well-defined vision statement supported by the most highly motivated change management team does not guarantee success. This can be viewed much like an architect and his or her responsibility in the development of a building. A good architect with all of the correct drafting tools is of little use if he or she cannot put to paper what exists in the mind. An architect must first develop a set of blueprints, then ensure that the construction foremen understands them, monitor the progress of the construction to make certain it is in conformance with the blueprints, and, once the building is complete, evaluate the final product.

Once the change management team has been identified, the following steps need to be completed by the team:

1. Develop a vision.
2. Develop goals and objectives.
3. Identify the method(s) of change.
4. Assess and select techniques for promoting change.

Although a vision statement serves as a road map, goals and objectives provide the details on the map. In essence, the goals and objectives serve as turn-by-turn directions of how to reach the final destination. The necessary characteristics of these goals and objectives can be remembered with the acronym SMART:

- Specific
- Measurable
- Attainable
- Realistic
- Timeline for completion

Four Methods of Change

As with any task, especially within the emergency services, there is more than one way to accomplish that task. The question is, what methods are best employed? The answer to that question is dictated by the nature of the desired change, which may necessitate a different process at different points in time. The four methods of change are:

- Technical
- Structural
- Managerial
- Personnel

Technical methods of change refer to the technology employed by the fire department to conduct business. An example would be the use of telemetry in the EMS field or the use of global positioning systems, geographic information systems, and automatic vehicle locators for dispatching emergency vehicles. The use of these technologies has drastically changed the manner in which emergency responders perform their jobs.

Structural methods of change refer to how a fire department is organized to perform their essential functions. That structure could be the division of work groups, number of shifts, number of employees, policies and procedures, standard operating guidelines, decision-making processes, and so on. A common organizational structure that exists within combination departments is the dichotomy between career employees, part-time employees, paid-on-call employees, and volunteer employees. Organizational charters, collective bargaining agreements, policies, and procedures may all play a part in defining the structure of an organization.

Managerial methods of change refer to those things that affect employees, which can include compensation packages, work environment, participation in decision-making processes, and so on. One such managerial issue commonly found in combination departments involves lines of authority between different groups within the organization. For example, does a paid-on-call or volunteer officer carry the same authority as their career personnel counterpart or vice versa? Are promotional and training standards equal and equitable for those different groups?

The personnel method of change refers to the interaction of individuals and working groups within an organization and their ability to work towards common goals thus facilitating the desired change. Combination departments inherently possess a multifaceted array of personalities with conflicting motivations regarding organizational philosophies, service delivery expectations, personnel deployment, and so on. These departments are undoubtedly labor intensive when it comes to managing personnel, which is the most valuable and often most scarce departmental resource. Oftentimes this area of change is the most essential and time-consuming part of the change management process, and is also the hardest to achieve while keeping everyone happy. Underestimating the importance of the personnel, their perspectives, and their vested interests will defeat any change management process, large or small.

Four Techniques of Change

It is well worth the time and effort of those on the change management team to champion their cause and successfully promote and market the change process and its benefits. The change management team will need to determine the best promotional technique, depending on the situation. Four techniques that can be utilized to promote or market the change are:

- Facilitative
- Informational
- Attitudinal
- Political

The facilitative technique involves the use of a specific sector of the affected groups that interacts with the change management team to create the desired change. This process draws a great deal of support because the affected personnel are part of the process, and so can give input about the change and how it affects them as individuals and in work groups. The same technique can be utilized for facility design, equipment purchases, training schedules, and so on. The fire chief or senior management personnel do not give up control; they set the parameters by which the committees operate. The facilitative process provides a conduit for personnel to have some control over their own work environment and circumstances.

Informational techniques involve keeping the affected personnel informed of the need for change, the change process, and the effects of the change. Clear and accurate communication is helpful in eradicating misinformation, as it is typical for personnel to make assumptions when there is a lack of clear, concise information. For example, a local fire department has recently experienced the effects of the economic downturn. As a result, they are in the process of implementing billing for all EMS transports. The fire department serves some of the transient population as a result of having a large retail base as well as serving a hospitality district with hotels and restaurants. This particular combination department has been historically supported by tax funds, but it is now forced to seek alternative funding sources to support the increased cost of providing services and the increase in service delivery needs. In this case, it is essential for the department's administration to communicate to the membership the need for this major shift in revenue streams and to explain the rationale for utilizing the alternative funding source, which in this case means billing for all emergency medical transports. If the department fails to properly educate and explain to the line personnel the issues involving this implementation, then they will give misinformation to the general public, and there will be a great deal of turmoil among staff members who are against alternative revenue services. Note that a change such as this cannot apply to some clients and not to others. There are Medicare laws dictating that once billing starts it must be consistent among all transports.

Attitudinal techniques involve the behaviors that members of the fire department have regarding their work or those attitudes that they bring to work with them. This technique involves changing those behaviors and thus creating the desired outcome. For example, the American Fire Service, through the National Fallen Firefighters Foundation, recently implemented the program "Everyone Goes Home" with the focus on core fire fighter safety initiatives. Many of these initiatives clash with the masculine attitudes of the American fire fighter. Use of full protective clothing, self-contained breathing apparatus, and seat belts are just a few examples of initiatives that are diametrically opposed to the attitude of some fire fighters. For many, the implementation of these safety initiatives involves finding a creative way to educate all employees about the need for change and the benefits for the entire fire department as well as every fire fighter.

Political techniques are negotiations between groups or the building of alliances within the fire department to affect the change. Often times the changes necessary for public safety agencies are more politically charged than one would think. One fire department, which has historically been a volunteer department, is facing daytime staffing issues as a result of a changing workforce and double-income families. The department is attempting to improve both internal and external political support for the addition of daytime career staff so that the department can reduce response times and meet customer expectations. Even though the ultimate goal of the community and the fire department is to respond to emergencies in a timely manner, the transition from an all-volunteer department to a combination department needs a great deal of political support to be successful.

Now that the appropriate tactics and strategies have been identified for instituting the change, it is time to execute the plan. One of the most important factors when leading organizational change is communication. One of the common pitfalls that stalls or sabotages plan implementation is a lack of communication, especially to internal customers. In the information age, we are saturated with ways to communicate. We have e-mail, Web sites, voice mail, memos, streaming video, alphanumeric pagers, voice pagers, message boards, face-to-face communication, and so on. The question is, which is the most appropriate method of communicating about the change to those that are affected.

The change management team should develop a communication plan that identifies the best communication methods, including when and how each will be utilized throughout the process. The team should identify the communication needs for the entire fire department.

Kickoff Meeting

One way to minimize resistance to the impending change is to involve the affected members of the organization in the implementation process from the beginning. This is easily accomplished through a kickoff meeting that serves to:

- Share the vision of change.
- Define the change management model.
- Define the change management timeline.
- Define the change management communication plan.

This initial meeting, if conducted properly, can be a positive catalyst that fuels the department through a successful change process. It also serves as an opportunity for members of the department to ask questions, develop a level of comfort with the change management model, and understand how the suggested changes will affect them. Remember that all of this takes time, so the leadership should be patient throughout the process.

Evaluation

The process of change is no different than any ongoing event with a defined outcome that must be monitored. The evaluation mechanisms employed are also indicators of needed modifications to the process to ensure that the desired outcome is achieved. The change management team should identify evaluation criteria and mechanisms prior to the implementation phase to ensure that they are capable of effectively measuring the change or lack thereof. Communication during this period allows for consistent understanding that change is imminent, ongoing, and is subject to modifications as necessary.

Emergency responders are accustomed and trained to evaluate. From the onset of emergency medical training, department members are taught to ensure safety on a scene; make a first impression; and constantly evaluate the signs, symptoms, and condition of the patient. From the onset of fire training, members are taught to conduct an immediate evaluation of an incident to identify hazards, ensure adequate emergency responders are on scene, initiate a mitigation plan, and then continually reevaluate the tactics and strategies to ensure they are working in conjunction with established benchmarks.

The change management process requires constant evaluation and modification to be successful. It is next to impossible to identify the obstacles that exist or may develop during the process. Oftentimes the issues identified by the change management team may have masked larger issues that require modifications in goals and objectives. The failure of the change management team to remain aware throughout the change process will limit their ability to meet goals and objectives and employ the necessary tactics and strategies to bring their vision to reality.

Summary

Many fire departments have mastered the four-step change management model. By implementing what is already in place for emergency situations, the change management model will assist in addressing nonemergency activities. The result will be a smooth change. It is wise to remember that the structure of the change management process should be fluid and flexible. Inevitable detours will occur and obstacles will materialize, which will require modifications. Don't underestimate the time and resources required to see the change process through. These processes are time and labor intensive. Properly structured and managed, the change process can be an exciting and worthwhile process with positive results.

CHAPTER

11 Improving Your Insurance Services Office Public Protection Classification

Dan Eggleston

Case Study

Everyone at the Maple Valley Fire Protection District was sad to hear about the retirement of Deputy Chief Ed Fisher. Chief Fisher served the department for 30 years, 13 of which were as deputy chief. Chief Fisher's leadership style was slow and calm. He was well respected for his fireground knowledge. It was going to be difficult to find a worthy replacement. Although Fire Chief Greeley would miss Ed's friendship, he knew that the department's operational division needed some focus and drive.

To improve operations, Chief Greeley needed a deputy chief with energy and knowledge, who also had the respect and confidence of the staff. The fire protection board gave Chief Greeley freedom on attempting to fill the position, allowing him to promote from within or seek an outside candidate. Although he was tempted to look externally and seek a person with a new perspective, Chief Greeley also had his eye on a talented battalion chief, Clarence Jones. Clarence had the necessary personal qualities, was highly motivated, and had led various successful projects to improve department operations.

After much thought, Chief Greely met with the board and received approval to promote Clarence Jones as Maple Valley Fire Department's new deputy chief. The announcement went out immediately. Everyone was elated over the news.

The following Monday, Clarence was busy setting up his new office. Clarence was excited about his new promotion. He had worked hard to attend college while holding down a full-time job, and he had taken on numerous projects throughout his career. Clarence had a number of ideas on how he could improve the level of service provided to Maple Valley and he was anxious to get started.

Just as Clarence was finishing hanging the last picture in his new office, Chief Greeley walked in with a thick folder and large three-ring binder. As Chief Greely approached Clarence's desk, he said, "Clarence, I have a project for you that is a top priority. The board and I have been discussing for some time that we need to take a proactive role in improving our ISO classification. I have been delaying the project but now that you're here, we need to begin work." Clarence was somewhat surprised. Although Clarence was involved in the department's last ISO survey, he knew little about the rating process. In fact, Clarence was under the impression that a split rating of 6/9 was not bad considering the adjacent fire protection district was 7/9.

Not wanting to shake the Chief's confidence in his abilities, Clarence quickly responded with "Chief, I'll take a look at the information and give you a draft plan within the next 30 days."

Chief Greeley responded, "Clarence, this is your first time at bat and its important to get a good solid hit on this one. The board has told me if they need to invest more capital funds to improve our rating, they are prepared to borrow the money. We just need to give them a solid recommendation backed by solid outcomes."

As the Chief left the office, Clarence began to look over the papers and charts in the file and binder. None of the figures and calculations made sense as he started to read the material, but Clarence did recall previous conversations about the need for additional large-diameter hose, specialized equipment for the ladder, and changes to the dispatch procedures handled by the sheriff's department. As Clarence's mind began to spin, he closed the folder and binder, pushed back the material, and leaned back in his chair.

"Where do I begin?"

Beginnings of Insurance Service Office's Public Protection Classification Program

As the United States began to quickly urbanize during the early 1900s, many cities experienced large growth and larger fires. Such great fires included the Baltimore fire of 1904, the fire resulting from the earthquake in San Francisco in 1906, and the Atlanta fire in 1917. The Atlanta fire alone destroyed over 2000 homes, businesses, and churches, and caused over $5.5 million dollars in damage (1917 dollars). In 2006, this loss is calculated at over $98.5 million dollars (http://www.westegg.com/inflation/).

The demand for fire insurance increased as home owners and business owners began to seek ways to cover their losses. However, the frequency and intensity of fires were still increasing and insurance companies were paying out many large claims. The large losses highlighted the weaknesses in the current structure of the fire insurance business. As a result, insurance companies realized that more information was needed to combat the problem.

The National Board of Fire Underwriters (NBFU) was established with the mission to prevent fires and educate the public about fire protection. The NBFU created a municipal inspection and grading system that provided useful feedback to cities about their fire loss potential. The Insurance Service Office (ISO) Public Protection Classification (PPC) service is an updated version of the earlier NBFU grading system. The PPC program gives insurers advisory data to help them develop fair and equitable fire insurance premiums that reflect a community's commitment to its public fire suppression services.

Established in 1971, ISO is an international company that provides data, analytics, and risk management services to private businesses and governmental agencies. ISO collects and analyzes information about public services as well as building information (including loss information), from which it prepares informative products and services to help their customers manage risk.

Public Protection Classification (PPC) Program

The ISO Public Protection Classification (PPC) program is an evaluation program that gauges a community's local fire suppression services for property-insurance rating purposes. The public fire protection of a county, city, town, or district is evaluated using ISO's Fire Suppression Rating Schedule (FSRS). The FSRS measures the following elements of an area's fire suppression system: fire alarm and communication systems, the fire department, and the water supply system. Once the elements are evaluated and measured, they are used to develop a Public Protection Classification (PPC) number on a relative class scale from 1 to 10. These terms are defined further in **TABLE 11-1**.

In some areas, a single classification is given to all properties within the given area. However, some areas are classified as split ratings such as a 6/9. Generally, the first class (6 in this example) applies to properties that are within 5 road miles (mileage

Table 11-1 Important Terms

Insurance Services Office (ISO)	An international company that provides data, analytics and risk management services to insurance, finance, and government agencies as well as other similar clients
Public Protection Classification (PPC)	A number on a relative class scale from 1 to 10. A Class 1 represents exemplary fire suppression services. A Class 9 applies to properties that are within 5 road miles of a fire station but are over 1000 feet from a water supply (i.e., a fire hydrant providing a minimum of 250 gpm for 2 hours). A Class 10 indicates that no formal fire protection is available or the fire suppression services do not meet the minimum criteria of the Fire Suppression Rating Schedule (FSRS). Class 10 will also apply to properties that are located over 5 road miles (in most states) from the responding, recognized fire station. There is an additional classification, Class 8B. Class 8B recognizes a superior level of fire protection in otherwise Class 9 areas. It is designed to represent a fire protection delivery system that is superior except for a lack of a water supply system capable of the minimum FSRS fire flow criteria of 250 gpm for 2 hours.
Fire Suppression Rating Schedule (FSRS)	An evaluation program that contains scoring criteria based on the following major elements of an area's fire suppression system: fire alarm and communication systems, fire department, and water supply system

may differ in a few states) of a fire station and within 1000 feet of a fire hydrant. The second class (9 in this example) applies to properties that are beyond 1000 feet of a hydrant, but within 5 road miles of a fire station. A Class 10 applies to properties located over 5 road miles of a responding, recognized fire station.

Most insurance companies use the PPC, in conjunction with historical and prospective loss costs, as a means for determining commercial and residential fire insurance premiums. Insurance rates vary from state to state. A drop in the PPC rating could result in significant monetary savings for the residents and business owners in a specific community.

Fire Suppression Rating Schedule (FSRS) Elements

ISO evaluates the public fire protection of a county, city, town, or district using a grading system that is outlined in the Fire Suppression Rating Schedule (FSRS). The FSRS is an objective evaluation system that contains scoring criteria. National consensus standards such as those of the National Fire Protection Association (NFPA) and the American Water Works Association (AWWA) are referenced in the FSRS. The FSRS measures the following weighted elements of an area's fire suppression system:

- **10%:** The fire alarm and communication systems, including telephone systems, telephone lines, staffing, and dispatching systems
- **50%:** The fire department, including apparatus, equipment, staffing, training, and the geographic distribution of fire companies
- **40%:** The water supply system, including the size, type, and installation of hydrants as well as condition and maintenance of hydrants, and an evaluation of the amount of water available compared to the amount needed to suppress fires

The FSRS contains a comprehensive number of items under each major category. Each item in the FSRS is assigned a value, and the values and various formulas add up to a total score between 0% and 100%. **TABLE 11-2** relates the total FSRS score to the final PPC rating.

Table 11-2 FSRS Score and the Final PPC Rating

PPC	Points
1	90.00 or more
2	80.00–89.99
3	70.00–79.99
4	60.00–69.99
5	50.00–59.99
6	40.00–49.99
7	30.00–39.99
8	20.00–29.99
9	10.00–19.99
10	00.00–09.99

Receiving and Handling Fire Alarms

This section of the FSRS is a review of the facility and personnel that receive emergency calls from the public and dispatch the fire service members to incidents. Ten percent of the credit is related to receiving and handling fire alarms. The grading is divided into three sections:

- Credit for telephone service: 2%
- Credit for operators: 3%
- Credit for dispatch circuits: 5%

There is an extensive list of criteria for receiving and handling fire alarms in the FSRS manual. This section is generally related to National Fire Protection Association (NFPA) 1221: *Standard for the Installation, Maintenance, and Use of Emergency Services Communications Systems*.

The telephone service evaluation is based on the number of reserved lines for receiving fire emergency calls and nonemergency business calls, respectively. The number of lines is based on the given population of the area served. The telephone service portion also evaluates how emergency calls are forwarded through the system, how the emergency and nonemergency telephone numbers are listed in the phone directory, and a review of the recording devices used to record and play back emergency telephone calls.

The credit of dispatchers is based on an evaluation of the number of dispatchers on duty for a given workload minus the time for sleep, if allowed. A detailed public service answering point (PSAP) call receipt and dispatch performance record as well as staffing data will permit a full and comprehensive evaluation which may allow for the maximum possible points to be credited.

The dispatch circuit portion is related to the evaluation of dispatch circuits, monitoring the integrity of circuit, dispatch recording facilities at the communications center, and emergency power supply. This section is covered extensively in the FSRS.

Fire Department

This section of the FSRS is a review of the engine and ladder/service companies, apparatus, equipment carried, response to fires, distribution of companies training, and available fire fighters. The grading is divided in eight sections for a total of 50%:

- Credit for company personnel: 15%
- Credit for engine companies: 10%
- Credit for training: 9%
- Credit for pump capacity: 5%
- Credit for ladder/service companies: 5%
- Credit for distribution: 4%
- Credit for reserve pumpers: 1%
- Credit for reserve ladder/service companies: 1%

The engine company section compares the number of in-service engine companies and the equipment carried with the number of needed engine companies and the equipment identified in the FSRS (or equivalency list—see the ISO Web site). The number of needed engine companies depends on the basic fire flow (BFF), the size of the area served, and the method of operation. The number of existing engine companies is evaluated based on the number of pumpers that are staffed on first alarms and an evaluation of the pumper's pump capacity, hose carried, equipment carried, automatic-aid plan (if applicable), and testing data of the pumper's pump and hose.

The reserve pumper section evaluates the adequacy of the reserve pumper program. The needed number of reserve pumpers is one reserve for eight needed engine companies. Reserve pumpers are reviewed for pump capacity, hose carried, and equipment in the same manner as outlined in the engine company section.

The pump capacity section evaluates the pump capacity of the in-service and reserve pumpers as well as pumpers supplied by automatic aid compared with the basic fire flow (BFF). The FSRS considers a maximum BFF of 3500 gallons per minute (gpm). The FSRS develops a separate Public Protection Classification that applies to specifically rated properties that have a needed fire flow (NFF) greater than 3500 gpm.

The ladder/service company section compares the number of in-service ladder or service companies and the equipment carried with the number of needed ladder and service companies and the equipment identified in the FSRS. The number and type of ladder/service companies depend on the height of the buildings, needed fire flow, and the size of the area served. In general, the FSRS will look for a ladder company if:

1. The response area has five buildings that are three stories or higher or 35 feet or more in height *or*
2. The response area has five buildings that have a needed fire flow (NFF) greater than 3500 gpm *or*
3. Any combination of the above criteria

The FSRS further states that areas not requiring a ladder company should have a service company. An evaluation of the ladder/service companies is based on equipment carried and frequency in which ladders were tested in accordance to the National Fire Protection (NFPA) 1911: *Standard for the Inspection, Maintenance, Testing and Retirement of In-Service Automotive Fire Apparatus.*

The reserve ladder/service company section evaluates the adequacy of the reserve ladder/service program. The needed number of reserve ladder/service companies is one reserve for eight needed ladder/service companies. Reserve ladder/service companies are reviewed in the same manner as outlined in the ladder/service company section.

The company distribution section is evaluated based on the existing number of engine companies and ladder/service companies within a given area measured in distance traveled along all-weather roads. In general, the FSRS provides maximum credit for a fully equipped first-due engine company within 1.5 miles of all built upon areas and a fully equipped first-due ladder/service company within 2.5 miles of all built upon areas. The final score is calculated based upon the percentage of built upon area within the 1.5 and 2.5 mile distance factored by the tools, equipment, and hose carried, respectively.

The company personnel section evaluates the personnel available for first alarm fire calls divided by the existing engine, ladder/service, and/or surplus companies. On-duty personnel (career or volunteer) must respond to first alarms to be credited. The credit is based on actual records of personnel responding to structure fire calls. For on-call and volunteers not normally in the fire station, the FSRS reduces the value of the responding members to reflect the delay due to decision, communication, and assembly. The FSRS then applies an upper limit for the credit for staffing, as it is impractical for a very large number of personnel to operate a single piece of apparatus. Personnel responding from needed automatic aid responses or on special apparatus are added to on-duty or on-call/volunteer numbers. Complete on-duty staffing and/or call or volunteer response records documentation are important to maximize credit for company personnel.

The training section evaluates training facilities; training at fire stations; training of fire officers, drivers, and recruits; hazardous materials training; and building familiarization and prefire planning inspections. Facilities and aids include drill towers, fire buildings, combustible liquid pit (or an instructional video on fighting flammable liquids fires), library and training manuals, slide and movie projectors and pump/hydrant cutaways, and training areas. Credit is given for day and night training on single- and multiple-company drills at the facilities. Credit is also provided for company training at the station and training provided for officers, drivers/operators, and recruits. Credit is also granted for training on radioactivity detection (or hazmat training) and for conducting building prefire planning inspections. Complete training records are important to maximize credit for training.

At the time of this writing, ISO was in the process of updating its FPRS. Please refer to the latest information available from ISO.

Water Supply

The water supply section of the FSRS is a review of the water supply system available for fire suppression for the area. This section applies to built areas within 1000 feet of a recognized water system capable of delivering a minimum of 250 gpm or more for a period of 2 hours, plus consumption at the maximum daily rate at a fire location (e.g., fire hydrant). Credit is also available for fire department supply delivered by fire department vehicles carrying or relaying at least 250 gpm within 5 minutes of arrival at the fire site. If any built area is not within 1000 feet of a recognized water system or a recognized fire department supply, that area may receive Class 9.

Forty percent of the overall evaluation process focuses on the water available for firefighting. This section of the grading process is divided in three sections:

- Credit for water supply: 35%
- Credit for hydrants: 2%
- Credit for inspection and condition of hydrants: 3%

The water supply section is an evaluation of whether you have sufficient water for fire suppression beyond your community's daily maximum consumption rate. The supply works, the water main capacity, and fire hydrant distribution all contribute to the available water supply. The FSRS states that the fire flow duration be 2 hours for needed fire flows (NFF) up to 2500 gpm and 3 hours for NFF of 3000 and 3500 gpm.

The hydrant section evaluates the size, type, and installation of hydrants. Credit is issued based on items such as size of branch lines, size and number of hose outlets, thread commonality, and cistern or suction points.

The hydrant inspection and condition section is an evaluation based on the Inspection American Water Works Association Manual M-17—*Installation, Maintenance, and Field Testing of Fire Hydrants*. The section evaluates the frequency hydrant inspections, the completeness of the inspections, and the condition of the hydrants (leaks, ability to open, location for fire apparatus, etc.).

Divergence

Adequate fire protection relies on an effective fire department combined with an ample water supply. If the fire department is effective, but lacks water (or vice versa), the overall affect is limited. Therefore, the FSRS factors in divergence points as a means to reconcile differences in water supply and fire department evaluations. Divergence is a reduction in credit to reflect a difference in the relative credits for fire department and water supply features. Mathematically, divergence is calculated using the following formula:

PPC = Public Protection Classification
CFA = Credit for fire alarm
CFD = Credit for fire department
CWS = Credit for water supply

$$PPC^* = 100 - \{(CFA + CFD + CWS) - 0.5[I(CWS) - 0.8(CFD)I]\}/10$$

Note: Raise any decimal to the next higher number (e.g., 5.12 = 6).

It is important to understand divergence before taking action towards changing a classification number. If the fire department score is relatively higher than the area's water supply relative score, the divergence formula will reduce the point values of each item in the fire department section while increasing those items in the water supply section. The opposite is true when the water supply relative score is greater than the fire department's relative score.

Improving Your PPC Rating

Now that you have an idea of the major components of the FSRS, it's time to begin formulating a plan to improve your PPC rating. This section will cover the four major steps: make a plan, form a team, gather your tools, and getting started. The authors agree that ISO improvement strategies be focused on service improvement and cost considerations. There are many instances where fire departments have invested huge sums to achieve a better ISO rating and have not necessarily improved their service level. The concept of service first, then ISO rating, is a prudent means of addressing PPC ratings.

The authors agree that ISO improvement strategies be focused on service improvement and cost considerations. There are many instances where fire departments have invested huge sums to achieve a better ISO rating and have not necessarily improved their service level. The concept of service first, then ISO will follow, is a prudent means of addressing PPC ratings.

■ Make a Plan

Like any project, to reduce your ISO rating you should consider scope, time, and cost. The scope, time, and cost relationship is the traditional triad in project management. Keep in mind that one element cannot be changed without impacting the other two.

Scope

The investment involved to lower your ISO rating can be substantial and depends on your overall goal. Therefore, it is important to define the objective of the project. That is, what is the project supposed to accomplish in specific terms? Reduce the rating by one classification number, two classification numbers, or achieve a rating of Class 1?

A detailed review of your last rating may help in determining your objective. For example, your department may have received a total credit of 48 points during your last ISO visit. A 48 translates to a PPC rating of Class 6. Just two more points on your overall credit would equate to a better PPC rating of a Class 5. Some of the easiest opportunities for two credit points have been better documentation on hose and ladder testing, training records, and prefire plans records.

You may initially set a goal to drop your PPC rating by one classification, and half way during the process, you realize that achieving a two or three classification improvement is within your aim. It is perfectly acceptable to reevaluate the scope somewhere in the middle of the project. However, a change in scope should always include a reevaluation of the time and costs required.

Time

How much time do you have to complete the evaluation and implement the changes? Is ISO visiting next month or 3 years from now? Of course, the more time you have to develop a plan, the more likely that the outcome will be successful.

The three major components of an ISO rating improvement process are the evaluation of the current system, the implementation of the recommendations, and the ISO evaluation visit. There will be a strong temptation to jump ahead and start doing something. However, a change without first understanding how the change will improve the rating could be a waste of money and effort. Therefore, it is important that the steps not be skipped. Much understanding can be achieved from visiting the ISO Web site dedicated to the FSRS program.

Cost

There is a cost to both the evaluation process and the implementation process. Staff time is the major element during the evaluation process. How much time will be allowed to work on the project? Do you have staff that can be dedicated to the project? What are the costs of reassignment? What's not going to get done while you take on this project?

Although staff time is a concern during the evaluation process, staff, equipment, apparatus, and infrastructure costs are the major issues during the implementation process. For example, during the evaluation process, you may find that out of the potential 15% rating for company personnel, your department only received 5%. Further investigation reveals that the FSRS will credit you 0.62% for every additional on-duty staff. This percentage is variable depending upon the number of existing companies and the existing staffing with those companies. A well-funded duty incentive program could cost the department a sizeable amount, but in turn could provide you the extra points that are needed to lower your rating.

■ Form a Team

Embarking on a project to lower your department's PPC rating is not a one-person show. The process takes times, involves detailed research and analysis, and requires commitment and determination from all team members.

The project to lower your PPC rating is more of an analytical process than a creative process. Therefore, when considering team members, it is important to clearly define the expectations and select those members who understand the process and have the will to see the project through completion. Many departments choose to involve a small number of staff as a core group. It is easier to manage and coordinate schedules of a smaller team than a team of 10 or more members.

Depending on your locality's organization, you may choose to involve associate team members outside of the fire department who may have special knowledge or skills. For example, many departments solicit assistance from their local water authority or public works staff to help assess the community's water supply. In addition, you may choose to involve other fire service leaders who have recently been through a PPC rate-reduction process or have special talents that are lacking in your team. ISO is also available to offer assistance either through their Web site or by calling their Public Protection Classification experts.

■ Gather Your Tools

Some fire service leaders fail to realize that the fire suppression rating process is almost like taking an open-book test. The criteria used in reviewing the firefighting capabilities are contained in the FSRS manual that all fire departments can obtain from ISO. ISO will provide, free of charge, a single copy of the FSRS manual to the local fire chief or community chief administrative official. It can be ordered at ISO's Web site.

In addition to the FSRS manual, it's also advantageous to obtain from ISO the most recent copy of the Classification Details, Improvement Statements, and the Needed Fire Flow (NFF) Report associated with the last evaluation of your community. The Fire Suppression Rating Schedule will provide the team with detailed information about the criteria used in the evaluation of the fire alarm and communication system, fire department, and water supply system. The Classification Details and Improvement Statements are reports of the conditions found in the last area survey. The statements provide feedback on how your department performed in certain FSRS categories and what is needed to receive full credit. The Needed Fire Flow (NFF) Report is a list of the fire flows for significant structures in your community that ISO has within their database.

There is also a wide array of training and educational opportunities available to learn more about the PPC rating process. ISO works with state fire chief's associations and other training institutes to conduct seminars on the Fire Suppression Rating Schedule. The course is entitled *Understanding the ISO Fire Suppression Rating Schedule*. Work with your state fire chiefs or state fire fighter associations to request an ISO course at your next state meeting.

Last but not least, ISO provides assistance through their toll-free customer service number. You can request ISO to meet with your team. When contacting ISO, be sure to use specific terms and address questions that relate to your FSRS. Keep in mind that ISO wants you to be successful at improving your classification, but it's important for your questions be specific to your area. ISO may also provide a presurvey meeting to help develop a more effective strategy.

■ Started

Now that you have made a plan, formed a team, and gathered your tools, you are ready to begin the evaluation process. Keep in mind that during this stage of the project, the team will be evaluating the current FSRS score and will identify opportunities for improvement. Although the team should solicit input from the department and outside agencies, watch out for the groups that may attempt to advance their agenda. The evaluation stage is an objective process and every suggestion can be evaluated for its return on investment. An important tool to assist in one area of this evaluation can be found on ISO's Web site. ISO has provided "Relative Value Tables" that identify the FSRS point value of each apparatus equipment item relative to the FSRS score before divergence is factored in. There is also an "FSRS Equivalency List" that contains a list of equipment that ISO considers equivalent to apparatus equipment items listed in the Fire Suppression Rating Schedule.

It's important to fully understand which areas ISO found to be deficient. At this point, a spreadsheet that calculates and totals the major items in the FSRS process would be worthwhile. The spreadsheet could calculate the actual FSRS schedule numbers and allow the team to develop "what if" scenarios by changing actual numbers and observing the final score. The spreadsheet may help to identify simple and low-cost solutions that would improve the overall FSRS score.

■ Example—East Overshoe Volunteer Fire Department

As an example, let's review the FSRS score of East Overshoe VFD. East Overshoe received a split PPC rating of 6/9 during their last ISO evaluation. East Overshoe's grading sheet is as follows in **TABLE 11-3**.

East Overshoe is 3.2% away from receiving a split classification of 5/9. Notice that the department received a –5.13% divergence, which means that the water supply was rated higher in relative credits than

Table 11-3

Feature	Credit Assigned	Maximum Credit
Receiving and handling fire alarms	4.84%	10.00%
Fire department	20.46%	50.00%
Water supply	26.63%	40.00%
Divergence	−5.13%	–
Total	**46.80%**	**100%**

the fire department. The department received less than half credit (20.46%) under the fire department section. An improvement in the fire department section will not only increase the section's rating, but will reduce the negative divergence points. A further breakdown of East Overshoe's fire department section is listed in **TABLE 11-4**.

Using an FSRS spreadsheet as a tool, you can adjust the credits and determine that a 5-point increase in fire department credit will increase your FSRS score from a 46.8 to a 56.8 (divergence changes from a –5.13 to a –2.23) with room to spare. We will explore some relatively easy points to gain.

Credit for Engine Companies

After reviewing the improvement statement, it is noted that ISO declares that two engine companies are needed based for a basic fire flow of 2250 gpm. East Overshoe has two engine companies, but Engine 1 only received 67% credit due to insufficient equipment, inadequate hose testing, and inadequate pump testing. Engine 2 only received 40% credit due to insufficient equipment. The overall credit for engine companies is 5.42% out of a potential 10%.

Pump Testing

The FSRS will review the pump testing program using the National Fire Protection Association (NFPA) Standard 1911, *Standard for Inspection, Maintenance, Testing and Retirement of In-Service Automotive Fire Apparatus.* The overload test is not a necessary part of the FSRS review. To receive full credit for pump testing the average interval between the three most recent tests should be 1 year or less. If the interval is more than 1 year, a partial credit is granted (see the FSRS manual for more information). In the case of Engine 1, the last pump test was 3 years ago. Therefore, ISO only credited 50% of the maximum credit available. East Overshoe will need to demonstrate 3 consecutive years of pump testing to get full credit.

Hose Testing

The FSRS will review the fire hose testing program using the National Fire Protection Association (NFPA) Standard 1962, *Care, Maintenance and Use of Fire Hose.* As in the pump testing criteria, to receive full credit for hose testing the average interval between the three most recent tests should be 1 year or less. Therefore, the FSRS only credited a small percentage of the maximum credit available. East Overshoe will need to demonstrate 3 consecutive years of hose testing to get full credit.

Engine Equipment

In addition to the pump and hose testing, both engines lack a couple of nozzles, salvage covers, and other miscellaneous inexpensive equipment necessary to receive full credit (see the FSRS manual for a list of the equipment). An investment of about $2500 in equipment will help increase the credit for engine companies.

Credit for Company Personnel

East Overshoe received 4.36% credit out of a possible 15% credit for company personnel. A review of the improvement statement noted that an increase in one on-duty person will increase the fire department credit by 0.83%.

Table 11-4

Feature	Actual Credit	Maximum Credit
Credit for engine companies	5.42%	10%
Credit for reserve pumpers	0.48%	1%
Credit for pump capacity	5.00%	5%
Credit for ladder/service companies	0.55%	5%
Credit for reserve ladder/service companies	0.00%	1%
Credit for distribution	1.41%	4%
Credit for company personnel	4.36%	15%
Credit for training	3.24%	9%
Total	**20.46%**	**50%**

East Overshoe has a volunteer staffing schedule for nights and weekends and has a system to track volunteer staff hours worked. ISO credited East Overshoe with the daytime career staff and night/weekend volunteer staff based on the supplied documentation. However, often times, additional volunteers spend the night at the station or are around during the daytime and run calls with the career staff. In addition, a volunteer shift officer is on duty nights and weekends and responds from home. The problem is that East Overshoe has never documented the extra staffing and thus was never given credit for addition company personnel credited by ISO.

The solution is simple for East Overshoe. The department should establish a sign-in roster for those extra volunteers that choose to spend the night. The department could also explore the option of an incentive program to increase staffing at the station. In either case, let's assume a small 1% improvement from a simple documentation solution.

Credit for Training

East Overshoe received 3.24% credit out of a possible 9% credit for training. Like many departments, East Overshoe trains often, but does not do a good job with proper documentation. The problem with East Overshoe was that station training was being conducted while fire fighters were on duty, but the training was not documented. In addition, when the chief and company officers attend conferences and out-of-town training events, they did not document the time.

Training records should be kept on each individual fire fighter. In addition, each course should be well documented and show the material that was covered. A paper-based tracking system will work for most departments, but computer based programs are also available. Let's assume that through better efforts of documentation, East Overshoe is able to increase its training credit by 1% for a total credit of 4.24%.

Results

Now that we have made improvements in engine company, personnel, and training, let's review the results in **TABLE 11-5**.

The improvements in engine company, personnel, and training increased the credit from 20.46% to 27.71%.

Improvements to the fire department section (**TABLE 11-6**) increased the credit from 20.46% to 27.71%, an increase of over 7%. However, since there is a difference in the relative credits for fire department and water supply features, negative divergence points are added to determine the total credit. On the other hand, notice that the estimated increases in fire department credits actually decreases the negative divergence points, and therefore, an increase of more than 7% in credit equates to a net increase of more than 10%. With minor improvements, East Overshoe should be able to increase their overall score from a 45.80% to 56.95%. Referring to the PPC rating table (**TABLE 11-7**), a 56.95% score is a solid 5 PPC.

There are probably other low-cost opportunities to increase East Overshoe's ratings. The above example is just a snapshot of the process that should be used to evaluate the cost of improvement relative to the benefits from a lower PPC rating.

Table 11-5 Improvements in Engine Company, Personnel, and Training Section

Feature	Prior Credit	Estimated Credit	Maximum Credit
Credit for engine companies	5.42%	**9.6%**	10%
Credit for reserve pumpers	0.48%	0.48%	1%
Credit for pump capacity	5.00%	5.00%	5%
Credit for ladder/service companies	0.55%	0.55%	5%
Credit for reserve ladder/service companies	0.00%	0.00%	1%
Credit for distribution	1.41%	1.41%	4%
Credit for company personnel	4.36%	**5.36%**	15%
Credit for training	3.24%	**5.24%**	9%
Total	**20.46%**	**27.71%**	**50%**

Table 11-6 Improvements in Fire Department Section

Feature	Prior Credit	Estimated Credit	Maximum Credit
Receiving and handling fire alarms	4.84%	4.84%	10.00%
Fire department	20.46%	**27.71%**	50.00%
Water supply	26.63%	26.63%	40.00%
Divergence	–5.13%	**–2.23%**	
Total	**46.80%**	**56.95%**	**100.00%**

Table 11-7 PPC Rating Table

PPC	Points
1	90.00 or more
2	80.00–89.99
3	70.0–079.99
4	60.00–69.99
5	**50.00–59.99**
6	40.00–49.99
7	30.00–39.99
8	20.00–29.99
9	10.00–19.99
10	00.00–09.99

Summary

An investment in understanding the rating schedule, a through review of your last rating, and a motivated team with a clear goal will better your department's chances in improving your ISO classification. In some cases, major improvements in the fire alarm and communication system, fire department, and water supply system can be made without significant financial investment. As changes occur within your community it is important to notify ISO about the essential fire protection features in your community. For example, changes in your fire protection area, location of fire stations, enhancements to the water system, and the number of fire hydrants may help to lower your ISO rating, which may reduce your residential and commercial insurance premiums. The PPC program offers economic benefits for communities that invest in good fire protection. The program is a useful tool to help fire departments and other public officials as they plan for, budget, and justify improvements. The end result: Lowering your ISO rating will help to improve your community's fire protection systems.

CHAPTER

12 Leadership

Richard B. Gasaway

Case Study

For a week Chief Thomas was on a well-deserved vacation at the lake with his family. The time away from work has recharged his batteries. As he comes in to the office on Monday morning, he is smiling and cheerfully greets his staff. It has been a while since his administrative assistant has seen him so relaxed. She hesitates to tell him about the work that has piled up on his desk in his absence. But Chief Thomas is a veteran fire service leader. He knows what awaits him and is eager to return to what he loves most, leading the organization.

Chief Thomas has many admired qualities among his peers and subordinates. One is how effortlessly he leads his people and deals with problems. Today he is presented with numerous opportunities disguised as difficult challenges. Chief Thomas settles into his chair and logs on to his computer, revealing an inbox full of e-mail messages. The display on his telephone reveals dozens of voice messages.

With so much work to do, it would be easy for Chief Thomas to feel overwhelmed and frustrated. But he does not. Instead, he opens his desk drawer and pulls out the notepad to create a to-do list. He draws a line down the right side of the paper creating a two-inch margin. At the top, he writes Leadership Opportunities *above the margin. When he has worked his way through all his messages and the list is complete, he starts back at the top and uses the space in the margin to remind himself of the issues and tasks that await him* (**TABLE 12-1**).

Introduction

There is an abundance of material addressing the behaviors and traits of good leaders. The shelves of bookstores are loaded with books on leadership theories, how-to guides, and the latest instant success schemes. The number of books available on leadership may only be rivaled by the number of definitions and descriptions of leadership and leaders. The focus of this chapter is to evaluate some essential keys to successful leadership.

Leaders need to be resilient. Successful leaders are able to adjust quickly to the circumstances that they face and apply different tools to different challenges. The focus of this chapter is to evaluate principles of leadership. The authors acknowledge the existence and value of many more traits and qualities in addition to those chosen to be highlighted in this chapter.

Table 12-1 Chief Thomas's To-Do List

Item	Leadership Opportunities
✓ Resident John Jacobs called to complain about how long it took for the ambulance to arrive at his house.	1. 2. 3.
✓ Jerry Kingston, the union president, wants to meet to discuss the upcoming contract negotiations.	1. 2. 3.
✓ Neal Herling, the city manager, needs me to attend a hearing at the legislature next week for a new law that will impact city funding.	1. 2. 3.
✓ Paul Williams, a captain on the B-shift, called to complain about the mess that A-shift personnel keeps leaving in the station at shift change.	1. 2. 3.
✓ Jack Johnstone, a fire equipment vendor, called and wants to take me to an upcoming baseball game as a way to thank our department for our recent SCBA purchase.	1. 2. 3.
✓ Abe Watson, chief of the neighboring department, is having a disciplinary problem with an employee and wants some advice.	1. 2. 3.
✓ Kim Eslund, a resident and former council member, called and would like to have a presentation for her neighborhood block watch group on home fire safety.	1. 2. 3.
✓ Wes Robinson, the city finance director, sent an e-mail asking for input on a new and innovative way to administer the budget process.	1. 2. 3.
✓ Paul Williams, the B-shift captain, called again to inform me that his crew had a successful cardiac resuscitation on a 42-year-old male at the community fitness center.	1. 2. 3.
✓ Janice Baker, a police deputy, called to inform me that one of my engines was involved in a vehicle accident where one person received a minor injury. She also informed me that none of the crewmembers were wearing seat belts at the time of the accident, but she doesn't want to make a big deal out of it.	1. 2. 3.

Leaders Are Responsible

The fire service provides a critical role in public safety, a responsibility that must be taken seriously. Being responsible and holding others responsible are important components to a fire chief's success. With responsibility comes accountability. Leaders must be accountable for their own actions as well as the actions of their followers. Often when something goes wrong, the first person blamed is the leader. This is unfortunate, because the task of being responsible is not a task borne solely by the leader, but one that is shared by the followers. For example, in the fire service the responsibility for creating a positive, productive atmosphere is one that is shared throughout all levels of the fire department.

Sharing ownership and responsibility for the fire department can have several benefits. It can also present some challenges. In general, followers tend to act more responsibly when they have ownership in their work. To gain the commitment of followers, the leader should identify the benefits and detriments that the work holds. This will allow your followers to answer the question "What's in it for me?" The more the followers know about how they will benefit if a program is successful, the more likely they will be to take responsibility. Taking ownership for their work and wanting to do a good job is a direct result of motivation.

Leaders Understand Motivation

An elusive concept that leaders often find challenging is how to motivate followers. Many wonder, "How can I motivate people?" The answer is simple. You can't. There is no simple 3-step process to amazingly transform unmotivated followers into highly productive, highly motivated team players. It's just not that easy. Or is it?

The first step in the process to creating a highly motivated team of followers is to understand that motivation is an inside job. In other words, highly productive team players are internally motivated. It is possible for you to do certain things to motivate externally. For example, tell a company of fire fighters to wash the engine and when they're done, each of them will receive a one hundred dollar cash bonus. Chances are, they would do it, collect the hundred dollars, and feel very accomplished. However, it is unrealistic to think we could go around work all day long passing out hundred dollar bills to get our followers to do the things that need to be done. And even if we could afford to, a time would come when that too would cease to motivate them. Once they were able to collect enough money to live a comfortable life and have all the things they need and want, then their desire to do the same work for the same reward would diminish.

The second step is to understand what motivates people. This cannot be done by reading a book. Ask your followers what motivates them. Engage in real dialogue with your followers. Do not assume you know their motivations without asking first. Once you ask, brace yourself, because you will quite likely be surprised. The best way to gather this information is to meet with your followers one on one. This will allow you to understand the internal driver of each of your followers. This takes time, but it is time well spent. You may find that some followers are motivated just by the attention they receive by being asked.

An alternative to meeting each follower in person is to conduct a survey. The advantages of a survey are that you are able to gather information anonymously from your followers, and your followers may be more revealing with their feedback. For example, one of the things that may motivate a follower is a change in the leader's style. Some followers might not be comfortable saying that to a leader during a one-on-one session.

The leader could also engage a group of fire fighters in a discussion about what motivates them. This is also less threatening than a one-on-one session. If you conduct a group session, expect that the dominant member in the group will speak up and give the appearance to be representing the entire group where in fact he or she may simply be the most vocal person in the group. Regardless of how you obtain it, the only way to understand what motivates followers is to have them tell you.

The third step in developing motivated followers is to help them understand the mission of the organization and how they fit into accomplishing the mission. At your next staff meeting, ask your followers to write down on a piece of paper the mission of the fire department. Then ask them to write down what role they play in accomplishing the mission. Gather the responses. When the meeting is over, take them back to your office, close the door, sit down, and read them. Unless you have spent considerable time coaching your followers on the mission and the role they play, you are about to have an epiphany. You are probably going to find their responses are all over the board. Do not get too discouraged by this. If you find your followers are not all on the same page, this is an opportunity.

The diversity of the responses is an indication that your followers are not united in their understanding of the mission and the role they play. The secret to motivation, if there is one, is first to understand the needs of your followers; second, for them to understand the needs of the organization; and third, for you to find common ground where both sets of needs are met.

Leaders Empower

Successful leaders involve their followers in decisions that affect their workload, their work environment, and their well-being. Followers will give greater effort to work assignments when they are involved in the planning and decision process. Successful leaders establish generalized work goals and then challenge their followers to figure out how to accomplish the work in the best possible way. For example, at the company level, a good leader could empower their followers by involving crewmembers in the decision-making process when planning station duties and work assignments for the day. This does not seem like much, but most followers do not like being led by the hand. This is especially true of fire fighters who are, for the most part, self-starters and action-oriented problem solvers. Empowerment in decision making can go a long way toward a follower feeling pride and ownership.

Empowered followers are also motivated followers. Those who are given control over work decisions tend to be more personally involved and committed to the success of the work. Involving followers in strategic-level work can create even more internal motivation. Exceptional leaders make others feel strong. Empowered followers can accomplish amazing things. And, as they enjoy the success of their efforts, their confidence will improve and they will be more willing to take on greater responsibility.

A couple of cautionary notes are in order for leaders who desire to create an empowered work environment. Followers must be given the encouragement to take risks, and the leader must be willing to tolerate mistakes. Empowered followers are going to make mistakes as they learn how to be independent in their decision making. Mistakes are not a sign of failure (unless the same mistake is made over and over again). Mistakes are a sign of stretching and growing. A follower who is willing to innovate and make a mistake or two in the process shows a willingness to step outside the norm in an effort to take the fire department to a higher level.

Secondly, if you are a leader who desires to empower your followers, you must be willing to let go of the decision-making process and allow your followers the latitude to do the job as they best see fit. Some leaders are challenged by this. To make this a success, you must be willing to accept the fact there are many ways in which the work can get done. To empower followers is to say, "This is what I need done and when I need it done by, you figure out for yourself the best way to get it done." The leader then assumes the role of a coach, helping the followers only when they need help. Empowering leaders are macromanagers. They focus on the big picture of the mission and vision and leave the details to their competent, motivated followers.

Leaders Are Role Models

As a leader, it is important to understand that the followers are always watching. On the job or off the job, leaders are always modeling behavior for their followers. Successful leaders understand this principle and realize how influential their behavior is. The attitude of the leader is contagious and will be copied by followers. If a leader has followers with performance and attitude problems, the first place the leader should look is in the mirror.

Followers learn from the examples set by their leaders. Followers are always looking for cues on how to behave, especially during stressful situations. It's easy to be a good role model when things are going great. The role-modeling behavior of the leader is tested when things go wrong. For example, when a follower needs coaching, counseling, or discipline, being a good role model is very important. When dealing with a follower who behaves in a manner that is less than expected, the leader should be fair, firm, and friendly. Coaching a follower to better performance does not have to be a painful process. Being a positive role model of desired behavior while displaying a sincere willingness to listen, learn, and develop your followers are important to your success.

Active mentoring is a good way for leaders to develop and influence the behaviors of their followers. Working with followers to develop personal action plans for growth and development strengthens the bond between the leader and followers. It should be a goal for every leader to help the followers succeed. That, in turn, will make the leader more successful.

Leaders Are Predictable

Followers want to know that they can count on the leader and that the actions of the leader are, for the most part, predictable. Followers do not want to be surprised with unexpected behaviors from the leader. Consistent and predictable behaviors are important, especially the ability to remain calm, levelheaded, and focused even under stress. This is true on and off the fire ground. This can affect the behavior, good and bad, of followers. When followers feel their leader is consistent and predictable, it promotes self-discipline.

Predictable leaders make their expectations clear. They embrace the fire department's rules and follow the standard operating procedures. For followers to succeed, they need to know where their leader stands on important issues and their leader's philosophies on how work is to be completed. Thus, the leader owes it to followers to be consistent day in and day out. Also, being consistent and predictable makes the leader more approachable, leading to more interaction with followers.

Leaders Are Trustworthy

To earn the trust of another person is one of the most valued qualities of any human relationship. As trust deepens, so will the productivity of followers. It is very important for the leader to be as fair as possible, to avoid the appearance of favoritism, to display trust in followers, and to be honest.

Closely related to trust is the credibility of a leader. Trust and credibility are built into relationships over time by a display of mutual respect for knowledge and abilities and by a display of sincere loyalty. The ability to lead effectively is directly dependent on the leader's credibility with their subordinates and peers. Being trustworthy means being true to your word, delivering what you say you will, and never compromising your integrity. A good leader knows it is better to underpromise and overdeliver than to overpromise and underdeliver.

A critical component of being trustworthy is honesty. To always tell the truth when asked a question even if it means saying you just can't talk about it. Followers need to know that their leaders are honest. Numerous surveys of followers have revealed that honesty is the highest desired leadership characteristic. Clearly, it is one of the most important ingredients in the leader–follower relationship.

Leaders Embrace Change

Leading change is one of the most difficult challenges of a leader. It is also one of the most critical components toward building a successful organization. It has often been said that the only thing constant in life is change. This maxim is true regardless if you work for a large corporation, a nonprofit organization, or a government agency. An organization that is not learning, growing, and progressing is stagnating. To be successful, change must become the way of life.

With all change, comes risk, stress, loss, and gain. Successful leaders are proactive in communicating and justifying the need for change and building change into their vision of the future of the department. The challenge for the leader is to reduce risk, stress, and loss and to help followers realize the gain that comes from change. The relationship between leaders and followers can be strengthened or destroyed based on how a leader approaches change.

The best way for leaders to understand how followers feel about change is to ask them. When you ask your followers how they feel about change, expect the responses to be less than cordial and supportive. Remember, many changes bring about feelings of risk, stress, and loss. It is difficult to talk about these feelings in a positive way. But it is important to get these feelings out in the open where they can be discussed. Leaders do not shy away from making necessary changes simply because it may be unpopular with followers. They do, however, seek to understand how followers feel and work with the followers in developing ways to cope with the changes that are going to impact them.

Leaders Are Communicators

Good leaders understand the components of successful communications. Those components consist of a sender, a message, a medium for transmitting the message, and a receiver. The communication process can be degraded by many variables including external noise, gender, age, biases, strained relationship, and knowledge base. Good leaders are conscientious of their communications abilities and shortcomings and work to ensure that the message sent is the message received. Research has shown that miscommunications are one of the leading causes of conflict in organizations. Working on communications skills can help leaders and followers avoid personnel issues. In addition to being skillful orators, leaders are also active listeners. One way a leader can ensure that he or she is hearing the follower correctly is by paraphrasing what the follower has to say.

For example, a follower comes to the leader to vent. It takes him 20 minutes to get it all off of his chest. As a good listener, a leader must do four essential things. First, listen attentively to the entire message, working hard not to shut down if the message is unpleasant. Second, take brief notes about what is being said so you capture the most important points. Third, show compassion for the messenger. Finally, repeat back the most important parts of the message to ensure that what was said is what was heard.

Leaders understand that feedback is a powerful tool to build trust, establish clear and honest communications, and to show appreciation. Positive feedback is motivating. Constructive feedback is a developmental tool. Good leaders make positive and constructive feedback part of routine interaction with followers. This strengthens the leader–follower relationship.

The process of communicating effectively is often oversimplified. Some think that merely stating something once in a meeting yields clear communications, or posting a memo about an important operational change will result in complete understanding and total compliance. Effective communications is not that simple. It involves a process of repetition and using various mediums to appeal to the multiple senses of the receiver.

Leaders Are Team Oriented

Leaders who want to achieve their greatest success realize it requires the effort of the entire team. Thus, they promote teamwork and assume the role of coach, helping and encouraging team members to excel. As leaders progress upward in the organization, they learn to be more independent in their thinking and decision making. When this happens, they can lose sight of how important it is to rely on the other members of the team.

Unless you are self-employed and have no employees, leadership is not a solo act. Leaders must be willing to set ego aside and delegate responsibility, authority, and accountability. If the leader can let go of these things, he or she can begin to realize that the synergy of the team will propel the success of the leader.

The success of a leader is affected by their ability to create a team that accepts responsibility for the fire department's collective actions. When each member of the team accepts some ownership of the organization and feels responsible for its successes and failures, the team will be more cohesive and more supportive of each other. It is vitally important that each member of the team understand their role and the effect it has on the entire team. Letting your followers know what you expect of them and letting them tell you what they expect from you will lead to better teamwork, greater understanding of organizational expectations, and an appreciation for individual differences.

Leaders Are Driven by Goals and Vision

Great leaders establish clear objectives for themselves and for others. They engage followers in establishing mutually agreeable goals and expectations. For a team to work together, members must strive to accomplish a shared vision and goals. Followers must understand their responsibility in reaching the objectives and be committed to the goals. The good of the organization must supersede the self-interests of individual leaders and followers.

Successful leaders are visionary. They are able to conceptualize the future and anticipate what is on the horizon. Leaders then help their followers to envision the future success of the organization. The vision the leader is sharing can extend 3 to 5 years into the future. It is important for a leader to share this longer-term perspective to gain support in constructing the fire department's future design. Leaders ask for input into what the future should hold. They obtain buy-in from followers into the visioning process. This builds loyalty and ownership for the plan. A plan is much more likely to be followed if the leaders and followers work together to develop the fire department's direction. In other words, a process that involves followers in the development of goals and vision removes the blindfold and allows them to see all the challenges and opportunities that lie ahead.

Leaders Are Learners

Leaders are smart people. Sometimes their intelligence comes from formal education, and sometimes it comes from being self-taught. Nevertheless, it is indisputable that successful leaders are intellectually well rounded. Leaders *never* stop learning. Leaders read continually. Leaders are always taking classes. Moreover, leaders study other leaders to learn how other leaders handle personal and professional challenges and opportunities. However, it is not enough to simply read, take classes, and observe other leaders. Leaders must be able to glean the most salient lessons that will benefit them in times of need. That is what separates the smart leaders from everyone else.

A close relative of intelligence is job knowledge. Successful leaders are subject matter experts. They know their business inside and out. Most started in entry-level positions and worked their way up through the fire department, learning everything they could every step of the way. For those who did not start at the bottom, they have taken the time necessary to become intimate with the inner workings of the fire department. Expert knowledge allows leaders to gather facts and information quickly and accurately. They are able to organize that information into complex mental images of problems and opportunities they see on the horizon and formulate action plans to lead the fire department.

Leaders Have a Desire to Lead

Leaders have a tremendous desire to lead. This desire shows in everything a leader does. Successful leaders are ambitious people. The enthusiasm that they have for their work shows in everything that they do. Great leaders also have extremely high energy levels. They pour their heart and soul into the fire department. When you talk to a highly ambitious leader, you sense it instantly. The leader has an aura of excitement for what the fire department does and where it is going. That energy is contagious and draws in followers. Passionate leaders beget passionate followers. Leaders like to have fun. They genuinely enjoy the process of leading. When leaders enjoy what they do, it shows.

As many followers will attest, it is painful to work for a leader who is poorly suited for the position he is in or a leader who lacks the desire to lead. As much as a leader might be able to hide it, followers learn to read the leader. A leader who lacks desire will stand out like a neon sign. Leaders who lack desire become undesirable. Followers will struggle to follow the lead of an undesirable leader.

Leaders Have High Ethical Standards

Successful leaders have high ethical standards and conduct themselves at all times in a way that is morally unquestionable. A leader cannot instruct his followers to act a certain way that is incongruent with his or her own behavior. When this happens, followers can see the double standard, and it is confusing. The follower does not know whether to emulate the leader's words or actions. For example, a leader who is habitually tardy for work has no credibility in counseling a follower for the same offense. Although the rules for tardiness may be clearly articulated in policy, the leader cannot hold the follower to a standard that he is not willing to follow.

Some leaders fail to realize the scorecard for their ethical and moral behavior starts long before they are promoted into a leadership position. In fact, some would argue that the scorekeeping starts on the day that they are hired. Some leaders who did not realize this have found themselves in a very precarious situation once promoted. Perhaps you have known someone who spent many years as a follower, finding creative ways to bend the rules to their advantage. Then, this person gets promoted into a leadership role. When this happens, some of his or her coworkers, who have spent years watching them take liberties with the rules, are going to expect to get away with the same indiscretions. This leader will have a difficult time gaining the respect he needs to be an effective leader.

Leaders should also be aware that followers are always watching. Some leaders enjoy social relationships with their followers when away from work. Although there can be benefits to close working relationships, there can also be pitfalls. This is especially true if the leader believes that he can behave with high ethical standards while at work, while his off-duty behavior lacks moral fiber. It is rare that a follower can separate the behaviors they see in their leaders at work from the behaviors that they see away from work. Thus, a leader should always and in all ways, behave in a way that leaves nothing to doubt about his ethical standards.

Leaders Are Confident

To be successful, it is not enough for leaders to be smart and energetic. The leader must also possess a level of self-confidence that allows him to execute his leadership. There are few things as frustrating to followers as a leader who lacks the self-confidence to make a decision. Tied closely to self-confidence is the leader's self-esteem. Leaders with a healthy level of self-confidence and self-esteem do not let their desire to be popular or their fear of failing stand in the way of doing what is right for the fire department.

Followers enjoy a leader who is confident in his abilities. Do not, however, confuse confident with cocky. There is a fine line between being self-assured and being arrogant. Confident leaders are also humble leaders. They understand that leading is a privilege that is given by their followers. They understand that all of these traits are subject to the leader's humbleness. Overconfidence is a detriment that can destroy relationships that are built over time. The leader must be able to recognize the fine line between confidence and arrogance.

Leaders Are Self-Controlled

Leaders understand the power they have over people and the good and bad that can result from how they use that power. Successful leaders are socially well adjusted and are able to keep their ego in check. Leaders understand the nature of their power and the many ways they can influence followers. By the very nature of the position, leaders can create fear or excitation in followers. The person who possesses the authority to hire, terminate, promote, demote, reward, recognize, compensate, transfer, and assign work is powerful. Leaders have the power to influence the beliefs, attitudes, behaviors, and actions of their followers. This creates wonderful opportunities for leaders. However, a leader who lacks self-restraint can easily damage follower relationships.

Followers value leaders who are able to remain humble and control urges to use their power in unwise ways. Good leaders continually monitor their own performance. They pay very close attention to how they behave when faced with difficult challenges and exciting opportunities. Leaders are aware how they respond to unexpected events. They are resilient and able to change direction on short notice when the situation demands it. Much as a fire-ground commander continually monitors the impact of his decisions as the fire incident unfolds, so does a leader evaluate how his traits and abilities affect his followers and the success of the department.

Leaders Are Problem Solvers

Solving problems is an essential part of a leadership. In fact, if it were not for problems, it might be argued that leader's job would be far less valuable. Astute leaders know that problems to solve are the essence of their job security. Conversely, the job security of leaders with poor problem-solving skills may be in jeopardy. Successful leaders use a process in problem solving, as shown in the list below. The more complex the problem, the more important it is to use a problem-solving process similar to the following:

1. Define the problem.
2. Gather information about the problem.
3. Evaluate and analyze the information.
4. Generate action options.
5. Construct a best-fit solution.
6. Implement the solution.
7. Evaluate the results of the decision.
8. Change what does not work.

Problems are not always bad things to have. In fact, most of the time, problems give birth to wonderful opportunities. The most articulate leaders are those who can look at problems and resolve them in ways that the fire department emerges stronger and better prepared for the future. The leader who looks at problems as opportunities in disguise will soon find every challenge is a gift. It may not seem like it at the time. If you are facing a problem that seems to have no hidden opportunities, consult a mentor or close associate and ask him or her to take an unbiased, independent look at your problem. Sometimes the leader is too close to the problem or is too personally invested to be objective. A gifted mentor is especially helpful in generating action options (step 4 of the problem solving process). Sometimes your mentor's ideas will seem far fetched. Before you discount the input of an unbiased observer, take some time to reflect on what they are telling you. Sometimes the best advice comes from those unattached to the problem. A significant portion of the problem-solving process is to ensure that you are addressing the root cause. Failure to identify the root cause is in itself a grave problem in the process.

Leaders Value Relationships

Successful leaders realize they cannot be successful by working alone. They learn to value the relationships with other people. They know how to use social skills to help people work cooperatively toward a goal. They respect the differences that people have. Good leaders work very hard at establishing and maintaining a nurturing and trusting environment where those who get the work done are as important as the work itself.

People are complex beings. Time spent understanding and showing appreciation for each follower's contribution to the team will pay dividends. Leaders also spend plenty of time helping followers to establish and maintain relationships with each other. It is rare in the fire service for a person to be able to work alone on an assignment. Healthy relationships are as important, if not more, than any tool or equipment used to fight a fire. Friendship is similar to relationship. The best way to be a friend is to have a good relationship.

Leaders Share Information

Information is a powerful thing. Those who have it make better decisions. Yet some leaders withhold information from their followers, which prevents the followers from doing the best job they can. Clearly, there are some things that must remain private, such as personnel matters. However, most of the information held by a leader is not confidential. Some leaders fear that followers with too much information are a danger to them. Withholding information from followers is equivalent to expecting a company of fire fighters to attack a fire while withholding the water.

Successful leaders know that creative followers will obtain the information they need by one means or another. If the leader is withholding, followers will have little choice but to go around the leader to get the information from another source or the make up what they need to. Both of these situations are bad for the leader. The best way for a leader to influence followers is to share information in context. This is accomplished by providing followers with the information that they need to be successful coupled with dialogue to explain what the information means and how it will affect the work that the followers perform. Sharing information means sharing it in context in a way that is factual and unbiased.

Summary

Great leaders have many special talents or abilities. Understanding and applying a variety of traits, skills, styles, and behaviors helps leaders be their best. In the fire service, you know that every fire presents different challenges and requires you to apply different tools and techniques. The same is true for leadership.

On the fire ground, there are a handful of core principles you apply that help to ensure successful outcomes. Fire fighter safety is top priority. The focus is on saving lives and property and preserving the environment. Incident management and accountability systems are used. Fire fighters work in pairs or companies. Having a dependable water supply is important. What we have presented here are the principles to help you be a successful leader.

CHAPTER

13 The Value of Membership

Fred C. Windisch

Case Study

Chief Joe Hurley was working on a simple policy change for his department. He struggled for 2 weeks to produce the perfect seat belt rule, but he did not feel comfortable with his final draft. He did not want to be overly legislative in his approach, and his basic concern was for the membership to make the right decision. His gut feeling told him that legislating change or instituting a new rule was not going to work no matter how well the policy was laid out. He woke up at 2 a.m. and had a vision—maybe he could use some help.

Introduction

Fire chiefs should understand that there is a huge network of resources at their disposal. These formal and informal networks can share an array of value-driven resources—all you have to do is ask. Sometimes chief fire officers seem to forget that they cannot know everything. No man is an island.

The Big Picture

The International Association of Fire Chiefs (IAFC) is our portal to the world. Formed in 1873 to address hose standards, the IAFC continues to lead the fire service and has made great strides during the last decade in addressing national issues. Membership in this organization is inexpensive when weighed against the benefits to be gained. Being an active member allows the chief officer direct access to large amounts of information regarding leadership, strategies, and tactics for the fire service business that is so valued by our citizenry. Within the IAFC are specialty sections and committees that focus on the specific needs of emergency response agencies, and provide a forum called the Fire Service Leaders to enhance leadership abilities and brainstorm new ideas for dealing with today's fire service issues and demands.

This book is designed to assist the combination fire chief. With that in mind, there is one IAFC section specifically for educating the combination chief: the Volunteer and Combination Officers Section (VCOS). The mission is simple:

> The mission of the Volunteer and Combination Officers Section of the IAFC is to provide fire/rescue/EMS chief officers who manage volunteer/combination departments and affiliated support personnel with professional and leadership opportunities.

A Personal Experience

I can remember when I joined the IAFC's volunteer committee in 1989. A group of dedicated IAFC members accepted me and pulled the rope that provided many opportunities to interact with others as well as to improve the capabilities of the parent organization. Many years have passed since that fateful day, and the VCOS team still nurtures chief officers because the belief is that sharing expertise will benefit the whole.

■ Force in Numbers

It is too easy to sit back and let someone else do the job. The fact of the matter is that we have force in numbers. Today the IAFC's membership exceeds 12,500 members, and the VCOS has about 2500 members. Many of the IAFC's members have not added the VCOS to their membership credentials. The hope is that all chief officers who have volunteers in their systems will join the VCOS.

The United States Congress, your state's legislature, and your local government operate by the majority rule. It would seem apparent that "force in numbers" speaks loudly for issues that affect our organizations, so we must be a part of the national group to assure that our voice is heard. The opportunities to interact with one another are simply too great to ignore.

■ State, Regional, and Local Opportunities

Chief fire officers also need to consider and take steps towards becoming active in their respective state's fire chief associations. State fire chief's associations influence the direction of emergency response, and once again the strength in numbers will cause change to occur. The networking will also allow for leadership sharing that will result in time saved by addressing common issues among similar response systems. There is a growing trend of state chiefs' organizations having specific alignments with the IAFC. This formal network that unites fire chiefs is paying dividends that positively impact each of us in our daily work and on the national level.

Local and regional fire chief organizations have the same impact on affecting positive change and networking. Think about a flow chart where the locals interact with the regions, then the state, and then the national organization. Clearly, we can see that the direct linkages provide for a continuum of improvement for our agencies. There are far too many examples of failures within organizations that stem from the concept of "we can do it ourselves."

Summary

Chief Hurley's wake-up call morphed into a new policy development method that produced outstanding results. He contacted some friends, he researched the IAFC's resources, he included his members in the policy development, and now he can state with confidence that "everyone goes home." He knows that when he needs assistance that someone has already experienced the situation and has developed the needed procedures that we would otherwise struggle with on an individual basis.

Fire chiefs have great opportunities at their fingertips if they reach out to discuss issues that affect all of us. Take the time to join national, state, regional, and local organizations so that you can share experiences and make your system better every day. Join the IAFC and the VCOS by calling 703-273-0911 or via the Web.

APPENDIX

The Blue Ribbon Report

A Call for Action: Preserving and Improving the Future of the Volunteer Fire Service

Executive Summary

America's volunteer fire service has faithfully served our nation for more than 300 years. Volunteer firefighters serve their communities with dedication and enthusiasm. Volunteer fire departments save local communities approximately $37 billion[i] per year—money that can be reinvested to improve local infrastructure, social programs and minimize the local tax burden.

Since the terrorist attacks of September 11, 2001, America has learned that local emergency responders are the community's *first* line of response, regardless of the event. Community protection and well-being depends on the experience, expertise and tenure of local emergency service providers. The volunteer fire service faces significant challenges in overcoming a basic lack of resources—both financial and in human capital. Only by aggressively confronting both of these issues will we create the necessary atmosphere of stability that will allow volunteer fire and rescue departments to meet the new expectations and challenges of the 21st century.

While volunteer firefighters and emergency workers provide a tremendous contribution to our country, they are often under-funded and ill-equipped. Lacking cohesive national leadership, efforts to correct these problems are often fragmented and ineffective. Additionally, volunteer fire departments have a difficult time retaining volunteers. Ultimately, much of the blame for these problems can be attributed to poor leadership. Unfortunately, there are few programs at the local, state or national level to assist fire chiefs and volunteer managers in acquiring the knowledge and skills necessary for effective management.

Support from the local, state and federal levels is necessary to ensure that the volunteer fire service continues to be a full partner with all facets of homeland security response and effectively functions as the first line of defense within local communities. The International Association of Fire Chiefs believes that by enacting the specific recommendations outlined in the text of this report, we can enhance the role of the volunteer fire service in this critical mission. The most important of those recommendations are outlined below.

At the **local** level, we must:

1. Emphasize the importance of local support for this basic community service
2. Provide appropriate levels of funding for necessary safety gear and training
3. Engage in strategic planning that emphasizes volunteer retention
4. Use mutual aid to offset service and technical deficiencies
5. Use uniform incident management systems
6. Use performance measurement to measure and analyze response times, fire fighting effectiveness, training and retention rates of volunteer fire departments.

At the **state** level, we must:

1. Emphasize the importance of the state government in developing and promoting disaster planning
2. Certify fire and emergency medical services (EMS) personnel to comply with basic training standards
3. Promote regional service delivery where local capabilities and technical expertise are weak
4. Provide statewide volunteer benefit programs to protect both the firefighter and employer from the risks associated with volunteer fire service.

Finally, at the **federal** level, we must:

1. Work to produce a national climate encouraging individuals to volunteer within their local communities

Introduction

America's volunteer fire service is deeply woven into the basic fabric of our nation. According to the National Fire Protection Association (NFPA), there are close to 800,000 volunteer firefighters across the United States, and the majority of this nation's geographical area is protected by volunteer fire departments. Of all the fire departments in America, 73 percent are all-volunteer departments.

Firefighters, both career and volunteer, are extremely dedicated to public service. This trait explains why firefighters often take tremendous risks to save the lives of the citizens they are sworn to protect. Volunteer firefighters, because of their diverse educational and employment backgrounds, bring tremendous depth and diversity to any emergency scene based upon their regular jobs and expertise in their communities. In many cases, volunteer firefighters invest an enormous amount of time and dedication to fire fighting, moving the fire service forward through improved fire fighting techniques and technological innovations.

Volunteer firefighters provide an enormous economic benefit to our nation. It is estimated that volunteer firefighters save the American taxpayers $37 billion[ii] per year that can be reinvested in each community's infrastructure, social and other community programs, and/or a general reduction in local taxes.

Unfortunately, despite their tremendous contribution to American society, volunteer fire departments are often underfunded and ill-equipped, putting many in a position where they must raise their own operational funds to provide apparatus and safety equipment. In addition, the number of volunteer firefighters is declining across the country. During the mid-1980s, it was estimated the volunteer fire service was more than 880,000 members strong, but those numbers have dropped to less than 800,000 in recent years.[iii] Finally, the volunteer and combination fire service continues to be unorganized across the nation with no clear leadership representing the volunteers. It has no unified position on national legislative initiatives or research issues affecting their services. Volunteer departments and their managers will continue to struggle until local, state and national attention directs a concentrated effort to assist in preserving and improving the management of this long-standing American tradition.

The perception of the role of emergency services changed with the events of Sept. 11, 2001. These attacks against America changed the expectations of local emergency providers who are now clearly each community's first line of response, regardless of the event. As the country and local communities reevaluate their abilities to respond and handle new threats, such as weapons of mass destruction and biological incidents, the stability of the American volunteer fire service has become a significant issue. Community protection and well-being depends on the experience, expertise and longevity of local emergency service providers. It becomes imperative that local communities understand that the homeland is secure when the hometown is secure. Local communities and the leadership of those communities will look toward the volunteer fire department for answers to questions of terrorist threats and threat assessment. Enhancing the overall community safety is a new responsibility for local responders. Significant improvements in the volunteer fire service will be necessary to improve retention and create an atmosphere of stability, allowing local, volunteer fire departments to meet the new expectations and challenges of the 21st century.

Volunteer and Combination Fire Departments Across the United States: Examples of Value and Effectiveness

Campbell County, Wyo., is governed by a Joint Powers Fire Board and covers the City of Gillette, the Town of Wright and all of Campbell County, with a total response area of 5,000 square miles and a population of approximately 40,000 residents. One-third of the nation's coal supply is mined in this community. The combination fire department is composed of 19 career positions and 175 volunteers. The volunteer firefighters of Campbell County have saved local taxpayers more than $21 million in wages alone since 1996. The department's savings are calculated on the reduced need for full-time career staffing and the actual dollar savings for 226,243 donated hours[iv] during the study period. When assigned a value of $16.05 per volunteer hour[v] (used as a national mean), the volunteer contribution of $3,413,244 annually becomes a significant savings for the community.

The Campbell County Fire Department provides all of the normal city emergency services—fire suppression, emergency medical response, rescue, etc. It enjoys an above average working relationship with law enforcement and provides extensive industrial and wildland fire response expertise. The department offers full administrative services including building inspections, plan reviews, investigations, public education, vehicle and building maintenance, and an aggressive industrial fire training and hazardous materials training program to community businesses. All career employees provide both shift coverage and administrative duties. Tactical operations are considered fully integrated and all personnel, regardless of career or volunteer status, meet the same training and experience standards for the rank that they hold. The department retention rate for volunteers is 17 years per person.

The department is family based with yearly activities that support and promote a strong family unit. The department sponsors the Campbell County Cadet Program, which functions as a worksite for juvenile offenders and was chosen as the number one Junior Emergency Services program in the United States in 2000 by Volunteer Fire Insurance Services. Volunteers are active in a number of community events throughout the calendar year, including a community pancake feed serving more than 2,000 people on the Fourth of July and a number of fundraising projects to assist less fortunate families in the community.

The midwest village of **Tinley Park, Ill.** is protected by a 120-member paid on-call volunteer fire department. All firefighters are certified and tested under the state of Illinois certification program. Tinley Park provides coverage for hazardous materials incidents, and features a Combined Area Rescue Team (CART) that provides special services for building collapse and major structural incidents, as well as a Rapid Intervention Team (RIT). All department members are trained to the Hazardous Materials Awareness level, and members of CART and RIT are certified by the state of Illinois.

The department also employs two personnel specializing in public education, inspections, preplanning, and investigations, supporting the overall safety mission of the department and relieving these administrative duties from the volunteers. It is the largest volunteer fire department of this type in the state of Illinois, protecting a population of 56,000 residents and an estimated 100,000 daytime work population within the 17-square mile area. Full city services are provided from four fully equipped fire stations, and personnel are trained at a state-of-the-art training center. Tinley Park has an Insurance Services Office (ISO) rating of Class 3. The department averages 800 calls per year with a turnout rate of 30 firefighters per call.[vi] The department also assists the local EMS provider with incidents requiring extrication or reported entrapment. All fire department motorized equipment is secured through fund drives. For example, in 2004, community fund drives will finance and pay for the cost of one Class A pumper and a one combination Quint 95' aerial unit with a total cost $1,500,000.

The department boasts a retention rate of approximately six years per firefighter. Because of the volunteer coverage, the estimated yearly savings to the village exceeds $3,744,000 per year, deducting the direct volunteer expenses. This is one-quarter the cost of a full-time department.

In **German Township, Ind.**, the predominantly volunteer department (two paid personnel and 70 active volunteers) serves 11,000 residents and provides the community a direct savings in staffing costs of $441,000 per year. German Township Volunteer Fire Department responds with an average of 10 volunteer personnel per call. To replace the volunteers with an all paid staff would cost their residents more than $1.5 million dollars annually.

The community is a suburban bedroom community. Nearly 99 percent of the residents own their residences, and 50 percent of the population has moved into the community within the last 10 years. The fire protection challenges are significantly impacted by the availability of water. A large segment of the population and geographical area has a rural water system that does not provide hydrants every 500 or 1,000 feet. The other portion of the population is protected by a municipal water system that does provide hydrants in the normal configuration. The water or lack of it requires the department purchase apparatus with large water tanks.

The department historically has made a significant commitment to training its members. It has always been its goal that each member is highly trained and competent in all necessary skills. The department's training program is outcome-based and requires a significant investment of time and energy. The instructional staff has identified more than 70 basic skills, and written drills have been developed to allow members to train and measure their competency without attending every regular training session. These basic skills drills have significantly improved the members' competency. They know that when confronted with a dangerous situation, they will be able to perform the fire ground evolution safely and effectively.

Leadership development and certification are encouraged, and in most cases tuition reimbursement is available for course work. The promotional process for leadership positions is based upon a written test, experience, education, seniority and personal performance evaluation. It is not based upon an election or the buddy system. Officers maintain their ranks on a permanent basis provided they continue to receive satisfactory evaluations.

The **Ponderosa Volunteer Fire Department** is an ISO Class 3 rated, combination fire department in northern Harris County (Houston), Texas—the third most populous county in the nation. The department, formed in 1972 as population growth in the area exploded, currently serves a population of approximately 45,000 people in 13 square miles and provides the community a direct savings of $439,000 per year based on the hours donated by volunteers. To provide the same coverage with an all-career department would cost the taxpayers an additional $3,315,000 per year in personnel expenses.[vii]

The county lacks the tax base to provide the necessary funding to transition to full-time career positions. The Emergency Services District levies a tax of 6 cents per hundred dollars of evaluation, which equates to $60 per $100,000 of property value that fund all operating and capital expenditures. The 65 volunteers continually demonstrate their commitment to the community by their performance and by maintaining a very effective response system that includes fire, technical rescue, EMS first response, water rescue, hazardous materials response, public education and a host of other services. The cost of the high quality services is only $27 per resident, which compares to full career departments that are above $110 per resident.[viii]

The **Roseville, Minn. Fire Department** is staffed by two full-time career firefighters and 70 volunteer firefighters serving a first-ring suburb of Minneapolis and St. Paul. The volunteers represent a vast cross-section of the community, ranging in age from 18 to 55. The chosen full-time career fields of Roseville's volunteers include: police officers, accountants, software engineers, bankers, career firefighters, city employees, teachers and a dentist. More than half of the department's members have college degrees. In addition to their very demanding full-time jobs and family commitments, each volunteer contributes an average of 16 hours every week serving the community. Many say that serving as a volunteer firefighter completes their lives, giving them an opportunity to serve others during difficult times and gives them a reward and sense of fulfillment and teamwork they are unable to achieve in their full-time occupations.

Providing fire and rescue services from three stations, the volunteer department consumes only 3.8 percent of the city's $35 million budget, easily earning it the accolades of best value in town. In addition to an intense commitment to provide high quality service for the department's 700 annual emergency responses, members contribute thousands of hours supporting hundreds of community events each year. Throughout the summer, Roseville firefighters are frequent visitors to the city's 28 parks, giving hundreds of kids of all ages an opportunity to ride a fire engine and learn fire prevention tips. Firefighters will dress-up a parent in firefighter gear, using the opportunity to teach kids about the equipment firefighters use while educating the parents about the cost of a firefighter's ensemble. Kids are quizzed on fire safety and awarded prizes for correct answers. Roseville's firefighters attend more than 100 community block parties each year, regularly visit senior centers and pre-schools, hosts birthday parties in the fire stations and occasionally show up with a fire engine when the candles are lit on the cake during a celebration of a special senior resident's birthday.

Each and every one of these events represents an opportunity to educate, a role the department takes seriously. While some departments focus primarily on the response to emergency calls, in Roseville, the priority is prevention and education. It's no accident that first line of the department's mission statement reads "To continually strive for the prevention of fires, injuries and accidents. . ." When it comes to emergency responses, the department is well-trained, well-equipped and well-prepared. In 2001, the ISO scored the department with a 79.36 (ISO Rating 3).

Issues Confronting the Volunteer Fire Service

While there are many volunteer fire departments across the country that play a vibrant role in their community—as exemplified by the examples mentioned in the previous section—much of the volunteer fire service across the United States is currently in crisis. While many departments function at a very high level, many other departments struggle for their very existence. Particularly in rural areas, volunteer departments are closing their doors and shuttering their windows for two basic reasons: 1) lack of financial resources and 2) lack of volunteers. But this problem is not only found in rural America. Many volunteer departments in more populated areas are in a state of crisis and face a deep-seated struggle to provide adequate services. In order to ensure that we maintain a vibrant, capable volunteer fire service throughout the United States, we must confront both of these complex problems head on.

■ Lack of Resources

Few local governments understand the true value of their local volunteer fire department—both in financial terms as well as the social capital generated by the department. A number of departments are independent corporations that do not have direct attachment to their local government, yet they are the sole providers of emergency services. In addition, volunteer fire departments often serve as the social and communal hub of their towns. As detailed by the earlier examples, volunteer fire departments save local communities significant expenses. Unfortunately, most volunteer firefighters not only donate their time for this basic community service but also are required to spend a significant amount of time conducting fundraisers to generate revenue. In many communities, local governments take for granted the services provided by the volunteer fire department. They are not willing to assist with even the most basic expenses, such as appropriate safety gear, functional apparatus or station facilities.

Compounding this problem, the demands on volunteer fire departments have increased significantly over the past 20 years. Today, because of increasing call volumes, departments provide more and more traditional services (firefighting, EMS response, etc.). However, volunteer fire departments also are being asked to expand their role in order to address new problems, the most prominent of which are new duties surrounding homeland security. This increase in responses and responsibility, combined with the lack of resources noted above, means that many departments must make hard choices about the level of service they can provide. This is difficult in a mobile society, where urban dwellers often move to more remote locations and continue to expect the same level of service they were receiving previously. Often, they do not appreciate the funding constraints placed on rural communities.

In addition, the costs associated with new apparatus and equipment have increased exponentially. In 1972, a Class A pumper was about $25,000; today a new pumper can easily approach $350,000. Just a few years ago, a single self-contained breathing apparatus (SCBA) cost about $1,900; today an SCBA unit costs about $3,500. The cost for this basic equipment has increased over and above the funding levels available to many volunteer fire departments. As a result, many communities have had to reduce their capabilities by not purchasing needed apparatus, equipment and technology. Other communities have reacted by extending the life of their current equipment. Unfortunately, this decision can give rise to numerous safety related issues.

The following charts demonstrate the growth in emergency response calls in the United States. Total emergency calls in the United States have increased by an estimated 61 percent since 1983[ix] to nearly 18,000,000 responses per year.

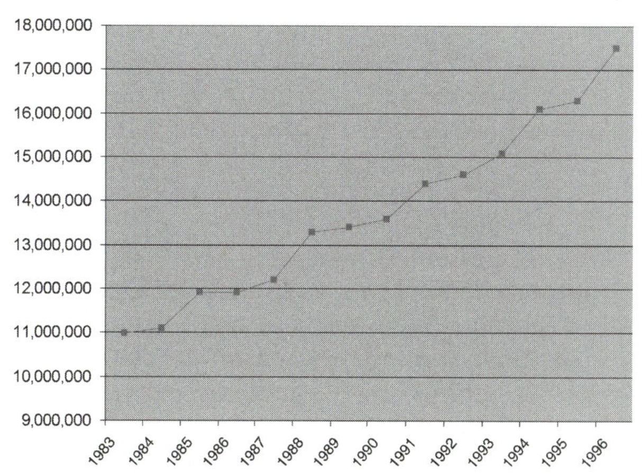

What is particularly interesting about these statistics is the change in the nature of emergency responses. While total responses have increased, the number of actual working fires has decreased 47 percent since 1977.[x] Residential fires have decreased from a reported 472,000 incidents in 1992 to 396,500 fires in 2001, a 16 percent reduction.[xi] Because departments are responding to fewer fires, managers are often concerned about the promotion of engine/command officers who lack sufficient experience actually fighting fires. This problem affects the safety of emergency operations and could lead to increased liability exposure for departments.

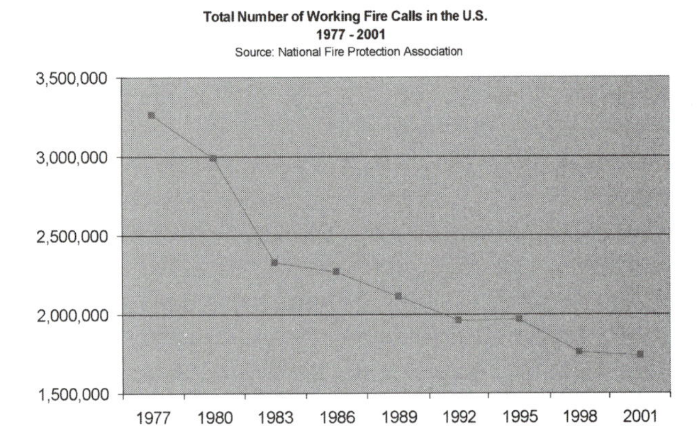

Detailing the Lack of Resources

As part of an effort to better understand the needs of the fire service, the Congress directed the National Fire Protection Association (NFPA) to conduct a Needs Assessment Study of the U.S. Fire Service for the United States Fire Administration (USFA). The study attempted to define problem areas in the nation's fire service as well as function as a guide for future planning to enhance the fire service and firefighter safety.[xii]

The following issues were outlined in the executive summary provided in the NFPA report.[xiii] While the report surveyed all types of fire departments, items selected for this report have the most impact on volunteer/combination departments. All of the problems documented below are a greater problem in smaller communities.

Concerns with Facilities, Apparatus and Equipment

- Roughly 15,500 fire stations (32 percent) are at least 40 years old and 27,500 fire stations (57 percent) have no backup electrical power.
- It is estimated that 60 to 75 percent of fire departments have too few fire stations to meet maximum response distance guidelines promulgated by the Insurance Services Office.
- Approximately half of all fire engines are at least 15 years old and more than one-third are over 20 years old.
- One-third of firefighters per response are not equipped with self-contained breathing apparatus (SCBA) and nearly half of SCBA units are at least 10 years old.
- Fifty percent of emergency responders per shift are not equipped with personal alert safety system (PASS) devices that assist in locating firefighters trapped in burning buildings.
- An estimated 57,000 firefighters lack even basic personal protective clothing, and an estimated one-third of personal protective clothing is at least 10 years old.

Communications and Communications Equipment

- Fire departments do not have enough portable radios to equip more than half of the emergency responders. This is a particular problem in small communities.
- Only 25 percent of fire departments can communicate on scene with all of their public safety partners at the local, state and federal level.
- Forty percent of all fire departments lack internet access.

Training Concerns

- An estimated 233,000 firefighters, most of whom are volunteers serving in small communities, lack formal training in structural firefighting—the most basic service the volunteer fire service provides. An additional 153,000 firefighters have received some training but lack certification in structural firefighting.
- An estimated 27 percent of fire department personnel involved in delivering EMS lack formal training in those duties. And in the majority of fire departments, EMS personnel are not certified to the level of Basic Life Support.
- An estimated 40 percent of fire department personnel involved in hazardous materials response lack formal training in those duties; the majority of them serve in smaller communities. In 80 percent of fire departments, personnel involved in hazardous materials response are not certified to the operational level.
- An estimated 41 percent of fire department personnel involved in wildland fire fighting lack

formal training in those duties; there are substantial training and certification needs in communities of all sizes.

Ability to Handle Unusually Challenging Incidents

- Only 11 percent of fire departments can handle a technical rescue with EMS at a structural collapse of a building involving 50 occupants with local trained personnel. Nearly half of all departments consider such an incident outside their scope.
- Only 13 percent of fire departments can handle a hazmat and EMS incident involving chemical and/or biological agents and 10 injuries with locally trained personnel. Forty percent of all departments consider such an incident outside their scope.
- Only 26 percent of fire departments can handle a wildland-urban interface fire affecting 500 acres with locally trained personnel. One-third of all departments consider such an incident outside their scope.
- Only 12 percent of fire departments can handle mitigation of a developing major flood with locally trained personnel. The majority of departments consider such an incident outside their scope.

■ Lack of Volunteers

Nationally, the number of volunteers has continued to drop since a high of 880,000 in 1984.[xiv] Today, the total number of volunteer firefighters has declined by about 10 percent, representing a reduction of approximately 90,000 individuals to 790,000.

The decline in the number of volunteers is a two-faceted problem. It stems both from difficulties in retaining current volunteers as well as problems with recruiting new volunteers.

Number of Volunteer Firefighters 1984-2001

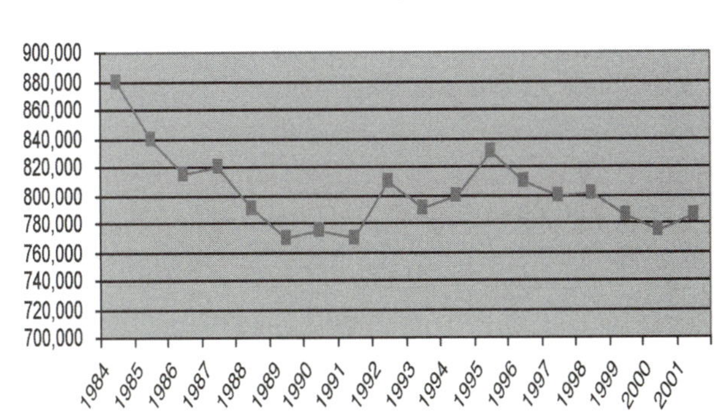

Retention

Retention of volunteer firefighters is a substantial concern for the fire service. It is estimated that the national retention average of volunteer firefighters is approximately four years[xv] per person, per department. When recruiting volunteer firefighters, the fire department will spend approximately $4,000 per person in orienting, equipping and training those recruits. While this figure would appear to be conservative in some jurisdictions, the cost to maintain one volunteer for the national retention rate average of four years is approximately $1,000 per year. This may not be an effective investment based on the return by the volunteer.

Retention of volunteer firefighters is a complex issue with a number of variables that can contribute to the lack of longevity. In 1993, the consulting firm Tri-Data, on behalf of the United States Fire Administration (USFA), conducted a national study titled *Retention and Recruitment in the Volunteer Fire Service, Problems and Solutions.* The study was assembled with input from volunteer departments across the country. The following areas were identified as major issues affecting retention of volunteer firefighters:

- Volunteers face increased demands from the fire department stemming from the increase in emergency response calls, the need for ongoing training and the increasing need to undertake specialized training.
- Demands on the volunteer's time are also increasing away from the fire department as families struggle to balance the career and family obligations of today's two-income families.
- Many of those who volunteer for the fire department do so in order to improve their employability. A volunteer fire department will provide training at no cost. This training can then be used to obtain a full-time position within the profession.
- The lack of a comprehensive benefit and incentive program. Benefits are necessary to protect the livelihood of the volunteer and his or her family in the event the volunteer suffers a significant injury or dies while on duty, while incentives are designed to recognize their personal achievements and to motivate them to improve their skills and participation.
- Finally, the lack of quality local leadership within the fire department is cited as the most

significant problem to retaining volunteer firefighters. Echoing the sentiment of that finding, it is the opinion of the contributing authors that ineffective leadership will doom an otherwise excellent organization. Sound management practices have the potential to significantly enhance retention rates.

The Value of Good Management

The following passage is taken directly from the Tri-Data report discussed above.

> The ability of a fire department to retain its people is directly related to its ability to manage those people. *It was unanimous among workshop attendees that poor management contributed heavily to people leaving the volunteer fire service.* The leadership issue was considered the most important; in one way or another, nearly all the other causes were either directly or indirectly traced back to the leadership problem. (emphasis added)[xvi]

The lack of quality leadership is the most critical issue confronting the volunteer and combination fire service. Few programs at the state or national level have been established to assist and provide fire chiefs and/or managers with the skills necessary for effective management. An example of how poor management can exacerbate a problem, such as an increase in call volume, is illustrated through the example below.

An increase in emergency service calls can significantly affect volunteer retention, so an effective manager will look at ways to minimize this intrusion on the daily life of a volunteer. A department that provides emergency medical services (EMS) will intrude on the life of a volunteer more often than those departments without EMS. EMS is an emergency response that can be reasonably predicted. As a result, staffing for EMS response is generally easier than staffing for activities that occur with a much lower frequency—such as structural fires. In addition, the number of staff required to respond to each call is relatively low. Three emergency care providers can handle the overwhelming majority of EMS calls. When a volunteer fire department providing EMS alerts a volunteer component of 20 members to an EMS call when only three members are needed, it can be damaging to a system. This intrusion into the life of the volunteer sets up a "cry wolf" syndrome where the pager is alerted but the volunteer is not needed. This increases the risk that the volunteer will not respond when actually needed.

The Challenges of Managing a Combination Department

Another difficult management challenge is the management of fire departments staffed with both career and volunteer personnel—combination departments. Combination fire departments are difficult to manage because career and volunteer firefighters often have different institutional interests. Administrative changes such as the transition from an all-volunteer department to a combination system may exacerbate the problem. The individual volunteer's sense of identity is important. Although the financial consequence of resigning a volunteer position is small, the psychological cost to an individual is extremely high because of the firefighter's great personal investment in the organization. The structural distrust the volunteer and career groups have for each other might be more tolerable if each group did not have to work with the other, but they usually do. Efficiency is a desirable goal; however, reaching that goal can be a tortuous path of management anxiety arising from personnel conflict between the two groups. The conflicts within a combination department can lead to unproductive involvement by the local government that sees itself as legally and often politically responsible for resolving the conflict.

A combination system will not work when it is based on prejudice or when either group of firefighters, volunteer or career, functions in a minority role and is perceived as subservient to the other. This situation often creates an atmosphere where the department is unable to tap the knowledge and expertise possessed by the individual. This can be perpetuated when we lose site of our basic mission—serving the public. The real test of a successful combination department is its ability to fully integrate tactical rank structure. The training and performance standards should be the same, regardless of the firefighter or officer status with parallel lines of authority, bringing personnel resources into harmony. The quantitative measure of that success is the retention rate of the minority group.

Nature of Volunteerism in the United States and Its Implications for the Volunteer Fire Service

In 2002, the Bureau of Labor Statistics of the U.S. Department of Labor released the Volunteer Service Indicator, a new national measurement of volunteer behavior developed by the Census Bureau, the Bureau of Labor Statistics and the USA Freedom Corps. The indicator provides a wealth of information relating to volunteerism in the United States. Those findings indicate that 27.6 percent of individuals (more than 59 million) over the age of 16 volunteered with a volunteer service organization between September 2001 and September 2002. The findings suggest that certain groups are more likely to volunteer, while others are more likely to volunteer more hours. The findings also offer information regarding what types of organizations and activities enjoy support from different segments of the population. Finally, they give information on how much time people are dedicating to volunteer work, including data that more than 34 percent of those who volunteered did so for more than 100 hours during the past year.

Perhaps the most striking statistic from the survey is that volunteers spent a median of 52 hours volunteering during the year. Volunteering for the fire service can be and most often is substantially more demanding. Depending on the call volume, designated shift coverage and the level of training standards mandated by the local community, an average volunteer could easily contribute in excess of 1,000 hours per year in community service. In Campbell County, Wyo., an active average volunteer can expect to donate 750–1,000 hours of service per year,[xvii] German Township, Ind., 500 hours per year[xviii] and in Ponderosa, Tex., 360 hours per year.[xix] In two communities with mandatory 24-hour volunteer shift coverage—Tinley Park, Ill. and Roseville, Minn.—an active volunteer will be required to provide 1,000 hours[xx] to maintain his or her volunteer membership. Both departments provide volunteers with a monetary stipend as part of the compensation for services provided.

The estimated 800,000 volunteer firefighters account for less than one fifty-ninth of the estimated number of individuals who volunteered, in some fashion, for their communities during the time of this study. The available personnel pool for volunteer firefighters may be more extensive than we realize, and a more detailed review of this study may provide insight into the recruiting strategies and diversification options that must be developed to fill open positions within our departments.

To be competitive, the volunteer fire service may need to refocus recruiting efforts, develop diversification strategies and design other volunteer opportunities within the organization that utilize skills outside of traditional recruitment considerations.

■ Recommendations–A Call for Action

The International Association of Fire Chiefs represents the leaders of America's fire service, both career and volunteer. Through the technical expertise and guidance of its Volunteer & Combination Officers Section (VCOS), the IAFC is well positioned to lead the volunteer fire service forward to confront the difficult issues detailed in this report. The recommendations that follow are broken down by the level of government that should address the solution. While most of these recommendations must be implemented at the local level, the IAFC will be active at the national level to secure the necessary resources and climate to make these important changes in the volunteer fire service.

Federal Responsibilities/Recommendations:

- Advance a Congressional Resolution supporting the American Volunteer Firefighters Bill of Rights.
- Create an Office of Volunteer and Combination Fire Service within the Department of Homeland Security.
- Develop a grading system for evaluating local emergency response capability.
- Create a national definition of allowable compensation for volunteer firefighters.
- Develop and support administrative changes to the Internal Revenue Code to clarify legislative issues related to length of service awards programs and allow "cafeteria style" benefit programs for volunteers.
- Create national job protection for volunteer firefighters.
- Fund the Assistance to Firefighters Grant Program (FIRE Act) at its full authorization, allowing the fire service to build a solid baseline of apparatus and safety equipment within its hometown communities.
- Appropriate funding for the Staffing for Adequate Fire and Emergency Response Firefighters Act (SAFER Act).
- Provide tax incentives for the installation of automatic fire suppression and alert systems.
- Provide national tax incentives for certified volunteer firefighters, reducing federal income tax by 3 percent annually.

State-Level Responsibilities/Recommendations:

- Develop community, regional and state disaster plans with specified review dates. Plans should include identified resources and certifiably trained personnel available for regional and statewide deployment.
- Develop methods for certifying fire and EMS personnel to enhance their professional commitment and achieve minimum training standards.
- Develop a benefits plan for all emergency responders to protect and provide for responders who are injured or killed in the line of duty.
- Develop a benefits plan that provides college tuition, including books, to the immediate family members of firefighters killed in the line of duty.
- Develop regional and statewide recruitment campaigns.
- Assure that volunteer liability protection is provided.
- Assure that statewide mutual aid places response liability on the responding agency versus the requesting agency.

Local Responsibilities/Recommendations:

- Volunteer firefighters, leadership within the volunteer fire department, community leaders, elected officials and citizens should expect that standards, rules and regulations be used, adopted and enforced at the local level that measure the effectiveness of services provided.
- Strategic planning must become institutionalized as an integral part of fire department operations and community resource allocation.
- Plan development should be performed in conjunction with the community to meet community expectations, growth and staffing requirements.
- Planning should be done in conjunction with surrounding communities affected by automatic mutual aid agreements. A multi-jurisdictional approach must be utilized to provide specialized services such as technical rescue, hazardous materials response and water rescue as well as covering training needs for these responses.
- The planning process should be developed with immediate, intermediate and long-range goals and have established review dates.
- An evaluation of the current volunteer response capabilities must be completed as part of the strategic planning process.
- A risk management policy must be instituted that clearly identifies the necessity of performing defensive fire suppression operations under noted conditions.
- A management structure must be developed and maintained. It should address business management operations, training, EMS, member benefits and operational leadership strategies.
- The number of calls significantly increases the business aspect of running a fire department. A department that responds to more than 750 calls per year, which is an average of two calls per day, should consider providing a compensated leadership position for developing and executing an organizational plan.
- If transition to paid personnel is necessary, the emergency service delivery system must prepare for an orderly transition from an all-volunteer to a partial career staff with identifiable funding options. Critical issues such as pay rates, job descriptions, duties, responsibilities, positions and status authority for career and volunteer personel must be examined. When the overall composition of the department is predominately volunteer, then career personnel serve to support the volunteer system.
- A funding plan for vehicle and equipment maintenance and replacement, as well as a plan to replace personal protective gear and accessory equipment in order to ensure adequate protection of emergency service personnel should be developed.
- Local, county, regional, state, federal and industrial resources that are available within the jurisdiction should be identified as part of a mutual aid agreement.
- The organization must develop a service delivery approach to meet the risks that are presented, consistent with what the community expects and can afford (standard of response cover).

Recruiting and retaining quality personnel continues to be the most important element in the overall success of a volunteer or combination fire department. Therefore, it is important to look at developing the following:

- Programs designed to certify and credential volunteer and career firefighters as well as officer positions at the state minimum level (NFPA Firefighter I/Fire Officer I or equivalent) to improve individual educational levels, emergency scene proficiency and safety.
- A diversification plan that maximizes individual talent and skill in order to enhance the overall efficiency, safety and effectiveness of the department. It should also guide the educational growth of the individual while maximizing his or her potential and enthusiasm in a specific discipline(s) within the organization.
- Ongoing educational opportunities that reinforce minimum training standards, enhance awareness and reinforce safety precautions dealing with local target hazards.
- Training that is measurable and emphasizes safety, command, multi-company drills, multi-agency drills and multi-jurisdictional responses.
- Benefit programs that encourage long-term participation from individual volunteers. Programs could include, but are not limited to, workers compensation; health, accident and life insurance; and coverage that will protect the livelihood of the individual volunteer against lost wages.
- A housing analysis to document housing availability and, if necessary, contingent housing alternatives for retaining reliable and well-trained volunteers within a community. Those options may include, but are not limited to, subsidized housing, dormitories, low or no-interest loans or relief on property taxes.
- Adequate liability coverage to protect an employer from costs associated with injuries that occur while performing duties. This consideration may extend to policies that provide the employer with overtime coverage to fill the position of the injured volunteer.
- A recruitment program that ensures adequate staffing and delivery of emergency services.
- Appropriate recognition and award programs to identify individuals or team members because of their performance or commitment to the department and community.
- A promotional process that ensures fairness for all members within the existing rank structure. Promotional systems should replace the traditional method of electing officer positions. It should be based upon merit with appropriate performance, education, training, skills and experience.
- Partnerships with other community emergency entities working to maximize resources.
- Partnerships with civic organizations and local businesses to integrate the fire department within the local community.
- Training programs that provide all new recruits with basic firefighting skills and First Responder level training before they are allowed to respond to and perform on fire, medical or rescue emergencies.
- A physical assessment program designed to evaluate each member's physical ability to perform the activities and tasks required for every job description within the organization. This assessment should be performed at least annually.
- A written policy prohibiting drug and alcohol use with specific enforcement, discipline and follow-up procedures.
- An "Emergency Vehicle Operational Policy" to qualify each member as a driver/operator of fire and rescue apparatus.
- A process to check the status of each member's driver's license annually.
- Criminal background checks on all prospective members.

Community Support Services are necessary elements to the overall image and success of the department and the well-being of the community. Departments should develop the following:

- Fire prevention and education programs to educate at risk groups as identified by the USFA. Programs should direct educational, awareness, prevention and support groups to assist in reducing concerns.
- Safety and accident prevention programs beyond the normal scope of fire prevention to augment identified needs of the community. Those programs could include, but are not limited to, drowning prevention; bike, rollerblade and car safety; and sponsorship of SAFE KIDS projects.
- Practices that would prevent fire loss, injury or death based upon occupancy, construction, apparatus, water supply, available personnel, communication abilities and response capabilities.

- An annual evaluation of water systems that affect local operations, including county, industrial and/or private delivery. Evaluations should include the capability of the water supply to deliver the required fire flows based upon existing occupancy as well as planned growth. Ensure that appropriate steps and procedures are in place to properly maintain supply.
- Customer service programs that provide community feedback and satisfaction ratings.
- The capability to complete investigations in an efficient and reliable manner involving police agencies where applicable.
- Appropriate pre-plan documents, including target hazards, to provide timely and accurate information to incident commanders.
- A partnership with the Local Emergency Planning Committee to work for a fire safe community.

The volunteer fire service is at a critical juncture in the United States. On one hand we have a positive can-do spirit, on the other hand we have forces that are creating ever-increasing challenges that attack that spirit. The needs and realities of the volunteer fire service appear to be moving in divergent directions, so when the spirit dies, all that remains is historic fact. It is imperative that local, state and federal government understand the challenges listed in this document, develop a problem solving attitude and be proactive in creating a new pathway that will allow the volunteer fire service to survive and flourish. The IAFC stands ready to work with all partners to lead this charge. This great country cannot afford to lose the rich legacy of the volunteer fire service.

End Notes

i *Fire Protection in Rural America: A Challenge for the Future.* National Association of State Foresters, 1993.

ii *Fire Protection in Rural America: A Challenge for the Future.* National Association of State Foresters, 1993.

iii *U.S. Fire Department Profile Through 2000.* National Fire Protection Association, Quincy, MA, December 2001.

iv *Coal Bed Methane Exploration, Campbell County Fire Department Partners in Progress*, Impact Study Prepared for the Campbell County Commissioners January 2002, Addendum Report 2003. Campbell County Fire Department managed by a Joint Powers Fire Board responsible for the City of Gillette, WY, Town of Wright, WY, and unincorporated areas of Campbell County, WY.

v Figure of $16.05 per hour provided by the National Volunteer Center as a national means for calculating time donated by volunteers.

vi Tinley Park Village Fire Department volunteer firefighter staffing agreement per 24-hour shift.

vii Ponderosa VFD Response and Training Statistics 2002, Ponderosa, TX.

viii *Firehouse* magazine, Fire Department Annual Statistics.

ix *U.S. Fire Department Profile Through 2000.* National Fire Protection Association, Quincy, MA, December 2001.

x *U.S. Fire Problem 1977–2001.* National Fire Protection Association, Fire Analysis and Research Division, Quincy, MA, March 28, 2003.

xi *U.S. Residential Fire Data 1992 – 2001.* National Fire Protection Association, Fire Analysis and Research Division, Quincy, MA, March 28, 2003.

xii *A Needs Assessment of the U.S. Fire Service*, A Cooperative Study Authorized by U.S. Public Law 106-398, FA-240/December 2002.

xiii *ibid.*, pages iii–ix.

xiv *U.S. Fire Department Profile Through 2000.* National Fire Protection Association, Quincy, MA, December 2001.

xv Figure is estimated based on the experiences of the authors. No formal studies have been developed to accurately define this figure.

xvi Retention and Recruitment in the Volunteer Fire Service, Problems and Solutions, National Volunteer Fire Council and The U.S. Fire Administration, August 1993, pg 1.

xvii *Coal Bed Methane Exploration, Campbell County Fire Department Partners in Progress*, Impact Study Prepared for the Campbell County Commissioners January 2002, Addendum Report 2003. Campbell County Fire Department managed by a Joint Powers Fire Board responsible for the City of Gillette, WY, Town of Wright, WY, and unincorporated areas of Campbell County, WY. Volunteers actively participate in numerous wildland campaigns each summer.

xviii Calculation based on an average of three hours of training and seven hours of emergency response each.

xix Ponderosa VFD Response and Training Statistics 2002, Ponderosa, TX.

xx Based on an average of 16 hours of shift coverage and three hours of training per week.

APPENDIX

B The Red Ribbon Report

Lighting the Path of Evolution: Leading the Transition in Volunteer and Combination Fire Departments

Introduction

Nearly 300 million people live in the United States today and the number keeps growing. Many areas of the country that traditionally have relied on citizen volunteers to provide fire protection and emergency medical services are finding fewer people available or willing to carry on the honorable tradition. The demand for service grows and the number of providers declines. How are communities' needs to be met? Finding the answer to that question is one of the most daunting challenges facing local governments and fire service leaders all across the country. What is the appropriate level and menu of emergency services to be offered in the community? How do we assure that those services are delivered reliably? If not by volunteers, then by whom?

It's an issue of considerable national and local importance. As the March 2004 Blue Ribbon Report by the Volunteer and Combination Officers Section of the International Association of Fire Chiefs noted, of the 26,354 fire departments in the country, about three-quarters of them that serve 19,224 communities are staffed by volunteers. The balance—and these numbers have been rising as more departments are unable to provide adequate services using only volunteers—includes 4,892 departments that operate with a combination of compensated and volunteer staffing and 2,238 that are fully staffed by paid personnel. The 800,000 volunteer firefighters who today protect large areas of America number ten percent fewer than 20 years ago. Why the decline?

The answer lies in a combination of factors that reflect our society's evolution. The growth in population has meant an increase in the numbers of calls for service just about everywhere in the country, putting added pressure on the volunteer staffing component and systems. There's the matter of rising expectations by citizens in most communities that have led to demands for increasingly sophisticated services. External drivers, such as legislative mandates, legal considerations, and the need to deal with the potential threats of terrorism, have all had an impact on volunteers. So have family considerations: two-job parents, two-earner households, and more competition for personal and family time . . . they all factor into the decline in the number of volunteers on the front lines. What does this mean for fire service and community leaders?

The fire service is evolving as well; in fact, it always has been. As demand for services outstrips resources, there has been in many areas of the country a natural progression from departments fully staffed by volunteers, to some form of combination system, to a fully paid service. The pace of that change is different from place to place, as are the problems encountered along the way. It depends largely on how successfully deficiencies, at all levels in organizations, are identified and resolved by the chief officers and the extent to which appropriate services are delivered successfully.

An evolving and progressive volunteer fire department will encounter a number of service delivery options before actually migrating into the combination fire service arena. This evolution process, if managed, can be systematic and prolonged. There is no cookie-cutter approach to staffing an evolving department, but there are a variety of approaches that have been successful in many communities. This report will share several of those successful models. As the accompanying chart illustrates, there is a progression from a purely volunteer-staffed organization to one that is primarily staffed by paid personnel.

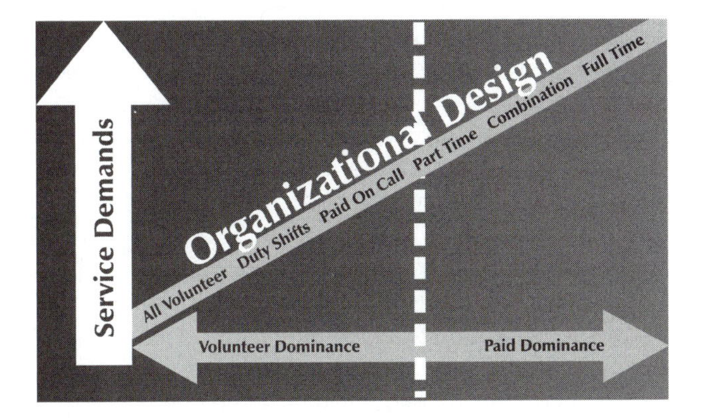

The goal of the fire service is to protect life and property by delivering the highest possible level of service consistent with need at the lowest possible cost consistent with safety. The goal of this report is to call attention to some of the strategies and options available to fire service and community leaders who are looking to do just that.

Signals of Change

Longtime volunteers often look back on the "way it used to be." They recall a time when training was much less demanding and time consuming and the local fire department had fewer responsibilities. Fires and accidents were pretty much the game. Attendance and training standards were achievable. There were fewer calls but each was an event that required the assistance of neighbors, who took great pride in their membership in the local department. The community appreciated their neighbors' help, local businesses supported the volunteer fire department, and the call volume was small enough so as not to interfere with the requirements of the members' jobs. The system was manageable, the emergencies were mitigated, and it was fun to be a member.

The reality today is that in many communities, to be a contributing, effective firefighter, a person has to meet significantly higher standards physically, in terms of training, and in terms of time "on the job" gaining experience. Not everyone has the luxury of time, or in some cases the inclination, to meet those requirements in today's hectic environment. Anymore, the fire department is not just a group of people trained to suppress fire and render first aid. It has become the premiere provider of choice for different levels of emergency medical services and in many cases transportation, as well as the provider of just about every other service that is not provided by the police department—hazardous materials response, high-rise and below-grade rescue, inspections, prevention and education, and community emergency planning and management, to name a few.

This is not to say that volunteers can't handle the job, for their abilities and successes are demonstrated daily in many places from coast to coast and border to border. But where they can not, community and fire leaders are challenged to meet their community's needs. In some cases, they will find ways to reinvigorate the volunteer members of their departments and improve their performance. In others, they will recognize the need for another type of change, moving to some form of partial or fully paid department, and they will set out to make it happen.

■ Indicators for Change

A natural evolution for a volunteer department is the growth in services and added responsibilities as the demographics of the community change. When the system develops problems, people generally know about them long before they are willing to admit that they need serious attention. For fire department managers and local government leaders, it is critical that they recognize the signs of problems ahead and prepare for change before it is forced on them by external circumstances. It is helpful when they recognize these pointers to change:

Community Growth

Emergency services are directly impacted by community growth—more people, more businesses, more emergencies. The larger a community, the higher level of service people expect. In many areas people moving to "suburbs" assume wrongly that emergency services are delivered in the same way they are provided in the more established cities and towns. A history of community growth and projected increases in demand can help managers forecast and plan for changes in the delivery of emergency services. In some cases, population growth projections might even help a department determine to limit its services based on available staffing.

Community Aging

A fire department's ability to recruit new members in part depends on the supply of new, younger people who can be tapped for service. A community's age profile can be an indicator of problems ahead. The age factor in your community is revealed by data showing who are moving in and moving out. If the younger people are moving away, or if schools are showing or expecting declining enrollment, the fire department may have a difficult time maintaining appropriate levels of service in the future.

Missed Calls

When an emergency call goes unanswered—a "scratch" on the East Coast or in other communities a "did not respond"—the fire department has a serious problem, not just because life and property are at stake, but also because it is a failure highly visible to the public. Equally serious is a department's overreliance on mutual aid for coverage and the lack of adequate personnel to handle subsequent calls when primary units are on an assignment.

Extended Response Times

When units regularly fail to get out of the fire station in a timely manner because of inadequate staffing resources, the community is endangered and fire department managers have a reliability problem. Response time is a critical factor for any fire department determined to provide appropriate service to the public. It is especially critical for medical calls when the first-due company fails to respond for whatever reason and an EMS unit responds but fails to meet the response-time standard, a common occurrence even when mutual aid is not involved.

Reduced Staffing

Units responding with fewer than the required number of people needed to perform that unit's functions pose a serious problem for the safety of citizens and the responders. This is another indicator of reduced service capability.

All of these situations indicate an inconsistency in a department's ability to provide necessary service, though not all are necessarily caused by a shortage of volunteer members. Staffing deficits can be related to other factors, such as changes in local business and industry policies regarding employees leaving the workplace, the number of volunteers who are employed outside their response areas, a lack of understanding on the part of new corporate managers of the community's needs, a tight labor market driven by rapid community growth, or even members' apathy. Where workforce restrictions are at play in the community, they typically lead to daytime response shortages and a significant challenge for the department.

Other Considerations

While employment issues tend to be the major factor in volunteer staffing shortages, other factors also contribute. Decreased interest among members who fail to participate could be the result of unreasonable community expectations, some problem with the fire department's internal requirements, or other organizational issues, such as:

- *Responsibilities outpace capabilities.* Mandated and selected responsibilities and response commitments exceed the department's capability to manage outcomes properly. Mandated responsibilities may have their basis in state statutes or local resolutions, proclamations and ordinances. Selected responsibilities are response categories that result from self-imposed obligations to provide a service.
- *Inability to raise funds.* Growth in the department as it faces new demands outpaces the volunteers' ability to raise capital and operational funds.
- *Waning political support.* A once-supportive political climate begins to falter and less emphasis is placed on the volunteer-staffed fire company. This becomes noticeable when apparatus is not replaced, new purchases are postponed, or local government wants the volunteer company to operate less expensively. The volunteer-staffed fire company needs to be a vital, supportive and healthy part of the local governmental infrastructure.
- *Internal conflict.* A department has internal struggles over its mission in the community and that conflict involves the preservation of the system as a fraternal organization rather than a service-delivery system.

- *Officers filling lower operational positions.* Staffing shortages that result in the fire chief driving the fire truck or fulfilling the responsibilities of other line firefighters is another sign of a serious staffing problem.
- *Mission creep.* When first-responder programs that once managed to provide essential services and also extra staffing for critical events and rescues become subject to all kinds of other assignments: or to policies that dictate that fire units respond every time an ambulance is dispatched, chronic staffing shortages can be a problem.
- *Controversy.* When internal controversy becomes the focal point and public image of the department, its effectiveness is impaired. Controversy can be inflamed by a poorly managed emergency, an event that exceeds the capabilities of the volunteers, or public criticism that home response is no longer adequate for the number of emergency calls handled by the department. The problems are exacerbated when the volunteers are unable to reorganize and meet the increased demands, or when the news begins to publicly question the effectiveness of the service. Few volunteers join the department to fail or be exposed to a community philosophy that "they tried hard, but they are just volunteers."
- *Too many jobs, too little time.* Another indicator: The department cannot provide fire prevention, public education or inspection responsibilities because of training and response demands occupy the time volunteers have to commit.
- *Kingdoms come first.* Some jurisdictions consider their response areas their "kingdoms." Boundary disputes can occur when department leaders fail to understand that the public does not care what color or name is on the fire truck. The "kingdom" attitude also leads to contentious working environments with neighboring agencies.
- *Lack of budget support.* Failure by elected officials to approve budgets that include capital expenditures for the department is an ominous sign.
- *Missed deadlines.* When critical administrative deadlines, such as daily response reports, training records, and legally required documentation are not completed or budget deadlines are not met, the department's effectiveness is compromised.
- *Catastrophic losses.* Catastrophic events, such as the loss of a firefighter or a civilian fatality, focus great attention on the department, and perhaps its problems and deficits, which can discourage members.
- *Volunteers priced out of the community.* In many communities the price of homes and property taxes makes it difficult for the children of current volunteers or others who have time to volunteer to live in the community, thus reducing the pool of potential members.
- *Demographic changes.* Shifts in the community that drive decisions by current members to purchase homes outside the fire district are a detriment to member retention.

When the Time for Change Has Come

Once a department recognizes there is a need for change, it must examine carefully both the organization and the options available to it. It is essential that all members of the organization identify the department's mission and core values. Whether in the end the change is a revitalized volunteer organization or a move to some type of paid or part-paid organization, a careful articulation of core values is critical to the success of the organization. Those core values must be incorporated and reinforced as employee strategies in new career positions and the core values must be carried throughout the evolution process. If the members expect the organization to be a mirror of what it once was, everyone must believe in and apply its core values. If you expect to maintain big city services with small town pride, the organization must maintain the focus on their core values and reinforce those values at every opportunity.

Once it is clear that change is necessary to preserve the department's ability to engage in its core mission, creating a paid staff is not necessarily the first option to consider. Having the answers to a number of key questions may help resolve a department's staffing issues.

Does the department have the right leadership? An initial examination of problems should always include a review of the fire department's leadership. The lack of dynamic, adequately prepared leaders has long been identified as a significant issue for the volunteer fire service. Poor leadership has a significant impact on the retention rate of volunteers, on a department's desire and ability to meet new levels of service demand, and on the quality of the service provided.

Does the department offer benefits and incentives? Benefits are safeguards provided by the community or the department to protect firefighters and their families against unexpected financial strain should the firefighter be injured, disabled or killed while on the job. As demands for service increase, so do the chances that firefighters will be injured or worse at the emergency scene. Departments need to provide protection—such as insurance and retirement or wage supplement plans—to ensure that the health, welfare and financial stability of firefighters and their families are protected. Such benefits are essential to assure that members are treated as valuable assets.

Incentives can provide motivation for members to improve personal performance and participation. These are defined by personal or team recognition programs or awards. Young people today, the future lifeblood of all fire departments, are interested in immediate feedback and that includes benefits and incentives. It is more cost-effective to pay for benefits than it is to pay people.

It is imperative that the community be involved in determining the level of support for volunteer or part-time firefighters. How willingly the community provides benefits for them now may help department leaders gauge its willingness to sustain a combination system, if one is needed.

Are department membership standards appropriate? Fire department leaders should review membership standards to ensure that they are appropriate for the services provided. Do you need to increase requirements to ensure that volunteers have adequate skills to deal with the dominant types of calls to which the department responds? Does the department really need a requirement that all members have the expertise and the responsibility to respond to all types of calls?

Can you use diversification strategies? It is critical for department leaders to understand that not everyone is equal in skills or abilities. Diversification strategies—essentially, not everyone in the department has to be proficient in all the jobs in the department—can be helpful in attracting new members. Diversification strategies are fairly simple. Recruit subject-matter experts for the different disciplines within the department. You can take advantage of that to attract new members and take pressure off of a small group of dedicated responders. For example, you might recruit from a number of professions within the community that deal with hazardous materials. Attract and train those individuals as volunteers and use them when chemical emergencies are dispatched. By implementing diversification strategies, you may actually improve your volunteer base by reducing the demand on all your members and enhancing their subject-matter expertise.

Trim the non-essentials. Review your organization's mission and values and identify the essential functions and services it is required to deliver. A review can, in some cases, lead to reducing or eliminating nonessential services. Remember, you can't be all things to all people.

Transitioning from the Present System—Alternative Delivery Systems

When it is clear that the present system is not working well, departments can follow a progressive path that leads from a completely volunteer-staffed organization to one that is staffed by some combination of volunteer and paid personnel. A department can stop anywhere along the path when that step leads to a satisfactory resolution of the community's fire department problem. The stop may be transitory or it could be permanent. The incremental approach helps a community achieve the best possible resolution of its issues at the lowest cost. Here are steps along the progressive path:

- Divide volunteer members into on-call duty shifts to ensure adequate coverage.
- Develop a program for volunteers to provide 24-hour coverage. Shift coverage needs to be flexible to accommodate individual commitments of 4, 6, 8 or 24 hours. Allow flexibility of start times and lengths of shifts to accomplish the coverage.
- Convert all-volunteer members to pay-per-call members, financially rewarding their participation.
- Implement regional response coverage and develop station-specific expertise.
- Develop Standard Operating Procedures or Standard Operating Guidelines.
- Establish a paid-on-call system, allowing the chief the flexibility to actually schedule shift coverage with financial compensation, essentially setting up part-time employee contracts with the volunteers.

- Create paid-on-call positions for specific job functions such as training, public education, inspections or administrative duties.
- Consider part-time employees specifically hired to provide coverage for inconsistent and sporadic volunteer coverage.
- Establish full-time career positions for daily shift coverage and completing administrative duties while supporting and maintaining a predominantly volunteer system.
- Convert a predominantly volunteer department to a 50/50 split, or predominantly career department where volunteers assume the supportive role.

Any of these solutions requires a new level of commitment, planning and consideration.

Typically, paid personnel are brought in to take on administrative duties or provide coverage for specialty services such as hazardous materials or technical rescue at a county or regional level, freeing volunteer firefighters to provide core services. (When an organization begins to pay personnel an hourly wage, they are subject to the Fair Labor Standards Act.) Doing so also sets the stage for more cooperative efforts on a regional basis. Regionalization of services clearly has a future in the volunteer fire service, providing economic relief and maximizing, not competing for, available volunteers.

Transitioning to a Combination System

Communities need to understand the forces that drive departments to consider transitioning, which may include hiring paid firefighters. Doing so is critical to a successful transition. The community's expectations about services and what they should include must guide how the fire department deploys and seeks additional resources. Such expectations are best identified in the local government by finding the balance between expectations of service and what the available funding is will support. These expectations can be expressed in the form of response goals that provide the fire department a benchmark for success. It is important that goals are not set internally. Sometimes the fire service choose what they "think" is right and move forward with the goal. Community feedback is essential to determining the correct path to the future.

Sample performance goals include:

- Average time from dispatch to response.
- Average time from response to arrival.
- Average time on scene with basic or advanced life support.
- Number of certified/qualified firefighters assembled on scene within a defined time period (NFPA 1720).
- Generation of proper fire flow (as defined by locality/ISO expectations) within a defined time period.

Such data can provide "dashboards" (analogous to the array of gauges in a car) for the fire department and the local government to use in determining how the department's performance measures up to community expectations. Organizational dashboards provide a way to monitor in real time compliance with organizational goals. Translated to the fire service, the department can monitor response goals in real time and adjust response strategies accordingly. The system will be performing efficiently when the organization is in full compliance with the goals and expectations set by the community. When goals and expectations are not being met, the department needs to re-evaluate how it operates.

In volunteer and combination systems other dashboards may be used to monitor performance and progress in other areas besides response. Other benchmarks include:

- Average volunteer retention rate.
- Average annual recruitment and associated demographics.
- Average call per volunteer.
- Various fund-raising data.
- Less government taxing support.
- Controlling the cost of recruiting, hiring and training new personnel.

Such information can be used to monitor the health of the organization based on what is deemed important by the stakeholders, but it isn't determined in a vacuum. It takes a plan.

Strategic Planning

What Is a Strategic Plan?

The development of a strategic plan is an important aspect of the evolution process. Fire executives who adopt a strategic plan for transition are better able to predict and manage change successfully. Strategic planning for an evolving department requires a commitment from the department's leadership and members and also from elected officials and other leaders in the community. Developing a strategic plan without involving community partners and stakeholders lessens chances for success.

Fire chiefs, presidents, and elected officials are often so preoccupied with immediate issues that they lose sight of their ultimate goals. That's why a preparation of a strategic plan is a necessity. A plan is not a recipe for sure success, but without it a fire department is much more likely to fail. A sound plan should:

- Serve as a framework for decisions or for securing support/approval.
- Provide a basis for more detailed planning.
- Incorporate detailed plans that include timelines, assignments and evaluations.
- Explain the services provided to others in order to inform, motivate and involve.
- Assist bench marking and performance monitoring.
- Stimulate change and become the building block for next plan within established timelines.

Preparing a strategic plan involves a multi-step process addressing vision, mission, objectives, values, strategies, goals and programs. When you develop a strategic plan, you must involve all the stakeholders if you hope to be successful.

■ The Vision

Your first step is to develop a realistic *Vision* for the department. Present it as a picture of the community and the department in three or more years' time, stated in terms of the department's likely growth and development.

■ The Mission

Describe the nature of a fire department in terms of its *Mission*, which indicates the purpose. Some people confuse mission statements with value statements (see chart on next page)—the former should be very hard-nosed, while the latter can deal with "softer" issues surrounding the business.

■ The Values

This element expresses the *Values* governing the operation of the department and its conduct or relationships with society at large, employees, local community and other stakeholders.

■ The Objectives

It is essential that you state the fire department's business *Objectives* in terms of the results it needs or wants to achieve in the medium and/or long terms. Objectives should relate to the expectations and requirements of all the major stakeholders, including employees, and should reflect the underlying reasons for operating the department.

■ The Strategies

Strategies reflect the roles and guidelines by which the mission, objectives and the like may be achieved. They can be developed using a SWOT analysis to identify *strengths*, identify and resolve *weaknesses*, identify and exploit *opportunities*, and identify and avoid *threats*.

■ The Goals

Goals are specific interim or ultimate time-based measurements to be achieved by implementing strategies in pursuit of the objectives. Goals should be quantifiable, consistent, realistic and achievable.

■ The Programs

The final elements are the *Programs* that set out the implementation plans for the key strategies. These should cover resources, objectives, timescales, deadlines, budgets and performance targets.

Hard

What business is/does
Primary products/services
Key processes and technologies
Main customer groups
Primary markets/segments
Principal channels/outlets

Soft

Reason for existence
Competitive advantages
Unique/distinctive features
Important philosophical/social issues
Image, quality, style standards
Stakeholder concerns

Designing a Combination System

The transition from an all-volunteer department to a combination system works best when the system is developing through detailed communication and strategic planning, rather than blind evolution. Many departments have evolved into an awkward conglomerate of resources with little thought given to system design and functionality and the long-term effects such a transition may have on the future of the organization. In many cases the evolution process is made more difficult by a lack of stable leadership.

The revolving door process for selecting leaders within the volunteer fire service creates a difficult structure to overcome in developing long-range plans. In addition, the election of officers requires a constant political campaign, creating a significant strain on the organization's ability to evolve. This paper strongly recommends that the officers' selection process eliminates elections and focuses on credentialing with performance factors.

Casualties of Transition

As departments approach the task of transitioning from an all-volunteer organization to another form of deployment, they need to be aware of a variety of pitfalls. It is common for such transitions to be emotionally charged events for those closely involved, and emotions often lead to serious mistakes. When emotions are allowed to overtake rationality, departments should expect some limited attrition of volunteers. Casualties could be significant but the vast majority of the volunteer members, even though some may be skeptical and cautious, will be willing to work through the issues and contribute meaningfully to improve the department. The same dangers apply to paid personnel. Those who are unable to integrate effectively with volunteer firefighters will quickly become a liability to the system. They seldom last if the department leadership recognizes and addresses the issues.

Another common casualty of transition results from avoiding sensitive issues and dodging conflict. Some departments may deem themselves "combination" simply because they utilize both career and volunteer personnel, but closer examination may show they are organizations in which paid firefighters are segregated from volunteer firefighters and there is little cooperation and integration between the two. This type of system is best described as "dual" rather than combination. While some dual departments function successfully in the short term, their division makes issues between the two groups stand out even more, and they miss out on many of the advantages a combination system brings. Poorly managed "dual" systems often become "duel" systems that are destined to fail.

Some indicators of a dual system include:

- Volunteers operating in different quarters than paid staff.
- Volunteers riding on separate apparatus than paid staff.
- Separate rules and regulations used.
- One group receiving better equipment and apparatus than the other.
- Rank structures and supervision not integrated.
- No opportunity for social interaction.

Departments should work to ensure system fairness for all parties. Integrating personnel fosters relationships that help to sustain the system.

An effective indicator of transition casualties is the retention rate of the minority component of the organization. If the paid component of the organization is in the minority and the retention rates are less than two years, it is likely that issues exist that are driving these firefighters away. Likewise, if the volunteers serve as the minority and retention rates are declining, it is likely issues are present that have negative impacts on the organization. The key to avoiding these issues is to ensure that everyone fully understands the core values of the organization and is committed to its mission.

Basic Design Models

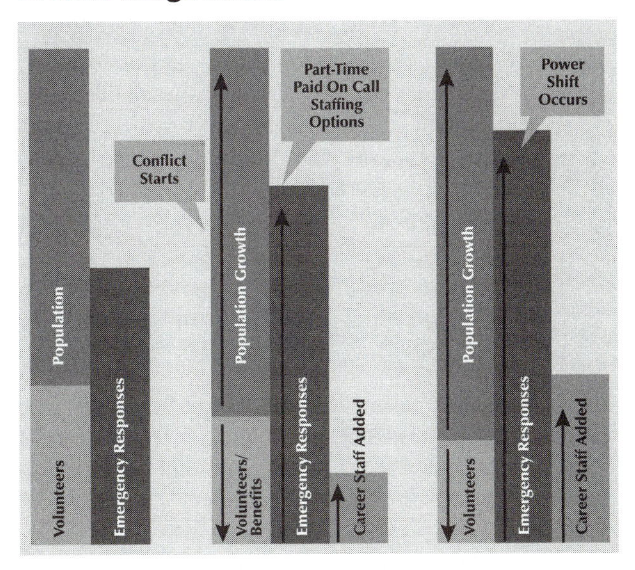

A department should conduct a cost/benefit analysis during system design to determine which model will function most efficiently for its locality. System design needs to recognize that volunteer/paid-on-call personnel are paid only for work performed. Career firefighters are paid for the POTENTIAL to be used. This does not mean that one is better than the other but it implies that department managers need to clearly understand the differences between the two as they relate to cost.

Some of the benefits/risks of the common system designs include:

All-Volunteer System

- Reduced labor costs.
- High-volume staffing during major emergencies such as natural disasters. Such influxes of manpower may be contingent on employers allowing volunteer employees to leave work during such events.
- Volunteers are willing or able to take off work to assist.
- Salary cost avoidance, which can be diverted to essential equipment and apparatus.
- Unpredictable response from volunteer staff.
- Volunteer systems can be more challenging to coordinate because of intermittent or sporadic participation from various members.
- Can rely on individual response rather than system response to meet call load, thus resulting in unpredictable service levels.

Combination System

- Can consist of any combination of career, volunteer, paid-on-call and part-time personnel.
- Enhanced staffing deployment as combination systems can capitalize on both the stability of a paid system and the manpower of the volunteer system during a major emergency, providing greater depth for staffing.
- Salary cost avoidance, which can free funds for essential equipment and apparatus.
- True integration of available resources and distribution of talent resulting in greater efficiency.

All-Paid System

- Consistent staffing providing predictable service level.
- Expensive due to increased salary and benefits requirements.
- Can lack depth during major emergencies because once multiple shifts have been deployed in a major incident, few resources are left to cover other service demands.

Another vital consideration when designing a combination system is identification of the stakeholders in the department. These stakeholders may include volunteers, employees, the fire department's management, local government interests, citizens and even the news media. A group of stakeholders should be convened early in the process to identify the obstacles to change and the processes to be used for overcoming them. The creation of a combination system can be challenging by its very nature and great care should be used to facilitate the change process.

■ Establishing Authority

One of the most controversial aspects of designing a combination system is establishing a clear line of authority and chain of command. Avoiding this challenge will breed animosity and mistrust over time, so it is critical that the lines be drawn early in the process. No matter if the chief will be paid or volunteer, the local government must empower the chief officer to lead the system as a whole. The local government—the "boss"—should be prepared to stand behind the chief as the transition progresses, even when political battles ensue, as they most likely will. Establishing local ordinances, resolutions or regulations that clearly define the authority of the fire chief and empower the position holder to effectively lead the organization is vital to success in the combination system.

Selecting a chief to lead a combination system is a delicate task. Leading a combination system should be approached as a specialty, and care should be taken to seek out candidates that have experience in this field. A qualified candidate should possess experience in dealing with both volunteer and paid personnel and have a leadership style that is conducive to conflict resolution and facilitation. Failure to select a candidate for chief with the appropriate experience and background can challenge the ability of the combination system to be successful. There is nothing that requires a chief in a combination system to be paid. The selection process for a chief in a combination system should not include term limits or an election, but should be based on common hiring practices. Equally dangerous is the philosophy that a chief from a fully career department automatically has the ability to motivate and supervise volunteers.

Subordinate officer selection can also be challenging. The selection of Assistant Chiefs and Deputy Chiefs should follow the same model outlined for the selection of the Chief. As captains, engine company officers, and other supervisory positions are created and people selected, the system's success is very much dependant on their enthusiasm and commitment to a combination system. They have to recognize and support the overall philosophy of a combination system and clearly understand their role in making it productive and successful. A promotional process should be in place that takes into account experience, education, service, testing and evaluation. Detrimental to this philosophy would be a career officer's inflated sense of rank based merely on wage compensation. Conversely, a volunteer officer's assumptions that the majority of "mopping up" duties are to be left to paid personnel defeats the cooperative nature essential to the mission of the department.

Substantial benefits can be derived from educating all the department members on the strengths and weaknesses of the system and clearly outlining the expectations of all the firefighters. It is important to ensure that career staff members understand they become informal leaders regardless of their rank because of their frequent exposure to emergency calls and the expertise they develop. Thus, every career firefighter is potentially a mentor who is expected to help others, including volunteers, capitalize on opportunities to improve, excel, and build confidence.

Local officials who believe that a system can always operate more cheaply have affected more than one well-organized and productive combination fire system. Sometimes their lack of knowledge about your department's history, the significant events that have formed it, its struggle for change, and the acceptance of its services by the public seem inconsequential when the time comes to balance budgets. Combination systems have a difficult challenge showing their cost effectiveness because there is no rebate of the monies saved or refund being returned to the funding entities.

It is imperative that local officials understand their obligation in this kind of system. Reducing the need for career positions saves the community substantial amounts of money that can be reinvested in other critical infrastructure. It is their responsibility to ensure that all firefighters have good equipment, apparatus, sufficient funds for overtime pay to allow career and volunteer members to train together, and capital to invest in new technology. Those issues make the system complete and they ensure that all the stakeholders have a vested interest in success.

Communication and Policy Development

Communication is essential in a successful combination system. The fire department leadership should constantly facilitate communication between paid and volunteer personnel and work tirelessly to manage information and dispel rumors. Leadership must constantly maintain open communications with elected officials and government authorities. Including stakeholders in the development of policies and procedures will help to develop ownership in the combination system and create transparency that will help avoid unnecessary conflicts. Both volunteer and paid members of the organization have special considerations that should be taken into account when developing policies. Time is often a major issue with volunteers who must juggle other jobs and family obligations, while paid members may be more interested in working conditions and wages. Each perspective should be considered valid and accommodations reached that allow both groups to succeed within their own abilities.

Job Descriptions and Expectations

It is imperative that all members of the organization understand their responsibilities and expectations. People often join an organization expecting one thing and then experience something quite different and over time they develop negative attitudes.

The job description must identify the specific roles and responsibilities of each member of the organization. Remember, paid personnel are being compensated for the work they do, but this does not mean that any less is expected from the volunteers.

The expectations of leadership are the same whether career or volunteer members hold top positions. These expectations are the same for firefighters as well. Being trained and competent is not determined by a paycheck but by the level of commitment. Everyone should be expected to be trained and competent; a paycheck or lack of one is not an excuse for incompetence.

Focus is important to any organization. If leaders fail to provide a focus, the members will develop their own, and their focus most likely will be centered on themselves and not the organization. Job descriptions and expectations help keep the members focused.

One specific responsibility of the career firefighter should be to help mentor the volunteers. The mentoring process should be spelled out in the Standard Operating Procedure (SOP) manual. This helps the volunteers grow and develop in their abilities and skills. Ultimately the volunteers will be able to function at a much higher level as a result of mentoring.

In addition, the career firefighter should help identify and correct problems within the fire department. It is no longer someone else's responsibility to solve problems.

Well thought out job descriptions and expectations can do much to keep the organization running smoothly. They help all members stay on the same page.

Clearly defined expectations, along with an evaluation system, will provide an excellent tool for managing the transition. The expectation model must include an analysis and evaluation of each individual's ability to function as a team member. Department leaders must enforce the model once it is set, but they should modify it when circumstances invalidate expectations. Maintaining expectations that no longer apply to the organizational structure can lead to conflict. The change model can be learned and implemented as long as management continues to understand that change is inevitable and most likely will produce improved service levels.

■ Sustaining a Combination System

Ensuring that a combination system stays focused requires constant maintenance and leadership. Local government and fire department leaders must embrace the combination philosophy and be prepared to endure intense scrutiny and political pressure. Leaders within the local government and fire department must regularly recommit to the combination mission and keep both paid and volunteer firefighters focused on service delivery.

The underlying philosophy of a combination system is improved service at a reduced cost. A combination department needs to be recognized for its value as a cost avoidance tool that reduces the need for employing full time career firefighters. Local officials should support the department with adequate funds committed to maintaining modern apparatus, protective gear and other equipment, and investment in improved technology.

Adequate allowance for overtime pay is necessary to ensure that all firefighters can train together, most likely on a schedule that ensures maximum participation by volunteers.

Monitoring the retention rate and/or general treatment of the minority group becomes a good indicator of how the combination system is performing and provides a mechanism for making internal changes.

■ Tactical Equality

One of the secrets of successful combination fire departments is full integration of career and volunteer firefighters at a tactical level. The concept is simple although it can be challenging to implement if you are already a combination department that is not so integrated. Tactical integration pays big dividends for the department and the community by improving emergency scene operations and increasing volunteer longevity.

Tactical equality recognizes that all positions, from firefighter through senior fire officer, require formal training and education to meet the expected performance level. Members of an evolving system must understand the complex issues facing the department and community and the serious nature of the service.

Officer promotions should be based on certification, tenure, experience and proficiencies in technical skills as well as soft skills, such as interpersonal communication.

Tactical equality is achievable if your department provides performance-based, certifiable training and the opportunity for the volunteer and career firefighters to train at the same time. This establishes a respect for the rank and the achievement to obtain the position and less emphasis on whether it is filled by a volunteer or career member. Position and rank are not affected by time of day or the day of the week. The attitude toward rank for everyone should be, "You earned it, you own it."

The importance of joint training and interaction means that it should be supported by adequate overtime funds to ensure training with the volunteers is a formal expectation of the job for the career employee.

One obstacle to tactical equality can be a requirement that volunteers meet training standards that are common in paid settings. The time commitment required to complete the training is an issue with many volunteers. Fire service professional standards and accreditation organizations should develop ways to incorporate performance-based training concepts into certification courses that allow students to gain qualification based on actual abilities rather than classroom hours.

Implementing this philosophy while the department is still a volunteer-staffed organization can improve the quality of officers available for promotion. Of equal importance, it fosters a cultural change that will dictate the value and respect that are placed on volunteer personnel long after career firefighters are incorporated into the system.

■ Resource Deployment Options

When looking at resource deployment of your department make sure that you consider all the options, thinking outside the framework of normal deployment systems and keeping all your search avenues open. It is imperative that leadership does not compare and try to create a system emulating large departments. Focus on what your department should be and create a model that is effective and efficient for your community.

Deployment can be a very simple or an extensive and costly process. As we look at the objectives of deployment (NFPA 1710/1720) we need to take into consideration items such as:

- Proper number of personnel.
- Time for response and delivery.
- Apparatus.

Deployment should take into account a means of getting the proper staffing, needed tools, and required resources to a predetermined location to effectively and efficiently mitigate the emergency. There should never be a differential between adequate "hardware" resources and payroll.

Some personnel deployment options include paging by radio, pager, or cell phone. Other items to consider are duty assignments allowing for the best utilization of staff, or working out a system that will provide required staffing without the traditional "everyone respond" or the use of several pieces of apparatus just to gain necessary staffing.

The fire service often utilizes apparatus as "expensive" taxis to bring together numbers of personnel rather than calling the proper apparatus for the performance needed. Departments should identify the required level of apparatus and staffing based on type of call, and then look at alternate means of transporting personnel, such as utility vehicles or command vehicles. It is not always prudent to buy bigger apparatus just to carry personnel. Another option is to manage personal vehicle deployment, but this method requires significant discipline to assure accountability and safety considerations.

Departments should have a method of monitoring the number of personnel responding to incidents so management can determine if sufficient staffing is going to be available to deliver the required service.

Staffing management is key to assuring the proper number of personnel on an incident scene. Assignment of duty hours or days or shifts can ensure proper staffing and help avoid overstaffing. Having each member assigned a time slot and/or a service function can regulate the number of members who respond to a call type so that the result is needed deployment but not overdeployment. This may also provide a more reliable time commitment for the volunteer.

Note: When attempting to define the timing element of your deployment system, you must consider the time for a response and the actual time required for the delivery of the service, which could be vastly different. A good reference for assistance in determining response and delivery times is *NFPA 1720: Standard for the Organization and the Deployment*, or a copy of the *Fire Service Standards of Cover*.

Looking at some nontraditional means of deployment could help solve overstaffing or understaffing concerns that often burden your department's service deployment. Reviewing all deployment opportunities will save the embarrassment of lack of proper resource development during the time of need.

Regional Response and Mutual Aid

It may be time to ask, "Can we be all things to all people?" Can two or three departments provide the needed services for the community versus parallel systems for each department?

The concept of regional response can help reduce the service-delivery burden felt by many fire departments today. Using this concept, fire departments can stop duplicating resources and services. This saves both time and money for the local community and addresses gaps in specialized response.

A relevant question is, "Do each of two fire departments that are located three miles from each other need two tankers and an aerial? Isn't it possible for one department to have two tankers and the other department an aerial?" Look at the possible savings that an agreement like this could mean for both communities. Although ISO ratings may affect these considerations, the fire service leaders must determine the best deal for the dollar and the best way to provide service to the community. Never allow ISO to be the driving factor. If your system is effective, your ISO rating will improve.

During the day two departments have a total of five people on staff. There are two people at one department and three in another. Could we place all five in one station to enhance the staffing? Now the first responding truck has a total of five instead of the possibility of only two. However resources are deployed, a minimum of four people per apparatus dispatched should be assembled on the fire ground. With the needs of the community as the primary driver, fire department leaders can develop many possible solutions. Never design a system to handle the worst-case scenario. Design it to properly address the vast majority of your responses. It may be that one department will no longer be the expert in all areas. Instead, each department in an area may have a specialty and its services can be offered to the region in exchange for specialized services from other departments. Regionalization of services can reduce the burden on many of the nation's fire departments.

The concept of mutual aid is sometimes abused. Departments that are unable to answer an initial call will rely on mutual aid to cover the alarm. This increases the burden on the other department. True mutual aid should be utilized when back-to-back calls are received, or when an incident is bigger than the resources that are on hand and additional people or equipment is needed.

The Impact of Emergency Medical Services (EMS)

EMS calls have created a strain for many fire/EMS systems as a result of increases in call volume. It is much easier to get people to volunteer for 150 fire calls than it is for 400 calls. The higher the call volume, the more strain that is placed on the personnel.

Even departments that don't provide EMS patient transports but only first response care are feeling the impact of higher run volumes due primarily to national issues related to health care conditions. To compound this, many stand-alone EMS systems are expecting and relying on fire departments to assist them on various EMS calls. In addition, EMS certifications have become a significant burden for volunteer and career members that results in additional costs and time commitments.

The EMS system benefits from this arrangement with the fire department's first responders in several ways. First, they can reduce the number of EMS units needed to cover a specific area. Fire departments are usually strategically located throughout the community. This allows for quick response and timely coverage, so in many cases the fire department will be on-scene before the EMS unit.

Some EMS systems will rely on the first responder to assist with lifting, CPR and other labor-intensive tasks. By doing this the EMS system reduces its cost of doing business, since the fire department is doing what additional EMS personnel would normally do, thus increasing the run volume.

Another aspect to consider is that some fire departments have consolidated fire and EMS operations. Although run volumes are increased substantially, additional revenue can be generated through EMS billing and additional services are provided for the community.

This additional service is good for the public image of the department and the additional revenue can provide money to help support a part-time, paid-on-call or combination system.

EMS can create many challenges for the local fire department. New methods for dealing with the challenges of EMS growth must be identified for the specific community. How the department deals with the EMS issue will ultimately determine its success. People expect to summon an ambulance for any reason at any time and be confident that someone will show up in a short period of time to transport them where they need to go. This public expectation becomes a huge burden when you staff with volunteers via home response. Even in a combination system this can create situations where nonessential EMS calls are taking up so much of the paid staff's time that other duties are not being completed. Most volunteers join for the excitement and the thrill associated with saving lives, but most EMS calls are not full of excitement and thrills. If peoples' expectations for service are to be met, they need to be aware that there are costs associated with its delivery. The cost of providing ambulance service in most cases must include career/part-time paid staffing.

Part-Time Staffing as an Alternative

There are alternatives to operating a combination department. One is transitioning from an all-volunteer system to one of all part-time firefighters. The part-time employee program can be designed around the specific needs of the department. It is dynamic in nature and can grow as the needs of the department change.

Under this system the volunteers are hired as part-time employees. Stations are staffed with part-time members around the clock or during peak call times. This allows for staffing that is comparable to that of the career department at a fraction of the cost to the community.

The administrators of the department can decide, based on run volume, the number of staff members needed on each shift. Shifts can vary in any degree of time blocks. Members can work their normal full-time job and sign up for shifts when they are available. Managers must be capable of making creative shift assignments. If a call requires more personnel than are on duty, members responding from home or work are paid from the time the call comes in until they are placed back in service and all equipment is made ready for the next run.

The pay scale for this system can be flexible. One example would be to pay those who are trained as firefighters in one pay range while paying basic EMTs and paramedics another range. This makes sense especially in those departments with high EMS call volumes.

Example:

Firefighter only	$ 7.00 per hour
Firefighter/EMT	$10.00 per hour
Firefighter/paramedic	$13.00 per hour

Under this system the members are paid more for education. A firefighter wanting to earn more money can return to train for a higher paying position and be paid at the level upon achieving certification. In addition to education, this system pays for performance compensating employees for what they do.

Scheduling must be monitored to prevent overtime and creating Fair Labor Standards issues.

The cost of the part-time system compared to the full-time system is greatly reduced. The need for many benefits is reduced when part-time employees are working full time at another career. For example, a department with 50 part-time members would save nearly $400,000 by not providing healthcare benefits. (50 employees × $8,000 per employee = $400,000)

There are intrinsic benefits to this system as well. The volunteer is now a paid employee. This can increase pride and he or she may feel more valued by the organization.

■ Leadership Selection

To ensure a healthy organization, it is imperative that strong leaders are selected for officer positions. Just because an individual is a good firefighter does not guarantee that he or she will perform well as an officer. Sometimes, technical skills are confused with leadership skills. Both are important but remember you are selecting a leader, not a "head firefighter." This means that special considerations must be taken in selecting those who have the ability to lead.

Leadership is a skill that can be learned and developed, but the leader must also have certain characteristics to ensure success. First, *integrity* gives the officer credibility. If the officer compromises his integrity, credibility is lost, and when credibility is lost his effectiveness is lost as well.

Leaders cannot lead where they cannot see. Therefore the officer must have a clear and distinct *vision*. Where does he see the organization moving? What will the organization look like in the future? Even a frontline officer must be able to see where he would like the people under his command to move. Once this vision is created, the officer must communicate it to the people and get them excited about it. People must buy into it if the vision is to become reality.

The officer must maintain integrity even when it hurts. In other words, the officer chooses to do what is best for the people and the organization even when another choice would benefit the officer.

Ask employees what they want from their bosses, and most often the answer is character and fairness. People want to be led by someone they can *trust*. And when trust is high, overall performance will increase.

Next, the officer must have a good attitude, be *optimistic* about the future, and focus on the positive more than the negative without avoiding problems.

An effective officer will also demonstrate a *caring attitude* for the firefighters, the organization and the community in which he serves. To put it bluntly, "If you don't care, then get out of the organization or at a minimum, get out of the position of leadership and influence." The officer sets the tone for the organization. If he is negative and constantly complaining, he creates a negative work environment that acts as a cancer spreading throughout the department. If an officer has a "no" attitude, or a "we cannot do that" attitude, the members will adopt the same philosophy. Eventually this will head into a downward spiral of defeat.

The officer must have self-discipline. *Self-discipline* is defined as "knowing what I need to do, not wanting to do it, but doing it anyway." Self-discipline demonstrates the officer's ability to stay cool under pressure. He tends to do what is right for the organization instead of what is popular.

Courage is an important trait for any officer at any level. Courage allows the officer to make tough decisions even when he knows he will be challenged. Courage allows the officer to show a healthy confidence in doing the job.

Another characteristic important to the officer is *humility*. A good officer is not driven by ego but by a value system that recognizes right from wrong. The humble leader will admit when he is wrong instead of pointing the blame elsewhere (self-esteem is managed internally) and will work to correct mistakes. Humble leaders are in their positions for the right reasons. They are transparent with nothing to hide and nothing to prove. They are there to serve the people they lead.

Effective officers *seek excellence* in all that they do. They constantly look for better ways of doing things. They learn from their mistakes and educate themselves on a continual basis. They strive to do the best they can in all situations. At the same time, successful officers understand that decision-making is a constant process of assessing risk versus gain or cost. So they incorporate into their decision process a "reality check" that helps to give their decisions a real-world perspective. They recognize that not every decision will be the perfect solution.

Leadership is a privilege. To lead and influence people is one of the highest honors an individual can be accorded. A leader must never abuse his or her authority or influence, no matter how tempting it may be. The organization must recognize this and have systems in place to ensure that high performers are rewarded for their successes. This provides an incentive for good people to strive for leadership positions.

When selecting an officer, look for people who are *good communicators*, people who are able to articulate information in a timely and accurate manner. They must believe in "open-book management," which means that information is not guarded but freely distributed within the organization.

Remember, the organization will never progress beyond the abilities of the officer. If the officer's ability to lead is limited, the organization will be limited as well. Strong leaders make strong organizations. Review your current officer selection process and determine how it can be improved.

Feedback from firefighters is often helpful for the command officer. An example of a firefighter evaluation of the command officer is included in *Addendum B*. Feedback from the bottom up is a very important part of the process of having great officers. Some fire officers fear this type of evaluation, but this method clearly demonstrates commitment and leadership.

Assessment Center–Leadership Selection

An assessment center consists of a standardized evaluation of behavior based on multiple evaluations, including job-related simulations, interviews and/or psychological tests. Job simulations are used to evaluate candidates on behaviors relevant to the most critical aspects (or competencies) of the job.

Assessment Center Exercises

An assessment center can be defined as "a variety of testing techniques designed to allow candidates to demonstrate, under standardized conditions, the skills and abilities that are most essential for success in a given job." Assessment centers allow candidates to demonstrate more of their skills through a number of job-relevant situations. The term assessment center is really a catchall for an assessment process that can consist of some or all of a variety of exercises. While assessment centers vary in the number and type of exercises included, two of the most common exercises are the in-basket and the oral exercise. Other possibilities include counseling simulations, problem-analysis exercises, interview simulations, role-play exercises, written report/analysis exercises and leaderless group exercises.

In-basket exercise. In a traditional in-basket exercise, candidates are given time to review the material and initiate in writing whatever actions they believe to be most appropriate in relation to each in-basket item. When time is called for the exercise, the in-basket materials and any notes, letters, memos or other correspondence written by the candidate are collected for review by one or more assessors. Often the candidates are then interviewed to ensure that the assessor(s) understand actions taken by the candidate and the rationale for the actions. If an interview is not possible, it is also quite common to have the candidate complete a summary sheet (i.e., a questionnaire). A more recent trend over the past 10 years has been the development of selection procedures that are based on the assessment center model, but which can be turned into low-fidelity simulations. Some low-fidelity simulations involve having an applicant read about a work situation. The applicant then responds to the situation by choosing one of five alternative answers. Some procedures have the applicant choose the response he/she would most likely make in a situation and the response that he/she would least likely make. These samples of hypothetical work behavior have been found to be valid predictors of job performance.

Recently, the in-basket has become a focus of interest because of its usefulness in selection across a wide variety of jobs. A variety of techniques have been used to develop in-baskets. Quite often information on an in-basket's development is not available for review because the reports do not contain the critical information. It is not uncommon for armchair methods to be used or for in-baskets to be taken off the shelf. A recent review indicated that nearly 50 percent of the studies do not describe how the in-basket was constructed. There is also a great deal of variation among the ways in which the in-basket is scored, with some scoring systems utilizing almost entirely subjective judgment, while others utilize a purely objective approach. The in-basket exercise may be thought of as an approach that assesses a candidate's "practical thinking" ability, by having a candidate engage in implicit problem solving for a job-relevant task.

It is now well recognized that a content-valid approach to constructing an in-basket is one that is professionally accepted as a technique that has passed legal scrutiny. However, despite the acceptance by the courts and practitioners, the reporting basis for content validity is often deficient. Schippmann, Prien and Katz in a 1990 report point out that all the studies they reviewed failed to establish a link between the task portion and the knowledge, skill and ability portion of the job analysis in order to provide a firm foundation for the construction of the in-basket. Often there has been no procedure for translating the job analysis information into development or choice of the test.

Oral exercises. Like all assessment center exercises, oral exercises can take many forms depending on the work behaviors or factors of the job being simulated. Common forms of oral exercises include press conference exercises, formal presentations and informal presentations (briefing exercise). In oral presentation exercises, candidates are given a brief period of time in which to plan/organize their thoughts, make notes, etc., for the presentation/briefing. Traditionally, the audience is played by the assessor(s), who observes the presentation and makes ratings. Assessors may also ask candidates a series of questions following their briefing/presentation. The questions may or may not relate directly to the topic of the presentation.

Leaderless group discussion. The leaderless group discussion is a type of assessment center exercise where groups of applicants meet together to discuss an actual job-related problem. As the meeting proceeds, the behavior of the candidates is observed to see how they interact and what leadership and communications skills each person displays.

Role playing. Role playing is a type of assessment center exercise in which the candidate assumes the role of the incumbent of the position and must deal with another person in a job-related situation. A trained role player is used and responds "in character" to the actions of the candidate. Performance is assessed by observers.

Several trained observers and techniques are used. Judgments about behavior are made and recorded. The discussion results in evaluations of the performance of the candidates on the dimensions or other variables.

Agencies should not utilize the assessment center as the only Pass/Fail portion of selection. Consider a "piece of the pie" attitude when utilizing assessment centers. Balance is the key objective.

■ Hiring Practices

Hiring career personnel is not only an important management/leadership decision, it is also a large monetary and professional investment for the organization. Depending on the size of the agency, it is estimated that approximately 70 to 80 percent of an operating budget for a combination department is dedicated to salaries, wages and fringe benefits for staff. Personnel are truly the most valuable resource for any organization, both in a monetary and asset sense.

With that said, organizations should ensure that their recruitment and hiring processes are designed appropriately to facilitate the hiring of qualified staff that meets the organizational needs and that the candidates are screened to appropriately identify strengths and weaknesses so that solid hiring decisions can be made. Once again, it is not only about firefighting and EMS skills—people skills are very important.

There are many models that identify the hiring techniques utilized by different organizations for the recruitment, hiring and appointment of volunteer, paid-on-call, part-time and career personnel. While specific criteria may vary with regard to experience, training, education and certification, several very important aspects should remain constant. Two of the most important aspects that come to mind are attitude and personality.

■ Attitudes

While hard to measure, attitude can be a driving force that overcomes many obstacles and can be the resolve that carries a person to a higher level of achievement. Southwest Airlines has embraced the motto "Hire for attitude—train for skills." Attitude is an individual trait that should be measured to ensure that potential candidates possess a level of commitment that will blend with and be accepted within the organizational culture.

The use of Likert scales or Semantic Differential Scales can be useful in developing appropriate evaluation mechanisms unique to the organization and can greatly assist managers in assessing a candidate's attitude. The following Web address provides an overview of the Likert scale and how to develop an evaluation mechanism: *www.socialresearchmethods.net/kb/scallik.htm.*

For more information regarding levels of measurement and scaling, visit: *www.fao.org/docrep/W3241E/w3241e04.htm.*

The following Web address provides an overview of the do's-and-don'ts of survey design: *www.unf.edu/dept/cirt/workshops/survey/polland_handout.pdf.*

■ Personality

As important as attitude is personality. The ability of an employee to survive and operate within an organizational culture may very well depend on his ability to fit in. Notwithstanding the fact that measuring a candidate's personality is subjective, it is extremely important to identify whether or not a potential candidate has the necessary interpersonal skills to connect with peers and supervisors in the organization.

A widely utilized mechanism for identifying personality traits is the Myers-Briggs Type Indicator® instrument, which provides a useful way of describing people's personalities by looking at their preferences on four scales (extraversion vs. introversion, sensing vs. intuition, thinking vs. feeling, and judging vs. perceiving).

Paladin Associates was formed as a nonprofit organization for the purpose of promoting the benefits of the Myers-Briggs Type Indicator® instrument. Their Web site is *www.paladinexec.com* and it provides a great deal of information and resources for personal and professional development. The Myers-Briggs Type Indicator® instrument is available free of charge at *www.paladinexec.com/mtbionlinetest.htm.*

The psychological assessment is based on the psychologist's knowledge of the requirements of firefighting duties. These requirements are based on a job analysis with identification of the psychological variables that are relevant to the knowledge, skills and abilities needed to be an effective firefighter. In addition to the variables that are more or less common to all or most fire departments, the psychologist can also customize the focus on those variables that are valued or required by a specific department.

The assessment procedures may include an individual interview with the candidate and a series of paper-and-pencil psychological tests. The comprehensive interview is primarily focused on work and career-related issues. The psychologist may also explore areas such as family history, education, interest in the pursuit of a fire service career, the individual's strengths and developmental needs, mental health history, legal history, exploration of the use of mood-altering drugs and chemicals, and overall adjustment.

The paper-and-pencil psychological tests may include the verbal comprehension, numerical reasoning and verbal reasoning subtests of the Employee Aptitude Survey® series; the Minnesota Multiphasic Personality Inventory-II®; the California Psychological Inventory®; the Myers-Briggs Type Indicator®; and/or a writing sample.

The three intellectual-based tests (verbal comprehension, numerical reasoning and verbal reasoning) provide an estimate of the individual's vocabulary knowledge and inductive and deductive reasoning capabilities. They identify the candidate's ability to analyze situations as well as deal with matters of a more conceptual nature. They provide an indication of how quickly the individual will learn what he or she needs to know to be successful on the job. The candidate's scores on the aptitude tests are compared to a sample of firefighter candidates' scores.

The Minnesota Multiphasic Personality Indicator-II® is a clinical screening instrument designed primarily to detect the presence of abnormal functioning, and to screen out clinically significant pathology that may impair an individual's ability to perform the duties of a firefighter. The California Psychological Inventory® is a general personality inventory designed primarily to differentiate among essentially normal individuals on a number of dimensions, including dominance, independence, responsibility, self-control, etc. Both of these inventories are used extensively in the selection of firefighters in the United States, and there are numerous research studies attesting to their validity and utility in the selection process.

The Myers-Briggs Type Indicator® provides useful information related to work style, including how people relate to each other, organizational skills and what information is relevant to them in making decisions (e.g., facts versus feelings).

The writing sample consists of having candidates write about a conflict situation. They are evaluated on the basis of the content as well as grammatical accuracy.

The conclusions of the psychologist regarding the candidate are based on all of the information gathered from the assessment processes described above. They represent the overall best judgment of the psychologist, taking into account not only the test results but also impressions gained from the interview. In addition to providing an overall description of the candidate in the report, the psychologist may also make a recommendation about hiring. For example, the candidate may be recommended unconditionally, recommended with reservations, or not recommended for hiring. When the psychologist recommends a candidate with reservations, the reservations may not be significant enough to disqualify the candidate but may cause some difficulty or be problematic. In some cases candidates may not be recommended for hire because the psychologist feels the candidate would not be a solid match or does not possess the characteristics that are particularly valued or required by a specific department. The agency must establish the benchmarks or it most likely will be saddled with a low performing employee (or volunteer).

Physical Abilities

One of the key elements of an organization's assessment of employees is determining their ability to perform the essential functions of the jobs that are detailed in position descriptions. Regardless of whether an employee is volunteer, paid-on-call, part-time or career, it is essential that the organization evaluate his or her physical abilities prior to appointment to the organization and thereafter on a periodic basis, to ensure capability of performing the essential functions of whatever position the employee fills.

A principal concern is the cardiovascular fitness of firefighters. The American fire service continues to see an increase in both injuries and cardiac-related on-duty deaths, which in turn leads to higher insurance premiums and increased workers' compensation costs.

The International Association of Fire Chiefs (IAFC) and the International Association of Fire Fighters (IAFF), through the Joint Labor Management Task Force, developed The Fire Service Joint Labor Management Wellness Fitness Initiative. The Guide to Implementing IAFC/IAFF Fire Service Joint Labor Management Wellness/Fitness Initiative is available via electronic format through the IAFC at no charge.

The manual includes information on these topics:

- Fitness evaluation.
- Medical evaluation.
- Rehabilitation.
- Behavioral health.
- Data collection.

In addition to these programs, the Joint Labor Management Task Force developed the Peer Fitness Training Certification Program, which is designed to train personnel to implement fitness programs, improve the wellness of personnel and assist in the physical training of new recruits. More information is available from the IAFF Web site, *www.iaff.org/safe/content/wellness/peer.htm*.

The IAFC, in conjunction with the IAFF, developed and adopted the Candidate Physical Agility Test (CPAT), as an entry-level physical ability test for measuring the physical capabilities of a firefighter candidate to perform firefighting functions. The Candidate Physical Agility Tests Manual is available through the IAFC and the IAFF. Additional information can be obtained from the IAFF Web site at *www.iaff.org/safe/wellness/cpat.html* and the IAFC Web site *www.iafc.org* (member only access).

It is essential that organizations utilize evaluation mechanisms that have been approved by their political entities and legal counsel to ensure compliance with local, state and federal legislation, such as Americans with Disabilities (ADA). More information can be obtained by visiting the ADA Web site at *www.ada.gov*.

Background Investigations

Many changes in the way we conduct business have come about as a result of an increased awareness of global terrorism and the new role of the nation's fire service as it relates to homeland security. Prior to September 11, 2001, many organizations were obligated under their state statutes to complete background investigations for health care providers with regard to offenses such as domestic violence, theft and drug abuse.

With the heightened level of security and the integral role that the nation's fire service now has at the local, state and federal level with homeland security, it is imperative that organizations perform a comprehensive background investigation on all candidates.

Some of the more common aspects of formal background investigations include:

- Employment history and verification.
- Reference checks and verification.
- Credit history.
- Criminal case history (*www.howtoinvestigate.com*).
- Certification/training verification.
- Polygraph (*www.polygraph.org*).
- Drivers license checks (current, tickets, suspensions, etc.).

There are numerous examples of potential candidates misrepresenting their training, education and previous employment and/or criminal record. By utilizing simple technology and/or services, organizations can quickly verify these areas thus confirming the validity of the information provided by a potential candidate.

Medical Evaluations

According to the National Fire Protection Association (NFPA), data show that in the ten years from 1995 to 2004, 307 of the 440 firefighters who suffered sudden cardiac death were volunteers.

NFPA 1582 Standard on Comprehensive Occupational Medical Program for Fire Departments should serve as a guideline for the medical evaluations of fire/EMS personnel. Organizations should be cognizant of specific requirements imposed by their state.

Tobacco/Drug/Alcohol-Free Workplace

A tobacco, drug and alcohol-free workplace should be a requirement of all emergency service organizations, regardless of their composition of volunteer, paid-on-call, part-time and career members.

If your organization is intending to apply for a FIRE ACT grant or is a recipient of the grant in prior years, it is required to be a drug-free workplace. Below is the language from the grant guidelines:

> As required by the Drug-Free Workplace Act of 1988, and implemented at 44CFR Part 17, Subpart F, for grantees, as defined at 44 CFR part 17, Sections 17.615 and 17.620:
>
> The applicant certifies that it will continue to provide a drug-free workplace by:
>
> (a) Publishing a statement notifying employees that the unlawful manufacture, distribution, dispensing, possession, or use of a controlled substance is prohibited in the grantee's workplace and specifying the actions that will be taken against employees for violation of such prohibition.
>
> (b) Establishing an ongoing drug-free awareness program to inform employees about:
>
> (1) The dangers of drug abuse in the workplace
>
> (2) The grantee's policy of maintaining a drug-free workplace
>
> (3) Any available drug counseling, rehabilitation and employee assistance programs
>
> (4) The penalties that may be imposed upon employees for drug abuse violations occurring in the workplace
>
> (c) Making it a requirement that each employee to be engaged in the performance of the grant to be given a copy of the statement required by paragraph (a).
>
> (d) Notifying the employee in the statement required by paragraph (a) that, as a condition of employment under the grant, the employee will:
>
> (1) Abide by the terms of the statement and
>
> (2) Notify the employer in writing of his or her conviction for a violation of a criminal drug statute occurring in the workplace no later than five calendar days after such conviction.
>
> (e) Notifying the agency, in writing within 10 calendar days after receiving notice under subparagraph (d2) from an employee or otherwise receiving actual notice of such conviction. Employers of convicted employees must provide notice, including position title, to the applicable DHS awarding office, i.e. regional office or DHS office.
>
> (f) Taking one of the following actions against such an employee within 30 calendar days of receiving notice under subparagraph (d2), with respect to any employee who is so convicted:
>
> (1) Taking appropriate personnel action against such an employee, up to and including termination, consistent with the requirements of the Rehabilitation Act of 1973, as amended; or
>
> (2) Requiring such employee to participate satisfactorily in a drug abuse assistance or rehabilitation program approved for such purposes by a federal, state or local health, law enforcement or other appropriate agency.
>
> (g) Making a good faith effort to continue to maintain a drug free workplace through implementation of paragraphs (a), (b), (c), (d), (e), and (f).

■ Training and Certification

The experience and training level for a particular recruit will most likely vary and is dependant on organizational needs.

It is not uncommon for smaller volunteer and combination departments to recruit personnel and then train them or assist in training them to the desired level. Other organizations recruit those who have obtained a minimum level of training and/or certification. Each organization will have to evaluate its hiring decisions based on:

- Immediate need of trained/certified personnel.
- Training/certification infrastructure of the organization.
- Available funds for training/certification programs.
- Cost versus benefit of training versus certification.

The International Fire Service Accreditation Congress (*www.ifsac.org*) and the National Board of Fire Service Professional Qualifications (*www.theproboard.org*) are two organizations that accredit training and education. Their respective Web sites can provide additional information regarding each organization.

■ Reverse Transitioning: Is It Too Late to Turn Back?

Just as it is appropriate to consider transitioning from an organization staffed completely by volunteers to a combination or fully paid department, there also may be situations in which it is appropriate to look at reversing the transition—moving from a fully career department to a combination system that incorporates volunteers.

The clues that reverse transitioning may be an option are clearly visible in career systems. Departments in which training opportunities are restricted, worn-out apparatus is not being replaced, building improvements are not made, or building and apparatus maintenance are deferred because of a shortage of funds, or which face staff downsizing and reduced minimum staffing levels, are candidates for reverse transitioning.

When there are serious budgetary shortfalls reverse transitioning from a fully career department to a combination system could, over a period of time, allow for much-needed tax dollars to be reinvested in a physically failing essential service. However, any reallocation of funds must not be at the expense of service to the community. Staffing alternatives of this kind should never diminish the need for qualified, well-trained and experienced emergency service providers.

Introducing volunteers to offset staffing shortages and career staff reduction through attrition is a subject that requires a great deal of department and community coordination before a switch can be made. While the number of communities that may have to consider this option is growing, organized efforts to make this switch will classify your department as a pioneer in rediscovering volunteerism and a trendsetter for others to follow. You will be recognized as an organization that planned and prepared the department and the community for the change with successful results.

While the needs for reverse transitioning may be obvious, a move to a combination system will require a great deal of planning and consensus building within the community. Community surveys may be useful in determining the practicality of such a move with some insight into the supportive population base. Solicited information should include available time commitments, average population age, types of industries and shift schedules, percentage of single-parent families, average income levels, local cost of living trends, and the involvement of local youth programs. All of these will provide clues as to potential availability of local residents. A key element of a successful effort is to include the union component in all discussions.

Other indications of community support may be obtained from a study of the activity levels of other civic groups, which may lend additional clues as to the available population to volunteer. Strong civic organizations with lots of activities and time commitment would most likely indicate an interest of the public to support volunteer functions. The opposite may be true if long-term civic events are cancelled because of a lack of volunteer assistance.

A task force encompassing a broad base of community interests and leaders may be useful in researching and documenting the success and effectiveness of similar-sized communities that operate with successful combination systems. This group may have a substantial impact on the decision to make the switch and provide a check and balance to the emotions that can be associated, real or perceived, with such a major change.

A timeline for this transition will have to include extended and multiple training opportunities for potential volunteers who have to maintain family obligations and full-time jobs. Career personnel must have the appropriate training to be successful mentors and guidelines for conduct to ensure success. Immediate and decisive disciplinary action may be necessary to curb willful attempts to derail the change.

Without proper planning and consensus building, claims of reduced or less than reliable service become a detriment and find their place in destructive rumors.

Examples of Model Combination Fire Departments

Department	Chief of Department	Web Address
Garden City FD (NY)	Edward Moran	www.gardencityny.net/fire_dept
Long Beach FD (NY)	Ralph Tuccillo	www.longbeachny.org
City of West Des Moines FD (IA)	Donald Cox	www.wdm-ia.com
Hanover Co. Fire/EMS (VA)	Fred Crosby	www.co.hanover.va.us/fire-ems
Vashon Island Fire & Rescue (WA)	Jim Wilson	www.vifr.org
Xenia Twp. FD (OH)	William T. Spradlin	www.xeniatownship.org
Jefferson Twp. FD (OH)	Keith Mayes	www.jeffersontownship.org/departments/fire
City of Vandalia FD (OH)	Chad E. Follick	www.ci.vandalia.oh.us/firedepartment.html
Montgomery Co. Fire/Rescue (MD)	Tom Carr	www.mocofiredepartment.com
Prince William Co. FD (MD)	Mary Beth Michos	www.co.prince-william.va.us
Bloomington FD (MN)	Ulysses Seal	www.ci.bloomington.mn.us/cityhall/dept/fire
Troy Fire Dept. (MI)	Bill Nelson	www.ci.troy.mi.us/fire
Clackamas County FD (OR)	Norm Whiteley	www.ccfd1.com
Hillsborough CO Fire/Rescue (FL)	Bill Nesmith	www.hillsboroughcounty.org/firerescue
Volusia County Fire/Rescue (FL)	James G. Tauber	http://volusia.org/fireservices
Marion County FD (FL)	Steward McElhaney	www.marioncountyfl.org
Ponderosa VFD (TX)	Fred C. Windisch	www.ponderosavfd.org
Kitsap County Fire/Rescue (WA)	Wayne Senter	www.kitsapfire7.org
Saginaw Charter Twp. FD (MI)	Richard Powell	www.stfd.com
Farmington Hills FD (MI)	Richard Marinucci	www.ci.farmington-hills.mi.us/services/fire
Evesham Fire/Rescue (NJ)	Ted Lowden	www.eveshamfire.org
Miami Township Div. Fire/EMS (OH)	David B. Fulmer	www.miamitownship.com
Clearcreek FPD (OH)	Bernie Becker	www.clearcreektownship.com/FD/findex.htm
Miami Township FD (OH)	James Witworth	www.miamitwp.org/fireems/fire.htm
German Township VFD (IN)	John M. Buckman	www.germanfiredept.org/
Tinley Park VFD (IL)	Kenneth Dunn	www.tinleyparkfire.org
Bath Twp. FD (OH)	Jim Paulette	www.bathtownship.org/fire/index.htm
City of Roseville FD (MN)	Rich Gassaway	www.ci.roseville.mn.us/fire
York County Fire/Life Safety (VA)	Steve P. Kopczynski	www.yorkcounty.gov/fls/index.html
Village of Savoy FD (IL)	Michael Forrest	www.village.savoy.il.us/index
City of Fitchburg FD (WI)	Randy Pickering	www.fitchburgfire.com

Addendum A: Employee Expectations

The following is a list of expectations that are not included in your job description. We feel it is extremely important for everyone to know what is expected of them. In order for the team to effectively operate all members must buy into these concepts outlined below. Please review the list and clarify any questions you may have. This list is intended to help you make an easy transition to our organization.

1. Maintain and promote a winning attitude

- Look at problems as opportunities. How can we improve?
- When you bring a concern to an officer, bring two possible solutions.
- Do not engage in chronic complaining. Be part of the solution, not part of the problem. Complaining does little to improve the organization. Help us work toward positive solutions.
- Don't accept negative attitudes in others. Bring negativity to their attention.
- Avoid negative thinking. Negative thinking is contagious and limits our potential.
- Remember . . . Attitude is a choice; choose to have a good one.
- Develop a "can do" attitude. You are in control of your potential.
- Focus on making a positive impact on others and the organization.
- Seek out opportunities and ways to implement them.
- Deal in FACTS not assumptions.

2. Practice the Golden Rule

- Treat others the way you wish to be treated.
- See value in others. Everyone has value.
- Care about the other members and help them succeed.
- Focus more on the positive attributes of others instead of the negatives. We will not ignore the negative, but we will emphasize the positive.
- Help energize others by being motivated yourself.

3. Be a team player

- Participate in meetings and trainings.
- Help your fellow members succeed.
- Remember . . . We win and we lose as a team, not individuals.
- Keep communications open.
- Always seek win-win solutions.
- Have fun. Enjoy working with the group.
- Make it a safe environment.
- Build relationships to improve trust and understanding.
- Allow mistakes. We will all make mistakes when we try new ideas.
- Learning must take place when we make mistakes.
- Poor performance is not tolerated.
- Recognize fellow members for a job well done.

4. Seek excellence

- Increase your education and skill level.
- Focus on helping move the organization forward for today and tomorrow.
- Finish what you start. Get help if you need it.
- Seek to improve everything we do.
- Think why we can, instead of why we can't.
- Be data driven.
- Understand our budget is limited. How can we make the biggest impact with what we have?

5. Do that which is right

- Everything you do must be done in a moral, ethical and legal manner.
- Contribute to the mission and vision of the organization.
- Help accomplish our goals.
- Always consider the internal and external customer.
- Be trustworthy and show integrity.

6. Stay focused

- Remember . . . You're here to help the organization succeed.
- Stay focused on contributing to the mission, vision and goals.
- Don't get distracted with personal agendas.
- You are our most valuable resource. . .We will support you through education, training, coaching and counseling.
- Every task that you engage in must be aligned with the mission.

7. Participate

- Participate in meetings, training, special details and emergency calls.
- Participate by communicating, asking questions and offering suggestions.

- Participate by helping the organization be better today than it was yesterday.

8. Capitalize on adversity

- We are constantly faced with adversity and problems. Don't let the problems pull you down. Our job is to adapt and overcome problems.
- Seek out opportunity any time you are confronted with adversity.
- Understand all of the facts when confronted with adversity.
- Help develop and implement the plan to overcome adversity.

I have reviewed and discussed the above list to clarify my understanding of the expectations. A copy has been provided to me for future reference.

____________________ ____________

Employee Date

____________________ ____________

Officer Date

Addendum B: Officer Evaluation

1. I do not interact with this officer enough to complete the survey. ❑

2. Do you personally get along with this officer?
❑ Yes ❑ No

3. How would you rate his/her ability to take charge of an incident?
❑ Excellent ❑ Above average
❑ Average ❑ Below average
❑ Needs definite improvement

4. How would you rate his/her ability to deal with personnel issues?
❑ Excellent ❑ Above average
❑ Average ❑ Below average
❑ Needs definite improvement

5. How would you rate his/her communication skills?
❑ Excellent ❑ Above average
❑ Average ❑ Below average
❑ Needs definite improvement

6. Do you believe that this officer has the appropriate leadership skills and experience to hold this position? ❑ Yes ❑ No
❑ Could, but needs improvement

7. Please rate this officer's abilities in the following areas. Rate on a scale of 1–5, 5 being the highest rating and 1 the lowest.
_____ Ability to adapt to change
_____ Level of personal motivation
_____ Ability to motivate others
_____ Ability to approach problems and issues in a logical fashion

8. Please rate this officer in overall performance with 5 being the highest rating and 1 the lowest. _____

9. Please rate his/her ability and experience to handle the following situations as a command officer. Rate each item on a scale of 1–5, 5 being the highest rating and 1 the lowest.
_____ Residential structural responses
_____ Commercial/industrial responses
_____ Hazardous material incidents
_____ Rescue operations
_____ Medical emergencies
_____ Station operations
_____ Interaction on mutual aid responses

10. Do you support this individual in his/her current position? ❑ Yes ❑ No

Addendum C: Sample Career Employee Evaluation

■ Interim Performance Appraisal

Employee Name: ______________________________

Title: ______________________________________

Reason for Review:

❑ Annual Performance Appraisal

❑ Interim Performance Evaluation

The purpose of this performance appraisal is to encourage and recognize the level of employee performance and the achievement of organizational objectives and accomplishments. The Interim Performance Evaluation is designed to solicit your opinion of your performance, combined with comments from your immediate supervisor and the Chief, and to develop a progressive plan to improve skills and performance. This form is designed to facilitate a mutual understanding of performance expectations and the performance appraisal process by encouraging the employee and the supervisor to participate in a meaningful dialogue.

Performance will be evaluated on the following rating levels:

5 = Outstanding
Performance consistently and significantly exceeds the requirements of the job and is beyond established standards. Employee achieves objectives at a superior level. The employee demonstrates exceptional skills and innovation in work performance.

4 = Commendable
Performance exceeds job requirements in all major areas. Employee displays leadership and initiative, produces quality work, and sets an example for others to follow.

3 = Effective
The employee consistently performs tasks at acceptable levels, produces the required amount of quality work, and makes effective use of resources (e.g.: materials, budget, time, guidelines, procedures, etc.)

2 = Needs Improvement
Performance is below job requirements in one or more important area(s) and immediate improvement is required. Employee fulfills some responsibilities; has difficulty completing others. Additional training or development is required to achieve performance expectations.

Each objective is scored with points assigned from 2–5; a score of 2 represents the lowest rating and 5 is the highest. Total points are divided by the number of scored questions. Employee comments to "Discussion Points" are to be typewritten and italicized in black and the Chief's, and/or supervisor's, comments will need to be italicized in blue.

1. Rate your personal performance regarding your specific administrative job assignments. _______ points

Discussion Points
List your contributions to the department during the past year.

Define your mission with each of the administrative assignments with which you have primary responsibility.

Prioritize your short-term (12–24 months) objectives.

Prioritize your general long-term goals.

What "cost saving" measures have you implemented or could be implementing within your area of responsibility?

2. Rate your personal performance as an emergency services provider. _________ points (as a volunteer) (for those assigned as emergency responders)

Discussion Points
What has been your best scene performance this past year and why?

What has been your least productive performance this past year and why?

Identify your technical strengths.

Identify your technical weaknesses and your plans for improvement.

What measures have you personally implemented to improve the safety of department operations?

3. Rate your performance as a team player with your co-workers. _________ points

Discussion Points
In what way have your actions, both as an individual and in the scope of your job responsibilities, contributed to building and enhancing the team effort with your co-workers?

In what way have your actions, both as an individual and in the scope of your job responsibilities, detracted from developing and/or enhancing the team effort with your co-workers?

4. Rate your performance as a team player with the volunteers. _________ points

Discussion Points
In what way have your actions, both as an individual and in the scope of your job responsibilities, contributed to building and enhancing the team effort within assigned duty shifts?

In what way have your actions, both as an individual and in the scope of your job responsibilities, detracted from developing and/or enhancing the team effort within your assigned duty shifts?

5. Rate your overall productivity this past year. _________ points

Discussion Points
Identify the critical elements/tasks of your job assignment(s).

Identify the non-critical elements/tasks of your job assignment(s), (those elements/tasks that could be transferred to someone else).

6. Rate your ability to effectively schedule your time. ________ points
7. Rate your ability to "self-motivate" and assume work without supervision. ________ points
8. Rate your ability to professionally resolve issues with co-workers and volunteers. ________ points
9. Rate your ability to "mentor" other co-workers and volunteers. ________ points
10. How would you rate your openness and approachability by co-workers? _________ points

General Discussion Topics
What single issue would you change/influence to improve the overall administrative operations of the department?

What single issue would you change/influence to improve the overall emergency services operations of the department?

What single item, within the work environment, serves as your most frustrating issue?

Rate how you feel you are compensated (i.e. wages and benefits) for work/duties performed as an employee.

What changes/adjustments would need to occur for you to reasonably perform your duties within your 80- or 86-hour allocation?

Overall Evaluation Score

Supervisor's Comments Provide a brief summary statement that characterizes the employee's overall performance and supports your rating. Supervisors should summarize performance strengths and indicate any performance improvement areas needed. Provide additional pages if necessary.

Employee's Comments Do you understand how your performance was evaluated? Provide additional pages as necessary.

NOTE: Employee signature does not necessarily signify the employee's agreement with the appraisal; it simply means the appraisal has been discussed with the employee.

________ /________________ = ____________

Total Score   Total Qualifying Questions   Performance Level

Performance Levels 2005–2006
Evaluation "Category"

4.0—Outstanding
3.0–3.9—Commendable
2.0–2.9—Effective
1.0–1.9—Need Improvement

Print Name		Signature	Date
Employee:	____________	____________	______
Supervisor:	____________	____________	______
Reviewed by:	____________	____________	______
Department Head:	____________	____________	______

APPENDIX

C The White Ribbon Report

Managing the Business of the Fire Department: Keeping the Lights On, the Trucks Running and the Volunteers Responding

A new fire chief usually will pay more attention to the quality of services delivered than the business management of the fire department because that is what is comfortable. However, as the department's leader, you are in charge of the responses, the oversight of the volunteers and the administrative duties. If the system that you inherited does not have the appropriate business practices in place, then you have some work to do.

The fire chief is the one usually held accountable and responsible by the public for mistakes, poor performance or slow response time. Leaders must make things happen.

Because you are the chief, you have the responsibility to set up a management system that takes care of the volunteers and their personal needs. Most fire departments dismiss this responsibility as something for career departments to dismiss do. However, volunteer firefighters are considered employees in most states and therefore have the same employer/employee relationship as any paid profession.[1]

1. Check with your city or county manager to verify the rules that apply to your department.

Mismanagement of volunteers and their personal needs will contribute to a reduced retention rate. Ignoring federal mandates can expose the department to serious legal issues. This is the time to show your ability to delegate and recruit. Ask for help if you need it. Consider appointing a volunteer business manager to help fix your issues and bring credibility to your department. Your goal is to be a good partner in local government.

Expectations

Your actions and personal conduct during your term as chief will impact your firefighters and the community's expectations of you. When you become a fire chief, it can be diffcult to get your hands around your responsibilities. No set of directions accompanies your badge. You will learn most of what you need to know from mentors and from experience.

Here are some of the **personal** values that will be expected of you as a chief officer:

- Honesty
- Integrity
- Dependability
- Commitment
- Knowledge and competence
- Respect for authority, your peers and your employees and volunteers

As the fire chief, you should know that others have expectations and trust in your abilities to manage the

*The eagle highlights important information throughout this report.

fire department. Here are some of their expectations of you and your volunteers.

The services provided by the fire department should reflect what the public desires and the taxpayers are willing to fund. In turn, the **community** should reasonably expect that:

- The taxpayers' money is spent in the best interests of firefighter and community safety
- The fire department will provide the services that are needed to keep citizens safe
- The fire department will respond in a timely manner
- The firefighters who respond to an emergency are trained and experienced
- The firefighters are physically fit
- The firefighters are not impaired by alcohol or drugs
- Services will change to meet growing demand because of an increase in population

The **local government** expects that:

- You will inform them of what their options are and what the consequences of their decisions will be
- The fire department is a partnership with local government in community protection
- You will manage the department in compliance with local, state and federal laws and regulations
- They will have the right to decide if you will be an all-hazards response agency or respond only to fires, or anything in between
- The apparatus and equipment purchased will meet the needs of the public and is not extravagant
- The fire department has negotiated mutual-aid agreements with other agencies and departments for those calls that require greater resources than you have on-hand
- The fire department is part of a regional response network for infrequent but important response situations such as hazardous materials response or technical rescues
- You are accountable for the money they give you

The **fire department** expects that:

- Its members are trained and proficient
- Its members are physically fit
- Its members share the response burden
- Its members will show up for calls
- You have a strategic plan for growth
- Personnel rules are in place to make the system stable

The **volunteers** expect that:

- You will provide a safe and professional working environment
- You will treat them fairly
- You will create an environment that encourages personal growth
- You will reinforce the importance of teamwork
- You are receptive to their opinions on major decisions
- You will use their time effectively and efficiently
- You will appreciate their service
- Officers will always put the good of the department first

The **volunteers' families** expect that:

- The department has—and strictly adheres to—a national standard of safety
- You will use their family members' time wisely
- You will take care of them if there is an injury or death
- You support their family bonds and responsibilities
- The department has a code of conduct regulating station behavior

Finally, as **fire chief** you should expect that:

- Your volunteers will honor their commitment to train
- Your volunteers will respond when required to do so
- Your volunteers will be honest with you about your performance
- You will have the opportunity to deal with internal issues before they become serious problems
- The local government will provide you with the resources to successfully run and manage the department

You may consider drafting and circulating a set of department conduct standards to which everyone must adhere. For a sample set of standards, please see Addendum A.

Vision and Planning

Every fire department needs a strategic plan. The overall objective of a strategic plan is to identify the risks to the community; determine the level of acceptable risk; and develop policy, plans and funding commitments to "buy down" the risk to an acceptable level. If a fire department does not have a strategic plan, it has no vision for the future and provides a high level of uncertainty to the volunteers and the community.

A strategic plan will help you identify your funding needs to the local government. Policymakers have three choices for the type of community safety and service that the fire department will provide: 1) prevention and early suppression; 2) prevention and response; or 3) response only. Ultimately, the level of funding and political support will make that decision for you. You will help shape the outcome, however, by interacting with the policymakers.

Often, the vision for the fire department is solely conceived by one person—the current chief—and changes each time a new chief takes office. Many volunteer fire departments are unstable because of a constant change in leadership. A strategic plan must be able to sustain the changes in leadership and become the guiding document for improved and anticipated changes in services.

Members of the fire department, community representatives and local government officials should work together to identify the immediate and long-range plans of the fire department and develop a strategic plan to meet those objectives. This kind of planning can best be facilitated as a workshop. The resulting plan should address facilities, equipment and apparatus needs, funding allocations based on planned replacement objectives and population milestones that may cause the department to reevaluate volunteer staffing and services. This process can change the working environment and the level of community support, obtain buy-in and understanding from local officials and solidify a financial commitment to accomplish the collectively adopted fire department mission.

Request that a government official be appointed to your department as a liaison to improve the communication between the fire department and the funding entity. This can be of tremendous benefit when it is necessary to gather support for large purchases or important changes. The liaison's inside experiences and views will help to promote, convey and convince that a need should be funded.

A strategic plan cannot discount the opinions of the volunteers and the services that they are willing to, and want to provide. They are the ones doing the work and donating their time, and therefore will have valuable insights and suggestions that must be balanced with the philosophies of the fireboard or commission and the funding agencies.

Your department board or fire commission should provide direction consistent with your strategic plan, establish overall governance and financial policy and oversee plan implementation. The board or commission should craft an evaluation tool to gauge the department's progress in meeting established policy planning goals and response objectives. This gives the chief direction—a basis to evaluate his or her performance and the ability of the department to meet the needs of the community. A factual checklist of objectives offers the department some protection from newly elected or appointed governing officials who may not understand the operations of your department or who have a personal agenda in reforming your department. This tool should be modified through official board or commission action.

Fire boards or commissions should not be involved in the daily operations of the department and should avoid active participation in personnel issues, preserving their ability to be the "impartial hearing board" at the conclusion of a disciplinary action by the chief.

■ Community Value of Volunteers

Fire departments staffed with volunteers provide a substantial cost savings to local governments across the country. Demonstrating this savings to local government officials is an important way to garner support for funding to keep your facilities, equipment and firefighters in shape to serve the community. The two main ways to demonstrate this benefit to government officials are "actual cost savings" and "cost avoidance."

■ Actual Cost Savings

The actual cost savings of volunteer firefighter departments is reflected in the amount of time that volunteer firefighters contribute to the community. To calculate this amount, you must keep meticulous records of the volunteers' time commitment, including responses, training, public education efforts, vehicle maintenance, station upkeep and any other contribution that volunteers make to the fire department.

To calculate the actual dollar amount of a volunteer's time, we suggest that you use the hourly figure

2. Independent Sector, *Value of* Volunteer *Time*, available at *www.independentsector.org/programs/research/volunteer_time.html*, viewed Aug. 4, 2006. Available on this Website is a history of hourly rates and specific state rates.

determined by the Independent Sector, which calculated a national average for volunteer time. The amount for 2005 is $18.04 per hour.[2]

Cost Avoidance

If people do not volunteer, then the community has two options: hire firefighters or provide a minimal-to-nonexistent fire service. "Cost avoidance" refers to the amount of money the community would have to spend on fully career fire services but avoids spending because of the use of volunteers.

The term "cost avoidance" is more accurate than savings because this comparative figure is never budgeted, meaning that the money is never exchanged and no cash carryover results from the payroll and benefit savings. If an exchange of money for the full service price occurred and that amount were rebated at the end of each fiscal period, then the savings would be a tangible amount of cash that would speak for itself.

Determine the amount of cost avoidance by calculating the number of career firefighters the community would have to hire if you did not provide firefighting services with volunteers. This figure should be based on local or state averages for career firefighters, including their benefits and support staff, to make the system functional. The staffing level should be based on communities with a population base similar to yours. For a sample cost avoidance calculation, please see Addendum B.

Also, calculate the amount of money your department saves citizens on their homeowners insurance because of your Insurance Service Office (ISO) rating.[3] Obtain the number of improved residential properties in your jurisdiction from the local tax assessor's office, the local planning office or the U.S. Census Bureau at *www.census.gov*. Then obtain the average home value for the area (a local realtor may be able to assist you). Finally, ask a local insurance agent to provide quotes on a standard homeowners policy on a home in your jurisdiction that is of average value. Ask what that homeowner would pay with a class-10 ISO rating and with whatever class you currently are or seek to be. For a sample calculation of a department ISO rating, please see Addendum C.

Asking local realtors and insurance agents for help is a good opportunity to use diversification strategies, as discussed later in this document. These professionals may be willing to donate their time to the fire department out of a sense of civic pride and responsibility.

Fire Prevention

Fire departments are responsible for fire prevention as well as response. This is another way to partner with local government in public safety. The following suggestions will help you work with the community in this area:

> *Implementing residential sprinkler codes is one way to sustain the delivery of emergency services with all volunteer or combination staffing.*

- Work with your local government to create and enforce a structured fire prevention program.[4]
- Look for opportunities to educate the public about fire safety. Ask to speak at local schools, civic centers and block parties, and bring your apparatus and gear to provide demonstrations. Consider including information in mailings and on Web sites. Build a relationship with your local media outlets (television, radio and newspapers). Fire chiefs should appoint a public information officer (PIO) to provide relevant information to the public on emergency responses and to coordinate public education opportunities.
- Work with your local government to develop building and fire codes, including potential requirements for sprinkler installation. Two major fire and building code development organizations exist to help you: the International Code Council (ICC) and the National Fire Protection Association (NFPA). The ICC and NFPA develop model codes to limit damage from fire and other natural hazards. If your local government has not yet adopted relevant fire codes, you should work with them to do so.[5]
- Work with your local government to enforce building and fire codes. Codes are of no use if they are not enforced. Help your local government establish a process for enforcing codes so

3. The Insurance Service Office assesses a community's risk under 10 categories. A community's risk rating determines its premium cost for building insurance. For more information, please visit *www.isomitigation.com*.

4. The U.S. Fire Administration (USFA) offers several programs to help localities establish such programs. For more information, please visit *www.usfa.dhs.gov/subjects/fireprev/*.

5. For more information on these groups, please visit their Web sites at *www.iccsafe.org* (ICC) and *www.nfpa.org* (NFPA).

you have an authority to cite when you find a hazardous situation. Pay particular attention to codes that require the installation of sprinklers. The U.S. Fire Administration (USFA) encourages the use of sprinklers as a significant way to prevent injury and death.[6]

Becoming a Part of the Community

Fire departments should serve their communities in more ways than traditional response. In order to embed your department more firmly within your community and strengthen your relationships with local policymakers, consider joining the local chamber of commerce and other civic organizations and participating in community events.

Regularly attending local government meetings gives you the opportunity to share information about the fire department and allows you to stay on top of changes within your community. In some cases, fire chiefs find out that new buildings or subdivisions are planned only when construction begins. Since any community development affects the fire department, the chief should know of any changes as soon as possible.

When attending any community meeting, always be as professional as possible and dress appropriately for the occasion. Be prepared with any information that policymakers might need, including concise written statements and other visual materials that are necessary to reinforce or explain your position.

Human Resource Management: Striving for Membership Longevity

The volunteer fire service is full of tradition that captivates, enchants and entices individuals to join this time-honored civic service. This tradition includes a personal feeling of importance, value and fulfillment of childhood dreams. It allows individuals to make a difference in the well-being of a community, regardless of whether a department is staffed by volunteers or a combination component. How we manage these strong emotions will make the difference in how long a person chooses to volunteer. The feeling of pride and the ambition to succeed is absorbed in the physical surroundings and the management's emphasis on success.

In departments staffed with volunteers, a widely accepted officer election system—minus personal qualifications and controlled with term limits—has created a situation where most of the officers' time and energy are spent on basic fire department functions such as keeping the trucks running, the station clean and training for and responding to calls. The election process requires a chief to be campaigning for reelection and keeping the members happy. Officer term limits in an organization without a strategic operational plan can create an unstable and unproductive environment because the goals and objectives change with the leadership.

Term limits for officers may create an unstable organizational environment. A strategic plan identifies the action required by the organization to reduce the rollercoaster effect of leadership that is constantly changing.

Successful volunteer managers understand the importance of fostering enthusiasm and focusing on opportunities to improve personal and team skill levels. Those departments understand the necessity of ensuring that volunteering is "hassle-free" regarding controllable issues. They structure their departments so that good service becomes the focus and mission of the organization.

A direct connection can be made between how we manage our human resources, the longevity of our personnel, the quality of services provided by our department and, ultimately, the safety of our emergency providers while operating at the emergency. That connection is directly related to a manager's ability to minimize conflict within the department, distribute prompt and fair discipline and provide an atmosphere that encourages and rewards substantive and positive improvements. Community protection and well-being depends on the experience, expertise and tenure of local emergency providers, whether they are volunteer or career.

The commitment to human-resource management provides the basis for the department's success. How we manage, motivate, mentor, design expectations, discipline and record all of these actions provides a basis for future individual and department success. What entices the volunteer is energy channeled toward a positive and productive outcome.

The position of human resource manager is equally important to chief officer positions. Regardless of the size of your department, the human resource functions are equally as important as the chief

6. For more information on USFA sprinkler data, please visit *www.usfa.dhs.gov/safety/sprinklers/*.

officer positions. If you do not take care of your firefighters, they will leave. While this role is behind the scenes, it covers all of the human aspects of the department, such as recruiting, hiring and terminating volunteers; dealing with personnel complaints and investigations; discipline; medical issues; managing long-term personnel objectives such as diversification and training plans; and record management. All of these duties are done with specific knowledge of federal laws to protect individual rights.

Hard work, done well, feels good, particularly when it is hassle-free and appreciated by organizational leadership.

This responsibility should not necessarily fall on the fire chief's shoulders. If someone in the organization does not have this kind of expertise, then outside assistance and guidance may be needed. Personnel management might be better transferred to individuals in local government who routinely deal with these kinds of job responsibilities.

Recruiting and Hiring Volunteers

Your challenge is to convince potential volunteers to: a) donate their time, and b) donate that time to your department. Americans spend about a third of their time at work and a third of their time asleep, which leaves only a third of their time for family, household activities and leisure, which includes volunteering. In fact, Americans spend only 5 percent of their time on leisure activities.[7]

Of the 65.4 million people who volunteered their time between September 2004 and September 2005, only 7.4 percent volunteered with fire departments, emergency medical services (EMS) or and related services.[8] A substantial pool of volunteers exists in our communities. We may have to change our recruitment philosophies to attract them to our departments.

The best recruiting program is a high retention rate.

For recruiting to be effective, you must understand the dynamics of your community and the reasons why people volunteer. The overriding reason is the self-imposed need to belong to something that makes one's community a better place. Few people join organizations to lower their social standing in the community; rather, a sense of achievement and increased responsibility are strong incentives for people to participate. Most individuals excel in organizations that have realistic and meaningful goals that improve both the department and the individual and allow the volunteer to balance his or her civic time with a personal life. Successful fire departments have found that their individual firefighter retention rate is higher when the department provides activities that include the entire family. Those include organized social events, special occasions and junior firefighter programs.

The best recruitment program that a department can have is a high retention rate of existing volunteers. That assumes that the department has mandates for training and response activity levels and is not merely a closed social club or fraternal organization. Few people want to join stagnant organizations or groups that have very limited opportunities for self-improvement or personal skill development. If volunteers think the department is disorganized, dysfunctional or offers little opportunity for self-improvement, they will most likely leave.

The length of time that a new volunteer will remain with the department will be determined in the first six months of membership. Actions taken by the department to make new members welcome, help them adjust, provide mentorship and minimize their discomfort will dictate how long they will stay.

One of the biggest hurdles to overcome is the local government's view of the community cost avoidance created by the volunteer service. These are deferred costs resulting from the actuality of being a volunteer entity. In theory, however, a portion of these savings should be invested in modern equipment, state-of-the-art protective gear and acceptable facilities that encourage volunteer participation.

To create a recruitment campaign, you should understand a number of factors:

- Traditionally, does your community depend on volunteer workers to provide various services?
- What is the level of community support for volunteers and how competitive is the volunteer market?

---

7. Bureau of Labor Statistics, *American Time Use Survey*, Sept. 20, 2005, available at *www.bls.gov/tus*.

8. Ibid, Table 5: Volunteer Activities for Main Organization for Which Activities Were Performed and Selected Characteristics, 2005, available at *www.bls.gov/news.release/volun.nr0.htm*, viewed Aug. 23, 2006.

- What are your community's specific needs based on demographics, population distribution, population age and employer support?
- Does a significant portion of the populace travel to neighboring communities for employment, creating a challenge for the department in providing daytime coverage?

In order to market your department and attract new members, you will have to be specific about their time commitment, training opportunities and your expectations for their involvement. Successful recruiting departments have a complete marketing division to attract new members. They use attractive brochures that outline the department's features as well as membership benefits and incentives programs. Other departments regularly advertise in local newspapers, on television, via Internet links, at movie theaters and on the radio. Some departments retain marketing firms to manage volunteer recruitment; the expenditure is more than justified by the amount of career salary savings. Other recruitment campaigns may be managed by a regional group of fire departments or by a state fire organization.

Recruitment efforts must be designed to attract individuals who have a solid sense of accomplishment and commitment and who meet the internal and community needs of the department. You must be sensitive that your department does not project an exclusive image that will discourage or exclude particular individuals from volunteering. The connection with the community is that a volunteer organization will use the talents of all who choose to donate their time.

Before recruiting efforts can be effective, you must ensure that the department is going to support the addition of new volunteer members. To do so, you must make sure that your selection process reinforces the values of your department, you have designed effective mentorship programs, and your benefit and incentive packages are attractive.

Your selection process should reflect your expectations of the traits and skills you consider important for the individual to be an asset to your department. The ultimate goal is to match these requirements to what the volunteer applicant can provide.

Membership Applications

Membership applications for your department should include a description of the job, necessary literacy skills and physical ability demands. This process should clearly outline the expectations for the volunteer position and distinguish between the levels and types of services that the department offers. Each level should have a list of skills, knowledge and abilities inherent to its specific needs. It is fundamentally important that you enumerate the physical attributes necessary to do the job since volunteers fill more than one role in a department. Different physical demands exist for the volunteer who offers to make meals for a significant wildland event and the volunteer assigned to the fire line.

Any questions that you ask should be in compliance with state and federal law. Simply because people choose to participate as volunteers in your department does not mean they waive their civil rights and allow you to act outside the scope of the law.

Applications that you use should be consistent with other governmental agencies in your area. The easiest way to ensure this is to modify an application that other local government agencies use. Most human resource professionals employed within your local government will be happy to help you modify the application to meet your department's specific needs.

The public expects that the members of your organization will be trustworthy and that they will meet higher standards than the general public.

Background Checks

Background checks help screen and eliminate individuals who should not be a part of your department because they do not share the department's values.

If a candidate has a criminal background, check your state laws to determine whether a previous conviction precludes that person from being a career or volunteer firefighter. Probable exclusions would be convictions for theft, narcotics possession or domestic violence. Law enforcement may not be able to reveal what types of convictions an individual has, however, they can compare the background of an applicant to your policy of acceptable conduct and advise you if a candidate is compliant or unacceptable based on that policy. Some law enforcement agencies will fingerprint a potential candidate. This process adds credibility to your staffing selections.

Since driving fire apparatus may be part of the job, the applicant should provide you with a certified copy of his or her driving record, which you should evaluate against the acceptable standards provided by your specific insurance carrier. An individual's past driving record may substantially affect their ability to volunteer and may disqualify them as a poor risk and uninsurable for this type of activity. Semiannual driver's records evaluation is recommended by insurance companies that insure emergency response agencies.

Written Exams

You may use written exams to eliminate or advance candidates for membership based on their level of literacy. If the position requires completion of reports, dealing with the media, correspondence or contributing to policies and procedures, the ability to compose narrative beyond a basic level is essential. Also, when you understand how an individual learns, you can implement training opportunities that are more valuable and comfortable to that learning style. Bear in mind that this information is personal. You must guard it to prevent any embarrassment to the volunteer.

Oral Interviews

Oral interviews allow several established department members to participate in the selection process. Depending on the level of position offered, you may wish to open the panel to representatives from other agencies in your community. This reinforces the importance of good working relations with other emergency agencies and provides the candidate some assurance that the oral board is not biased in any way. Oral board questions must be predetermined and compliant with state and federal law. Even during this process, the questions need to be directed to the job as a volunteer. Questions that address age, race, nationality, religious beliefs or sexual orientation are off-limits. The questions must be consistent for each member who applies. It is perfectly acceptable to rank the candidates based on their responses.

Physical Fitness Standards

Physical fitness is one of the most pressing issues in the fire service today. Physical fitness standards are a necessary part of being a line firefighter because we ask new members to expose themselves to an immediate onset of strenuous physical activity—for sustained periods of time—with elevated mental and emotional demands. Each department should have some type of physical assessment for potential members to ensure that they are capable of this kind of activity. If this is not practical to do as an independent department, consider coordinating with your local law enforcement agency or as part of a larger fire system in your area.

Following a tentative offer for membership, the department should offer some type of physical exam, including a drug screen. Entry physicals may be expensive, however, they may prevent substantial future costs by identifying physical impairments that would exclude a person from firefighting activities. Ways to defray the costs of these physicals may include special arrangements with local hospitals or recruiting the voluntary services of a physician.

"Popularity"

You must not validate a new member by membership vote. Few businesses, if any, and no other emergency service agency or organization hires personnel based on popularity. This type of action often leads to group discussions involving information that is protected and beyond the scope of "need-to-know."

Discipline and Termination

Discipline is one of the most sensitive jobs performed within a department. The degree of professionalism in dealing with each of these issues will substantially impact the retention rate of your organization and define you as either a quality department or one managed by vigilante justice.

Your goal should always be to provide positive direction to improve behavior and performance. Everyone has the right to expect that the chief will deal with any personnel issues in a fair and professional manner, regardless of whether the individual serves in a career or volunteer capacity. You must draft and circulate a disciplinary policy and you must follow it.

When implementing that policy, all officers need to know their line of responsibility and their maximum level of authority before they are compelled to advance an action. Also, all officers and firefighters should understand that the system protects the rights of the firefighter by providing the right to a hearing before a board of independent and uninvolved members at each stage of the process. Each department should establish its own hearing protocol.

Progressive Discipline

The most common disciplinary method within volunteer companies is "progressive discipline." Progressive discipline is meant to address repeat offenses of departmental values that do not rise to the level of criminal behavior. The most common application of progressive discipline may be infractions of safety policies at the emergency scene or around the station, driving infractions or interpersonal disputes between members. This system is not designed to deal with more serious allegations of inappropriate interpersonal behavior or alleged criminal actions.

Progressive discipline involves a three-step process: a verbal warning, a written warning requiring immediate corrective action and finally, significant disciplinary action.

A verbal warning is the first attempt to notify a member that their behavior is not acceptable. This usually includes some direction to correct that behavior. Although this step is verbal, the officer must keep notes documenting the conversation and the corrective action required. Some departments create an employee performance action plan that acts as an official document for all personnel actions. Please find a sample in Addendum D.

A written warning should be a formal document outlining the infraction and clearly noting when the verbal warning was issued, who issued it and the corrective action required. This document includes specific actions, definable action dates and a time period for re-evaluating the situation. This may involve more coaching efforts over a longer period of time and more officers helping to oversee the requested improvement. This phase should clearly outline the immediate disciplinary actions that will follow if any future infractions occur. Actions may include restrictions, suspension or termination.

In the final phase, implement the action(s) you specified in the written warning. Stand by your decision, no questions asked.

The progressive discipline policy must specifically address who has the authority to enforce it and to update it as necessary. As an example, some departments may allow the station officer to issue verbal warnings and written reprimands. After the written reprimand, the issue and the reprimand are advanced to a chief officer for follow-up. The station officer may or may not be involved in the follow-up, depending on the infraction and how close the officer is to the situation.

Investigations

Other personnel actions can be defined categorically as allegations of improper conduct or inappropriate interpersonal behavior, such as creating a hostile work environment or engaging in sexual harassment. Before engaging in discipline for these allegations, you must assemble investigative facts. These types of events may require a suspension from duty while the investigation is completed. Depending on how close you are to the situation and the individuals involved, you may choose to turn these issues over to an independent third party to ensure that the investigation and following actions are based on fact as opposed to emotion. That third party should have human resources expertise, such as a city or county administrator or human resources official.

Hostile Work Environment/Sexual Harassment

The chief and governing body must provide a work environment free from individual hostility and sexual harassment. Federal law in these areas applies to all departments regardless of whether they are volunteer, career or combination. Fire departments must implement a hostile work environment and sexual harassment policy with yearly staff training on its requirements. The policy should address the use of foul language, jokes or conversations that make light of ethnic or religious values as well as sexually explicit printed materials and figurines. The policy should apply both to spoken and written words, including e-mail.

A significant element in any alleged sexual harassment case is the message perceived by the complainant (as opposed to what the accused meant to say or imply). Tell your staff and members that if any possibility exists that a comment or action may be perceived to be harassing or hostile in nature, they should refrain from making that comment or committing that action.

Your members must be assured that any complaint of alleged improper conduct or inappropriate interpersonal behavior will receive immediate attention. Department officers need to understand their responsibility, in compliance with the policy, to immediately advance any complaint and the consequences for failure to act. These policies are easily obtainable from your local governing entity or drafted with assistance from legal counsel.

Poor Performers

Morale in your department is likely to drop if you retain poor performers for too long. This includes officers. The first step in a departmental review is for officers to be evaluated by their superiors as well as their subordinates. Then the officers should review firefighter performance.

Volunteers will have good days and bad days. You must be somewhat flexible and continue to communicate your expectations of their performance in a fair and impartial manner. Failing to correct poor performance will bring down department morale because members will think that you do not value a job done well.

When assessing a firefighter's performance, you must communicate clearly how you are conducting your evaluation. You should base this evaluation on quantifiable and objective standards for training attendance and response.

Consider setting aside time during each officers' meeting to discuss firefighters who are performing poorly. Company officers should maintain communications with those individuals to help them improve and to keep up their commitment. Try to help your members. Be sensitive to any personal problems they may be having. For example, if members are dealing with personal problems that are keeping them from honoring their commitment to the fire department, offer to place them on a leave of absence or probation rather than dismissing them. If a member's behavior is a problem, consider offering counseling. Be sure to treat each individual firmly and fairly.

Termination

Termination should be a last resort after you have made a full-fledged effort to improve the individual's behavior. If all else fails, however, you may have to terminate a volunteer.

In most fire departments, termination is the responsibility of the chief. Terminations must be done in the presence of at least one other individual who can verify the context of the conversation (which is a good role for the human resource manager) and should include a written document outlining the reason for the action. You may need to suspend the individual until the investigation and termination documents can be drafted.

As a safeguard, your system should include a written policy allowing volunteers to challenge the action and plead their position if they believe they have been wrongfully terminated. Such challenges should go to the governing board or fire commission rather than department officials to ensure an objective review.

To prepare yourself psychologically to fire a volunteer, you may want to visit the Web site of the National Court Appointed Special Advocate Association, which has a number of good tips: *www.casanet.org/program-management/volunteer-manage/fire.htm.*

Managing Medical Records and Related Issues

Federal mandates are specific regarding the confidentiality of a volunteer's medical information. The most recent legislation to impact medical records is the Health Insurance Portability and Accountability Act (HIPAA) of 1996 (Public Law 104-191). This federal law is very clear about what medical information may become public knowledge. You may not share volunteer or patient medical information with anyone else.

HIPAA mandates internal controls for record management, including both hardcopy and computerized information. Firefighter medical information may no longer be filed with a standard personnel file but requires a separate filing system. Medical information on patients that is contained in your response reports must be protected. You are responsible for creating security measures to keep these files private and to implement information policies regarding what information can be contained in a response record database.

You must document any on-the-job injuries in the individual's medical (and not personnel) file. When appropriate, you must complete workers' compensation forms. Reporting should always be in compliance with state statutes and administrative regulations. Follow-up is necessary on any treatable injury to ensure that the volunteer is following specified medical instructions and that the individual does not return to duty without medical clearance.

Retaining Volunteers

Retention of high-performing personnel is necessary to the success of any organization. This holds true particularly for volunteer fire departments where institutional knowledge can mean the difference between life and death for firefighters and the community at large. Community protection and well-being depends on the experience, expertise and tenure of local emergency providers. Volunteers bring tremendous depth and diversity to any emergency scene based on their regular jobs and their expertise in their communities. Weak retention rates often indicate a problem with an organization and diminish the level and quality of service to the public.

Not all attrition is bad. Many organizations use exit interviews to get honest reasons why people are leaving. You should consider these reasons carefully to determine whether your organization is experiencing positive or negative attrition. If your retention rates are low and the reasons why people leave are not negative (for example, they are being transferred out of state), then the organization probably is performing well. However, if people are leaving because they do not enjoy the work, they have conflicts with other members or they are concerned about safety, you are facing an organizational problem.

You should know the retention rate and average length of service of your department. Calculate your organization's retention rate for a given time period by taking the number of members at the end of the period and dividing it by the number of members at the beginning of the period:

Retention and Attrition Rate Calculation

1.	Total number of members	35
2.	Members who have left	5
3.	Total adjusted number of members	30
4.	Retention rate (#3 / #1)	86%
5.	Attrition rate (100% –#4)	14%

To calculate the average length of service (LOS) for your department, divide the total years of service by the total number of members:

Average Length of Service

1.	Total number of members	30
2.	Total years of service	300
3.	Average LOS	10

If your retention rates are low, consider implementing the following strategies.

Minimize Interpersonal Conflict

A very important factor in retaining volunteers is the level of conflict within the organization. This reinforces the notion that the single most important issue affecting retention is solid department leadership.[9] Leadership may suffer if popular elections are held with no requirements for promotions and officers are not trained to deal with personnel issues. Constant turmoil, a lack of discipline, improper management of personnel disputes, overly dramatic embellishments and immature conflict resolutions are often to blame for good volunteers leaving fire departments.

People continue to volunteer when they are liked and respected as people. The connection among members is critical to successful retention.

As the chief officer, you have the responsibility to minimize conflict and resolve interpersonal disputes in a predetermined, fair manner. Bear in mind that most individuals join community service organizations to provide a service, make new friends, learn new skills and have fun. At some point, you may have to ask difficult people to leave your department before they cause irreversible damage by driving good volunteers out. The health of the organization must prevail over the desires and ambitions of a single individual.

Show That You Value Your Volunteers' Time

The fire chief should create an environment in which people feel they are part of a group yet still are unique. When you task volunteers with specific jobs and give them the responsibility to complete them, you unleash tremendous motivational power and a desire to serve.

The role of a volunteer firefighter in a successful department is twofold. One aspect is emergency response—training and going to calls. The other is non-emergency response, such as finance, maintenance and human resources. A dedicated non-operational support staff—whose motto is, "Our job is to make your job easy"—can make the department hum by reducing the burden on operational volunteers. Non-operational volunteers can assist with training, logistics, administration and communications.

Do not waste your volunteers' time. Schedule non-emergency work far in advance and efficiently execute it. Similarly, make sure that routine tasks are routine. For example, do not take an entire day to replace minor equipment because too many people are involved in the process, or require firefighters to fill out forms in triplicate to obtain a new pair of gloves.

One way to effectively minimize volunteer inconveniences is to use a "one-stop" approach, whereby you deal with each of a volunteer's concerns during one visit to the department. While this takes a little more administrative time and organization, it clearly shows the volunteer that you value their commitment. The department can establish an appointment system so that you are aware of all the issues that need to be addressed when the volunteer arrives at the station.

Set an agenda for meetings. For example, if the department has one two-hour meeting each month, you might set aside 15 minutes for briefing or business, 45 minutes for training or work, a 10-minute break, 35 more minutes of training and 15 minutes for conclusion or cleanup. Determine a training and meeting schedule for the entire year and disseminate

9. See International Association of Fire Chiefs' Volunteer and Combination Officer's Section, Blue Ribbon Report—*A Call for Action: Preserving and Improving the Future of the Volunteer Fire Service*, March 2004, available at *www.vcos.org*.

it at the beginning of the year. This allows members to plan their fire department commitments in advance. A meeting will be more productive if your members have a chance to look over the agenda and any reading materials in advance of the meeting. This is easy to do with e-mail distributions.

Provide daily recognition for the contribution your volunteers make to the department and show your trust in them. For example:

- Say "thank you"
- Involve the volunteers in decisions that affect them
- Treat all volunteers equally
- Publish an internal newsletter to highlight your volunteers' important personal and professional milestones
- Show an interest in the volunteers' families
- Send a note of appreciation to the volunteers' families
- Allow volunteers to represent the department at community events
- Recommend deserving volunteers for promotion
- Remember the volunteers' birthdays
- Celebrate the volunteers' anniversary dates with the department

The occasional formal praise cannot take the place of daily informal interaction. Show your appreciation frequently, publicly and in a timely manner. You should be consistent and, most importantly, sincere. Finally, recognize the achievement, but praise the person who achieved it.

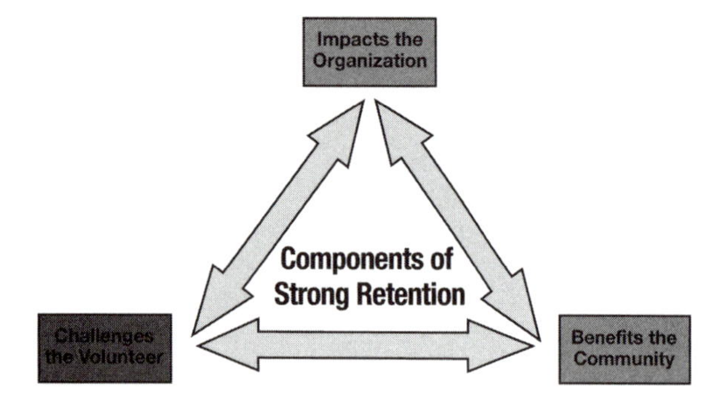

Offer Benefits and Incentives

Positive ways exist to retain volunteers. Among these are benefits and incentives. Benefits include the privileges and securities that are provided to you as a result of your membership. Incentives are rewards for improved performance.

Benefit programs should encourage long-term participation from the volunteer, clearly showing the department's commitment to the safety and security of the individual and his or her family. Those benefits should include workers' compensation; health, accident and life insurance; and coverage that will protect the livelihood of the individual in case of injury, such as wage-supplement insurance that adds to workers' compensation payments. Every volunteer has the right to expect adequate financial compensation in the event that they are injured in the line of duty. Every department has the responsibility to ensure that the volunteer and his or her family are financially protected should such an injury or death occur.

Non-monetary benefits may include using the fire department maintenance facility for personal vehicle repairs, using an empty apparatus bay to clean a personal vehicle and having controlled access to the Internet while providing station coverage. A number of departments are finding substantial value in organizing activities that include the entire family. Picnics, special showcase drills designed to demonstrate the kinds of tasks the volunteers perform or station fun nights are events that allow the families to interact.

Incentive programs should award individuals and team members for their performance and commitment to the department and community. An annual awards banquet provides an excellent opportunity to recognize many achievements. Most of these programs are acceptable expenditures within governmental accounting systems.

Awards should honor individuals as well as team members. They can be spread out over the course of the year and incorporated into other department activities. Here are some examples of awards that can be given during monthly business meetings or training and those that are best suited for the end of the year.

During monthly business meetings or training:

- Graduation ceremonies for individuals completing the fire academy or obtaining their initial firefighter certification. This is the first big step for most volunteers and, accordingly, should have a prominent ceremony that involves their families, the board of trustee, and chief and station officers.
- Recognition of individuals who have accomplished a state certification training level.
- "Certificates of Response" that are given when volunteers reach response milestones. The in-

crements should depend on your response volume. These awards promote call participation as well as acknowledge the individual for increased experience.
- Customer service awards that encourage individuals to go above and beyond the call of duty when dealing with the public and duly serve as a basis for improved community relations.
- "Life Saver" awards for special actions at an emergency scene.
- Team, crew and group recognition for extraordinary work on a firefighter call or rescue, such as a prolonged extrication or water rescue.

End of the year:

- Emergency Provider of the Year
- Rookie of the Year
- Medal of Valor
- Years of Service Awards
- Recognition of long-term projects such as funding drives and fire prevention activities

Consider recognizing other emergency providers as an effective way to improve interagency relations. These kinds of awards generally include the dispatcher, law enforcement officer and EMS provider of the year and are awarded in conjunction with a significant community event or if the individual has made a special effort to cooperate and improve relations with the fire department.

Invite former patients and representatives of businesses that you have helped to the awards ceremonies to improve community relations. Invite local, state and federal elected officials to solidify your political relationships. Do not invite every elected official to every event; rather, select events that you think your political leaders would be interested in attending. Consider inviting the local media, which has the dual benefit of improving the department's public image and providing increased incentive for public officials to attend.

All awards must be defined in a policy that clearly outlines the criteria for obtaining the recognition and the incentive that is provided for that accomplishment. Those incentives may be in the form of plaques, gift certificates for special events and/or dinners as well as jackets or caps. A department should not feel compelled to present an award simply because a category of recognition exists. Make every effort to ensure that the award is meaningful and maintains the level of prestige for which it is intended.

Financial Reimbursement and Tax Breaks

Financial reimbursement for volunteer time is becoming a popular method of attracting new members and retaining experienced members. Payment programs include a year-end bonus, monthly stipends, payment per call or hourly compensation for responses and station standby.

Departments that are looking to implement some type of financial reimbursement program are encouraged to consult with their legal counsel and their regional Internal Revenue Service (IRS) office to have the program validated. Departments that have financial payment programs should be prepared to withhold appropriate payroll taxes, Social Security and Medicare payments. Departments that provide hourly payments for services are most likely not volunteer companies and may be in a position to extend payment for overtime hours and appropriate employee benefits.

A number of states offer different types of tax breaks for volunteers.[10] However, please note that the IRS may record any form of compensation to firefighters as taxable income, including tax breaks or other benefits such as free water or reduced utilities.

Make the Department a Family Organization

Families of volunteer firefighters often experience a great deal of stress when the firefighter dedicates a substantial portion of time to the community, especially when that person misses family events or runs out of the house on a moment's notice.

You can mitigate this stress by making the entire family feel as if they are part of the department. Organize family events at the house and engage family members in tasks that are not necessarily firefighter related. They may want to assist with fundraisers, special event days, daily business operations or junior explorer or cadet programs. (If your department has or plans to have a junior explorer or cadet program, please see Addendum E for appropriate activity guidelines based on the age of the volunteer.)

Suggested family activities include:

- Potluck dinners
- Super Bowl parties
- Family picnics
- Nursing home visits
- Junior combat challenges
- Spouse recognition banquets

10. National Volunteer Fire Council, *State-By-State Comparison: Tax Benefits*, *www.nvfc.org/benefits/state-by-state.phpftype=Tax*, viewed Aug. 4, 2006.

Consider offering childcare for duty crews. Doing so would alleviate a significant cost and time burden for member families, as well as create a family atmosphere by allowing the children to bond with each other.

Finally, consider providing financial security for the volunteers' families through a length of service program. The Virginia Volunteer Firefighters' and Rescue Squad Workers' Length of Service Award Program is a good example. Information on this program is available at *www.nvfc.org/leg/leginfo_va.html.*

Avoid Motivational Traps

Look at what you may be doing to drive volunteers away. Actions that may discourage them include poor training, improper discipline or not enforcing rules uniformly, yelling and screaming to get your point across, lacking excitement in your job, failing to address problems and not following through on requests for help.

As the chief, you should make sure that:

- Volunteers have opportunities for promotion
- Training is interesting as well as educational

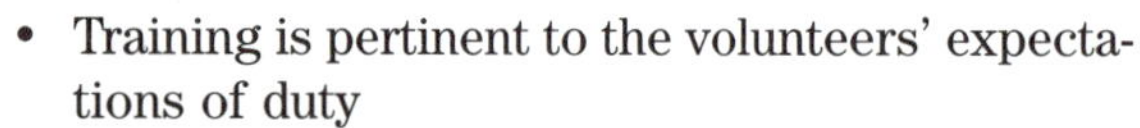

- Training is pertinent to the volunteers' expectations of duty
- Members share responsibility and accountability for important tasks
- Routine issues are dealt with quickly and efficiently
- Meetings serve a purpose and are run effectively
- Members are not subject to undue risk and they have the opportunity to voice their safety concerns
- You do not "sweat the small stuff," such as a truck that is not parked straight in the bay
- You give and earn respect

Diversification Strategies

The concept of diversification is based on the idea that a single individual cannot be an expert in all department operations. Diversification maximizes the talent and skill of the individual, which enhances the overall efficiency, safety and effectiveness of the department, while reducing the need for a single volunteer to respond to every incident.

The premise of diversification is to expand the number of volunteer positions and match individual talents and skills to a specific task. Restricting involvement to one or two tasks allows volunteers to become true specialists and reduces the amount of general training time. Introducing this principle should improve retention by reducing the dependence on a small group of individuals who must respond to all types of events. It will improve the general expertise of the department by developing service-specific experts, and open recruiting opportunities to fill task-specific functions.

To diversify the department, allow new firefighters to identify individual interests after they have fulfilled their baseline training commitment (which is typically Firefighter I certification). The fire department then should design and monitor training to give the firefighters opportunities to expand their knowledge and experience in several different tactical operations. These may include—but certainly are not limited to—apparatus operations, aerial operations, hazardous materials response, technical rescue and wildland fighting.

Consider nontraditional roles for volunteers. The challenge is developing department training and participation standards outside of traditional membership requirements. This philosophy means that not every member of the department must be a certified firefighter to maintain membership. As an example, a professional truck driver who wants volunteer may not be physically capable of functioning as a line firefighter but can contribute as a tender operator during wildland fire season. Should this individual meet all of the membership requirements that a line firefighter has to meet? In a number of departments the answer is emphatically yes. Those departments effectively reject individuals from volunteering who could simply reduce the number of hours a line firefighter is expected to contribute by doing routine, noncritical tasks. You should create an environment where these services are valued and firefighters are concentrating on safety, training and response.

This type of program can increase the vitality of any fire department that relies heavily on volunteers. It clearly delineates your understanding of the value of their time and the need to modify traditional systems to accommodate a wider variety of individual skills and distribute the workload more evenly.

Tactical Equality

One of the best ways to ensure that your system has parity and provides equal opportunities for each of your members is to base training and promotional systems on the process of tactical equality. This requires leadership to devise training programs that lead to some type of state or national certification. Those certifications, combined with specific years of service, are the basis for promotion within the ranks of an engine company and eventually to officer positions. Experienced-based training becomes a critical part of preparing an individual to go from firefighter to engineer or apparatus operator to engine company officer.

By developing standards for volunteers, you encourage them to be more active in training and provide an automatic incentive for personal improvement. This system forces trainers to make the most of the available training time and to expand the number of training opportunities to cover more specialized areas. Requiring firefighters to be certified at the tactical level at which they perform ultimately will force the entire system to select officers based on practical experience and an appropriate level of certification.

Finances and Budgets

Fire departments need money. Funding is necessary for equipment, training, facilities maintenance and staff payroll.

A fire department can obtain operating funds in three basic ways: fundraising, local budgeting or a combination of the two. The optimal funding situation is to have allocated funding and an agreement to use fundraisers as local matching grants for specialized equipment needs.

Smaller departments generally operate solely through fundraising. They hold annual fundraising events, solicit donations through mailing campaigns and operate community events such as weekly bingo. However, requiring volunteers to train for proficient service delivery, make responses and raise their own operational funds will likely hurt volunteer retention. If your volunteers see that the department is completely on its own for financial support, the value of the organization is already degraded in their minds because they believe that no one else thinks the department is important.

To obtain funding from a local government, a well-prepared budget document is a necessity. The key to budget negotiations is to realize that most elected officials like to see solid justifications for funding requests. The budget process often involves several rounds of meetings, justifications, additional justifications, requests for more information and the evaluation of needs with the remainder of community requests.

You must understand how the budget process works. Two common local budget processes exist: the base budget process and the line-item review. Federal sources of funding and other assistance also are available.

Budget Processes

The base budget process identifies those line items that are necessary to operate the fire department every year. Examples include fuel costs, insurance, utilities, operating supplies, training and public education. Once these core accounts are identified, they are funded each year at the same level. Budget discussions are confined to those line items that need to be increased, new line items to be added to the budget and those capital items that are designated to be purchased from special sales tax accounts, grants or other special revenue streams.[11]

In some cases, the local government provides limited support in return for a budget input with just a few line items. However, you still must develop a comprehensive internal budget to be effective. Know where the money is coming from and where it is going.

During the line-item review, you must justify each line item in the budget proposal.

The choice of the base budget or the line-item review is a local decision based on the legislature's view of good government and accountability for the taxpayers' money.

Federal Assistance

To supplement local government and community fundraising dollars, consider applying for federal assistance. The federal government operates a program called Fire Corps that allows community and business experts to assist local fire departments with non-emergency response functions. This may include assistance with budget development, fundraising, developing and maintaining the department's Web site,

11. Capital items are generally defined as major purchases that have a specified lifecycle and will most likely be carried as an asset in a formal inventory. They can be as large as a fire truck and as small as a nozzle. Local policy sets the dollar value for such purchases, which are generally funded from accounts, such as a local sales tax referendum.

12. For more information on Fire Corps, visit the program's Web site at *www.firecorps.org*; call the program's toll-free number, 1-888-FC-INFO1 (1-888-324-6361); or e-mail questions to *info@firecorps.org*.

providing rehabilitation services and doing preplanning.[12]

Another potential source of funding is federal grant money. Through the Assistance to Firefighters Grant (AFG) Program, commonly known as the FIRE Act, the Department of Homeland Security (DHS) provides grant money each year to fire departments for equipment and training needs. Departments must provide a portion of the funding, so be sure that you have an adequate amount of money in your budget.

The Staffing for Adequate Fire and Emergency Response Act, commonly referred to as "SAFER," sets aside federal grant money for staffing. Congress requires the DHS to set aside at least 20 percent of the funding for departments staffed by volunteers, who may use the funding for training, public education or hiring staff to assist the volunteers. Departments must provide an increasing portion of the cost over a four-year period, with the fifth and final year being the full responsibility of the department.[13]

However, do not rely on grants as a part of your base budget. Instead, view them as supplements since they probably will last for a limited time. In fact, the FIRE Act and SAFER require that any grant supplement—and not supplant—local budgets.

Drafting a Budget

We cannot stress enough how important a well-planned budget is to obtaining funding. To assist you with this process, we have included a budget template in Addendum F.

Consider including digital pictures with the budget submission. They can be a great visual aid that provides support for your budget justification. For example, they can show the need for facilities improvement better than a narrative description can.

Consider hosting a demonstration for lawmakers on donning personal protective equipment and using fire department apparatus. This will provide a better understanding than a written narrative of what is necessary to run a fire department and how much it costs.

You may want to consult the governing entity's administrator for help in aligning your proposal with other agencies' budgets. In the case of a special protection district, the governing board can provide this information.

Here is a simple way to look at the budget. Take the total amount of funded money and divide it by the number of runs. This gives you a basic cost per call for the services you provide. You can take this same concept for a specific service within the department, add up all the associated costs, and determine the cost per service. You should know the basic cost of a particular fire department service so that you can make informed decisions about its value to the taxpayer.

Smaller departments should take a look at their revenue and determine if they are able to provide adequate services for the community. A basic single engine fire department with 20 members providing basic fire services in a community of about 1,000 people will need a minimum of $50,000 per year to operate (see Addendum G).

Knowing the cost of the service allows the fire chief to be the chief and to protect the interests of the community. While all communities need basic fire suppression and EMS, specialized services such as heavy rescue and hazardous materials response may best be provided on a regional or shared basis. These services require additional training levels and their success depends on the experience of the firefighters. They require large capital investments for a service that may be of minimal use.

Far too many volunteer fire departments attempt to provide services that the community cannot afford, adequately staff or provide enough training and experience to make their firefighters successful when they handle these kinds of emergencies. Poor performance by volunteers in critical situations will cause them to leave because of personal and public blame for their lack of preparedness. Chiefs and local leaders may need to face the reality that not every department can or should be a "one-stop shop."

Finally, bear in mind that fire trucks and support apparatus are expensive. Two common philosophies apply to buying fire apparatus. The first is to purchase the vehicle and use existing equipment from another piece of apparatus. The second option is to purchase the vehicle with all new equipment and appliances. Several different funding options may exist for new apparatus, including purchasing the vehicle from one account and all the equipment and appliances from another. Lease purchase methods are also becoming popular.

Many fire departments find it difficult to sell older apparatus and equipment. Remember that every piece of rolling stock requires maintenance and in-

13. For more information on these programs, go to *www.firegrantsupport.com.*

surance. Many smaller departments are asking for these same items.

By balancing each of these considerations, you will be able to create a budget that serves the needs of the department and the community in an efficient and cost-effective manner.

Training: Leading and Learning

Each year, around 100 firefighters die and tens of thousands are injured in the line of duty.[14] Firefighters must receive the training necessary to do their jobs safely. Training should be considered a privilege to attend because it prepares firefighters to serve the public they protect, and to protect each other from serious injury or death.

The fire service is unique in that little enforceable fire service regulation exists regarding training. Typically, localities establish their own minimum training standards and qualifications. Such responsibility should not be taken lightly. Failure to set adequate standards can make the difference between a successful fire department and a social club that occasionally goes to fires.

Several standards-setting organizations exist at the state and national levels. A standard does not become law until adopted by a legislative body. However, state and national standards can be identified as a common practice. Professional standards can carry weight in civil court. The authors of this document firmly endorse the use of standards for all fire departments.

Training officers should look to their state firefighter training system for help in developing and delivering a regular training program. In addition, your insurance company may be able to provide training materials and other supporting information.

Areas of Training

The most common areas of training for departments staffed with volunteers include new member orientation, basic firefighter training, regular skills training and officer training. When new volunteers join a department, they should start with orientation and then participate in regular training to hone their skills. If they aspire to be leaders, they should complete officer training.

14. USFA, Statistics: Firefighter Fatalities, Preparedness Directorate, U.S. Department of Homeland Security, *www.usfa.dhs.gov/fatalities/statistics/history.shtm*, viewed Aug. 5, 2006.

An important component of any training program is experiential learning. This type of program provides lessons learned from your own department's or other departments' experiences in responding to large-scale events.

New Member Orientation

A common issue in volunteer organizations is how to make new members active as soon after joining as possible. While they wait for formal training programs to become available, they may lose interest and fade away. This is a valid concern that sometimes prompts volunteer leaders to put new members in dangerous positions for which they are not properly trained. This places personnel at unnecessary risk by putting them in situations where they may not be mentally or physically prepared for the trauma and stress of emergency services.

To combat these risks, develop a training program that gives new members the information and skills they need to operate in a support role, safely allowing them to be on the fireground early on. Tailor the program to the community's needs. Some organizations may recruit enough volunteers to support monthly training sessions. Another option is a one-on-one mentoring program in which new volunteers are assigned to experienced members to work through specific training objectives. The members fill in a worksheet according to the training they have undergone and submit it to the training officer when complete.

Whatever training program you implement, make sure it meets the state's and locality's minimum standards before allowing new members to run their first call. For sample program topics, along with the time commitment for each, please see Addendum H.

Basic Firefighter Training

Basic firefighter training should target nationally or state-recognized professional development standards directly. All interior structural firefighters must obtain this level of training and should be certified at the minimum level within one year. Some volunteers may have difficulty achieving this level of training because of work and family commitments. In such cases, the organization should recognize those members as "non-entry" and facilitate job functions that let the members give to the fullest potential for which they are qualified. The Fire Corps program works well in this situation.

Fire departments should establish formal training programs. Some states provide certification training while others defer to the localities. If the department must provide its own basic training, you should consider the format you wish to use. The most common method is for members to achieve the certification of Firefighter I on their own by seeking out and completing a class in-house or in the region. Then they may go back to obtain the Firefighter II certification.

Some departments have found it more beneficial to consolidate such training by forming "volunteer academies." These programs combine all of the basic training components into one program, usually lasting six months or less. By using this method, volunteers can complete all of their training at once. The benefit to this approach is that once the training is done, new members are finished with the basic training component.

All firefighters should be certified at the minimum level within one year. Officers should be certified at the Fire Officer I (or equivalent) level before promotion.

Departments should choose the delivery system that works best for them. They should consider the number of firefighters to be trained on a regular basis as well as the free time that each member has to offer. A common schedule for both academy and individual programs is to train two nights per week and every other Saturday. Most academy programs can easily complete the Firefighter I and II certifications in a six-month period using this template.

Regular Skills Training

Regular skills training, commonly called "drill night," is a training staple of a department staffed with volunteers. These sessions are usually held on a weeknight and feature a variety of topics that allow volunteers to come together to brush up on skills and/or techniques. More importantly, it gets the volunteers together at one location so they can work together and build the bonds that enhance volunteer retention.

Selecting topics for drill night is a common challenge for training officers. Coming up with a new and exciting topic month after month can be difficult and, as discussed in the recruitment and retention chapter, you should avoid wasting the volunteers' time.

To help prioritize training topics, break them into two categories: high frequency/low risk and low frequency/high risk. High frequency/low risk events occur on a regular basis and pose little risk to firefighter safety. Such topics include taking blood pressures, activating the fire alarm, responding to EMS calls and communicating over the radio. While local regulations may require occasional refresher training on these topics, the *members* certainly should not dominate a drill night. If you have to do training on these types of topics, comply with the regulation but make the training as quick and painless as possible.

Good training will motivate volunteers and make them more effective individuals and team members.

Much more important are events that happen rarely but pose a serious risk to firefighters. These events include fighting structural and vehicle fires, responding to hazardous materials incidents and specific tasks such as cutting vent holes in roofs or forcing doors. Because smaller departments staffed with volunteers seldom do these tasks on the fireground, they must compensate by practicing on the training ground. Drills on advancing hose lines, operating power equipment and throwing ladders should be in the regular drill schedule. Please see Addendum I for a sample list of a year's worth of drill topics.

Monthly training sessions should reflect the critical skills the firefighters carry out on the local fireground. The department's officers and firefighters should meet to create a list of these skills. Then,they should develop questionnaires to determine which skills need the most work. For example, you should ask how many times a firefighter has started the power saw on the truck, or changed the saw blade. You should ask how many times a firefighter has placed a 24-foot extension ladder, and whether he or she has removed the ladder from the side of the engine to the fireground. For a full list of potential questions, please see Addendum J for a sample experience assessment.

Finally, training should include a periodic review of standard operating procedures (SOPs) and standards of governance (SOGs) as well as any changes to the SOPs and SOGs that occur between scheduled review sessions.

Officer Training

Officer training is likely to be one of the most difficult areas to create. Often, instructors are in short supply and programs are complex. Also,

while state and national standards for firefighter training exist, none exist for leadership training. Professional development standards usually outline requirements for a fire officer, but the standards mainly target management functions. A successful pathway to officer development that includes educational milestones and performance expectations is the IAFC's Professional Development Handbook.

When departments lack the necessary resources to provide quality officer training, consider the idea of regionalizing the training among neighboring departments. If you pool your resources, you are likely to be able to meet local needs. You also will forge and strengthen relationships among neighboring departments.

Please see Addendum K for a sample officer training schedule.

Mentorship Programs

Mentorship programs recognize that simply donning turnout gear does not make an individual a firefighter. These programs involve a comprehensive effort to ensure that when new members arrive at emergency scenes, someone—usually a station officer—is available to explain to them what is going on and why. Many "micro-training" opportunities are available before, during and after responses to improve levels of understanding, define procedures and help develop skills for future assignments. This is a critical step. New firefighters do not learn anything sitting in the station because they missed a truck. If they miss enough trucks, they will not show up.

New members should receive a copy of their job description, an outline of duties they may be required to perform at the scene and around the station, and a copy of the rules that you expect them to follow, clearly stating any response or training attendance requirements. Mentorship reinforces your commitment to a volunteer's success by evaluating them and providing regular feedback during the probation period. The member needs to know "the good, the bad and the ugly" aspects of their performance, but presented in a constructive—and not destructive—manner. By nature, we want to do the best that we can. When we venture out of our personal comfort zones to volunteer, we expect honest feedback to improve our performance.

A comfortable environment in which to accomplish this is the "debrief session," which occurs when a response is complete and all the gear is again ready for service. Take the time to go over events in a constructive way, noting the good aspects of the response as well as the areas in need of improvement. Given the opportunity to criticize their own performance, people are often harder on themselves than you as a chief would be. However, the result generally is good feedback from the group and individual improvement the next time out.

Mentors should have a complete understanding of their role, authority and responsibilities. The program must be consistent from station to station, reinforcing the same values. Mentors need to have good people skills, the ability to function on-scene in the dual role of firefighter and mentor, and the ability and willingness to communicate with new members and share information.

■ Training as a Recruitment and Retention Tool

In addition to being necessary for safe departmental operations, training can be a solid volunteer retention tool. People like to volunteer for organizations that provide them with new skills and challenges and expand their abilities to learn and perform. However, as volunteers, their time can be limited. Departments with high retention rates have found several secrets to balancing training requirements without demanding more time of the volunteer:

- Decide on a baseline level of training that all members of your department need to have to provide good basic service.
- Each volunteer should select one or two areas in which to build expertise. This allows an individual to concentrate on specialty services that are of interest to them.
- Make sure that every formalized training opportunity that you provide puts a firefighter closer to meeting a training standard or certification. This allows an individual to break down certification requirements into achievable steps.
- Formalize training times and produce training schedules that allow the member to plan on specific times to commit to the fire department. Scheduling provides you with protected calendar space and promotes training nights as an essential part of the department, becoming a part of the rhythm of the organization.
- Provide constructive feedback to foster honest and open communication. All of us want to do a good job and feel good when we perform successfully.

Providing Balance

The job of being a volunteer brings a new level of personal stress that, if not managed, will reduce participation and ultimately cause members to drop out. The new physical demands and required training may resurface learning and physical disabilities that will be judged by an ultimate outcome of life or death.

Balance training with social opportunities to help manage the stress and build confidence and solid relationships with other department members. Recruiting a department chaplain can provide your firefighters and their families with an excellent resource to manage difficulties in their personal lives.

Leadership and Management: Leading and Following

Fire chiefs are the chief executive officers of their organizations. In the old days, the fire chief was the "best firefighter." Today, departments must focus on operational integrity and business excellence. "Business" is defined as the financial aspect of running a fire department. Fire chiefs must draft a budget, secure funding and put specific controls in place to ensure that spending is legal and appropriate.

Fire chiefs lead personnel. The strength and quality of an organization's leadership determines the cohesiveness, effectiveness and motivation of the units within that organization. Ineffective organizational leadership is a problem that multiplies and magnifies smaller problems and generally undermines the effective management of an organization. The organization decays from the inside, the mission does not get accomplished and the system fails.

Leaders should:

- Constantly observe and learn from their environment
- Observe the successes and failures of others and analyze the "how and why"
- Make mistakes, sometimes more often and with more consequences than others, simply because leaders are constantly in the spotlight
- Learn from their mistakes and use that experience to better themselves
- Set a positive example
- Surround themselves with subordinates that will hold themselves to the standard to which that leader adheres
- Demonstrate willingness to take on personal risk as they ask others to take on personal risk

A leader must be submersed in the organization. Two key ways to do this are to keep informed and to actively communicate with the members. At the forefront of every thought should be the fact that a leader is nothing without the people who work for the organization. If a leader is good to his or her staff, the staff will be good to the community.

"People skills" are key. The fire chief does not have to be the best firefighter but does have to know which firefighters are best for which jobs.

A Strong Leadership Foundation

Think of building your department as you would build a house. You start with the foundation and then add walls, the roof and drywall. Your challenge as the department leader is to ensure that the foundation is strong and that it is integrated seamlessly with other structural components.

To achieve this goal, chief officers must expect excellence of themselves first, always striving for success. Understand that you are always "on stage" and must set an example for the department. As a leader, you must subordinate yourself to the organization; your ego must be the lowest priority. The leader must be a genuine person who demonstrates openness and caring each day.

You must foster respect within the department. The position of leader demands respect, but the person in the leadership position must earn that respect. Treat others with integrity and respect and they will treat you the same way. Also, hold yourself accountable and responsible for your own actions.

You must build strong relationships with your members:

- Be sensitive to your members' personal issues and meet with them regularly to discuss their strengths, weaknesses and goals, both personal and professional
- Participate in social and family events with the members
- Delegate tasks as appropriate to show trust in your members' abilities

A genuine and caring individual must continually put the department and the public good first. A leader must separate personal wants and needs from what the community desires and what the agency can produce.

Continued Learning

A leader, whether elected or appointed, must never stop learning. This holds true even for four- or five-bugle chiefs. No fire chief school for community-based fire departments exists, so the drive for education and personal growth must come from within. The moment any chief thinks he or she knows all that is needed in this business is the moment that chief lets his or her people down and puts them at risk.

To make sure you continue to learn, subscribe to fire service periodicals that have officer and leadership development articles. Take community college courses in personal and professional development. Read books by fire service publishers and others that describe the traits and personal stories of successful leaders, both within the fire service and in the world at large.[15]

The National Fire Academy (NFA) provides free education for all members of the fire service, including the Volunteer Incentive Program (VIP) with several weeklong courses designed specifically for volunteers.[16]

In some departments, outdated election processes do not always ensure that the chief has the necessary leadership qualities. Most fire chiefs have achieved tactical excellence and can maintain those skills via continuing education. However, some elected chiefs have no formal training in leadership or accounting (to be able to maintain the department's budget). Departments that elect their chiefs should consider transitioning to a modern by-law system, whereby predetermined qualifications can help select a trained and educated chief.[17]

■ Conflict Resolution

People have differences of opinion, that is human nature. Those differences can escalate from warm to simmering to boiling. Your job as leader will be to resolve these conflicts, preferably before they reach the boiling point.

"Warm" issues tend to nag a person every day and continue to build unless someone intervenes. A leader should not ignore warm-level events. Instead, he or she must ask questions to find the root cause. Never jump to a conclusion before analyzing the answers to these questions. As with any conflict resolution, less damage will occur if you solve the problem early.

"Simmering" issues are more serious. They appear after warm problems continue without intervention, fueling hostility and resentment. At this level, a leader must actively seek root causes and focus on validating the problem. Is it a real problem or is it a perceived problem that began with a lack of communication? The leader should conduct interviews with the complainant and seek immediate "cooling" to get the issue back to the warm level. At that point, the leader can develop a resolution. Most people will react positively if direct communication demonstrates that you understand the problem and are developing corrective action.

"Boiling" issues are the most dangerous. At this level, people try to exact revenge, and nobody wins. You will need a third-party facilitator—such as the president of the board of directors—to intervene and try to reestablish relationships within the department. That third party should be trustworthy and credible and have no personal stake in the issue at hand. This method is very dynamic and requires that the facilitator be schooled in continuing negotiations. He or she will need to continually evaluate the facts, seek common ground and determine the best course of action.

The key to intervening at each level is to understand the root cause of the problem. This is imperative. Do not allow your emotions to take over or jump to conclusions before identifying the real issue.

Instead, follow this problem-solving guide:

- Identify each issue separately
- Ask why the issues have become problems—and continue to ask why until you get to the root cause(s)

- Identify various actions to address the problem(s)
- Pick the most effective and efficient resolution(s)
- Do not create a temporary patch—try to fix the problem(s) completely
- Evaluate the results and try new approaches until you have solved the problem(s)

■ Personal Integrity and Trust

Integrity builds trust. People will trust a leader they can believe in. Always hold yourself to the highest standard of personal integrity.

---

15. Other leadership programs that the authors of this report have found helpful include Ziglar True Performance (*www.zpgtraining.com*) and *The 360° Leader Comprehensive Assessment* by Character of Excellence, LLC (*www.characterofexcellence.com/index.html*).

16. For more information on VIP, contact your state training agency or visit the program Web site at *www.usfa.dhs.gov/training/nfa/resident/vip/*.

17. For a sample set of fire department by-laws, please visit the VCOS Web site at *www.vcos.org*.

You must show trust in your members. In short, you must delegate. Too often, leaders try to bear all responsibility and complete every associated task, which results in overload and trouble completing projects. Delegating—with clear guidance—provides ownership in a task that allows for excellence and personal pride.

Delegation can be a powerful tool, but you must define boundaries. As the chief, you will have the full picture of what must be done, while your members will be assigned tasks within the full structure. For example, the standard chain of command is followed easily on the emergency scene, but a multitude of tasks that are not related to emergency response are involved, such as shirts, gate locks and identification badges. A simple list will allow your members to understand who is doing what.

■ Conclusion

The VCOS and the IAFC stand ready to help you build a strong department. The authors of this report encourage you to visit the VCOS Web site at *www.vcos.org* for additional tools and advice.

Addendum A: Sample Department Conduct Standards

CONDUCT STANDARDS OF THE ANYTOWN VOLUNTEER FIRE DEPARTMENT, USA

The following list of directives represents the conduct standards for members for the ANYTOWN FIRE DEPARTMENT. The basis for these regulations is the following policy:

Every member of the ANYTOWN FIRE DEPARTMENT is expected to operate in a highly self-disciplined manner and is responsible to regulate his/her own conduct in a positive, productive and mature way. Failure to do so will result in disciplinary action ranging from counseling to termination.

ALL MEMBERS SHALL:

1. Follow operations, policy manuals and written directives of both the ANYTOWN FIRE DEPARTMENT and (YOUR LOCAL GOVERNMENT)
2. Use their training and capabilities to protect the public at all times both on and off duty [18]
3. Work competently in their positions to cause all department programs to operate effectively
4. Always conduct themselves to reflect credit on the department
5. Follow instructions in a positive and cooperate manner, and expect that supervisors will manage the department in an effective and considerate manner
6. Always conduct themselves in a manner that creates good order inside the department
7. Keep themselves trained and educated to do their jobs effectively
8. Be concerned and protective of each member's welfare
9. Operate safely and use good judgment
10. Obey the law
11. Maintain personal physical fitness

MEMBERS SHALL NOT:

1. Engage in any activity that is detrimental to the department
2. Engage in a conflict of interest with the department or use their position with the department for personal gain or infiuence
3. Engage in physical confrontations
4. Abuse their sick leave (career members)

MEMBERS SHALL BE TERMINATED IMMEDIATELY IF THEY:

1. Steal
2. Use alcoholic beverages, debilitating drugs or any substance that could impair their physical or mental capacities while on duty
3. Engage in any sexual activity while on duty
4. Wear fire department clothing/uniforms and/or equipment (such as radios and pagers) while consuming alcohol off-duty

18. For career staff, on-duty is defined as normal duty shift or when responding to an incident after normal work hours. On-duty for volunteer members is defined as time responding to incidents, shift coverage and training.

Addendum B Volunteer Fire Service Cost Savings Project

Step 1 Enter your fire department's data in the yellow boxes. Version 2.1

19,000	Population protected
7,400	Number of residences
$ 72,000	Current personnel expenses

Your Fire Department Data

	Number of Apparatus
4	Engines
0	Aerial Trucks
1	Tankers
1	Rescue Trucks
0	Utility Vehicles
3	Brush Trucks
0	Ambulances

Total Staffing	Minimum Staffing on apparatus	Company Officer	Driver/Operator	Volunteer Firefighter
3	Engines	1	1	1
0	Aerial Trucks	0	0	0
2	Tankers	1	1	0
3	Rescue Trucks	1	1	1
0	Utility Vehicles	0	0	0
3	Brush Trucks	0	1	2
0	Ambulances	0	0	0

Step 2 Go to "Career Costs" tab.

Career Cost Spreadsheet

Step 3 Starting Salary for career personnel in the vicinity (including benefits).
If there are no salaries paid for any position, use a nearby city costs for these entries. If there are no cities nearby that can be used, use police salaries for this portion.

Position	Salary
Chief	48,000$
Deputy Chief	44,000$
Assistant Chief	40,000$
Captain	37,000$
Lieutenants	33,000$
Drivers	27,000$
Medics	33,000$
Fire/Rescue Personnel	24,000$

Step 4 Staffing factor (positions needed to staff one full-time position) For 56 hour week use "3", for 42 hour week use "4". → 3

Step 5 Enter **your** department staffing for the "Chief" positions only.

Number and Cost of Career Personnel Necessary

Position	Number	Cost
Chief	1	$ 48,000
Deputy Chief	1	$ 44,000
Assistant Chief	4	$ 160,000
Officers	18	$ 630,000
Drivers	18	$ 486,000
Firefighters	15	$ 360,000
Medics	0	$ -
Career Chief Officer Costs		$ 252,000
Career Fire Fighter Costs		$ 1,476,000
Total Career Costs		$ 1,728,000

Step 6 Go to "Summary" tab.

Cost Savings to Your Community

$ 72,0000	Your Annual Personnel Costs
$ 1,728,000	Full-time Equivalent Fire Department Personnel Costs
$ 1,656,000	Annual Personnel Cost Savings
$ 223.78	Savings per Household
$ 87.16	Savings per Resident

Calculation tool provided by PivotPartners

Addendum C: Sample Fire Department ISO Rating Cheatham County (TN) Volunteer Firefighters Association

The following table was compiled in cooperation with local insurance agents and the property assessor's office. The property assessor's office provided the average home value in the fire district. Several local insurance agents provided a basic homeowners policy and its components.

Policy	
Coverage	**Amount of Coverage**
• Dwelling	$100,000
• Other Structures	$10,000
• Personal Property	$50,000
*Special Limits and Protection for:	
○ Jewelry	$1,000
○ Money	$200
○ Securities	$1,000
○ Silverware	$2,500
○ Guns	$2,000
• Loss of Use	$20,000
• Personal Liability	$300,000
• Medical Payments	$5,000

Policy	
* Deductible = $250	* Construction = Frame
Fire Protection Class	**Premium**
10	$813
9	$759
8	$504
7	$469
5	$428
* Deductible = $250	* Construction = Masonry
Fire Protection Class	**Premium**
10	$736
9	$670
8	$482
7	$425
5	$389

- The protection of the residence with a fire sprinkler system in compliance with National Fire Protection Association Standard 13D may result in an additional savings ranging from 10 percent to 30 percent.
- Each department should research the actual cost benefit of lowering the community's ISO rating. The rates for insurance premiums may be the same for residential units with a fire department class between three and seven. Commercial and industrial occupancies may realize significant savings in annual premiums if the fire department expends large sums of capital.

Addendum D: Sample Performance Improvement Plan

Use a performance improvement plan (PIP) when you are looking for ways to improve a member's performance. The PIP is a tool to monitor and measure the deficient work products, processes and/or behaviors of a particular firefighter in an effort to improve performance or modify behavior. Key items to remember:

1. Define the problem. This is the *deficiency statement.* Determine if the problem is a performance problem (firefighter has not been able to demonstrate mastery of skills/tasks) or a behavior problem (firefighter may perform the tasks but creates an environment that disrupts the workplace).
2. Define the duties or behaviors where improvement is required.
   - What are the *aspects of performance* required to successfully perform these duties?
   - Which skills need improvement?
   - What changes need to be made in the application of skills that a firefighter has already demonstrated?
   - What behaviors need to be modified?
3. Establish the *priorities* of the duties.
   - What are the possible consequences of errors associated with these duties?
   - How frequently are these duties performed?
   - How do they relate when compared with other duties?
4. Identify the *standards* upon which performance will be measured for each of the duties identified.
   - Are they reasonable and attainable?
5. Establish short-range and long-range goals and timetables for accomplishing change in performance/behavior with firefighter.
   - Are they reasonable and attainable?
6. Develop an *action plan.*
   - What will the officer/chief do to help the firefighter accomplish the goals within the desired time frame?
   - What will the firefighter do to facilitate improvement of the product or process?
   - Are the items reasonable?
   - Can the items be accomplished?
   - Are the items flexible?
7. Establish *periodic review* dates.
   - Are the firefighter and the officer/chief aware of what is reviewed at these meetings?
8. Measure actual performance against the standards to determine if expectations were:
   - not met
   - met
   - exceeded
9. Establish a PIP file for the firefighter.
   - Does the file contain documentation that identifies both improvements and/or continued deficiencies?
   - Is the firefighter encouraged to review this file periodically?
10. Put the PIP in writing.
    - Has plain and simple language been used?
    - Have specific references been used to identify areas of deficiency?
    - Have specific examples been used in periodic reviews which clearly identify accomplishments or continuing deficiencies?
    - Have you chosen an easy-to-read format such as a table or a duty-by-duty listing?
    - Have the *Terms of Agreement* been included in the PIP?

Addendum E: Sample Management Guidelines for a Junior Firefighter/Cadet Program

The following are management guidelines to consider if you are interested in developing cadet or explorer programs. The authors of this report highly encourage you to consult your state's office of occupational safety and health for the rules that apply to your department. You should consult your state's labor laws to determine whether a junior firefighter or cadet must obtain a work permit.

■ Cadets/Explorers/Junior Firefighters

The following activities for all firefighter trainees (cadets/explorers/junior firefighters) under the age of eighteen (18) years are prohibited:

1. Driving department vehicles greater than $^{3}/_{4}$ ton
2. Performing fire suppression involving structures, vehicles or wildland fires, except grass fires that are not in standing timber
3. Responding in a personal vehicle with blue lights
4. Performing fighting overhaul duties
5. Responding to hazardous materials fires, spills or other events
6. Performing any activity, except training performed by qualified personnel (after medical certification as required by 29 CFR 1910.134), involving the use of self-contained breathing apparatus
7. Performing traffic control duties
8. Using pneumatic/power driven saws, shears, Hurst-type tools or other power tools
9. Entering a confined space as defined in 29 CFR 1910.146
10. Entering a fire ("red") zone
11. Performing any duty that involves the risk of falling a distance of six feet or more, including the use of ladders
12. Filling air bottles
13. Operating pumps of any fire vehicles at the scene of a fire
14. Handling life nets (except in training)
15. Using cutting torches
16. Operating aerial ladders
17. Performing any duties involving the use of lines greater than 2 inches in diameter

■ 14- and 15-Year-Olds

Cadets/junior firefighters who are 14 or 15 may only perform the following duties. They may perform these duties only between the hours of 7 a.m. and 7 p.m., except from July 1st through Labor Day, when evening hours are extended to 9 p.m.

1. Responding to emergencies on fire department apparatus
2. Clean-up service at the scene of the fire, but only outside of the structure and only after the scene has been declared safe by the on-scene commander
3. Providing coffee/food service
4. Engaging in training that does not involve fire, smoke (except theatrical/latex smoke), toxic or noxious gas, or hazardous materials or substances
5. Receiving instruction
6. Attending meetings no later than 9 p.m. throughout the year
7. Observing firefighting activities while under supervision

■ 16- and 17-Year-Olds

Cadets/junior firefighters who are 16 or 17 may perform only the following duties. They may perform these duties between the hours of 6 a.m. and 10 p.m., except when there is no school the next day, when evening hours are extended to 11 p.m.

1. Attending and taking part in supervised training
2. Responding to emergencies on fire department apparatus
3. Participating in fire department functions within the rehabilitation area
4. Picking up hose and cleaning up at the fire scene after the on-scene commander has declared the area to be safe
5. Fighting grass fires not involving standing timber, with proper training
6. Performing search and rescue operations, not including structural firefighting

An alternative to standard recruiting is the Fire Corps program. Visit the program Web site at *www.firecorps.org*.

Addendum F: Sample Budget Proposal

■ ANYTOWN VOLUNTEER FIRE DEPARTMENT, USA Budget Proposal for Fiscal Year 2006–2007

Table of Contents

General Department Information

- Mission statement
- Department officials, including the fire department governing board and chief officers
- Organizational structure
- The department budget process and how the requests are assembled
- Special event coverage and the impact on the department budget
- Summary of existing federal, state, and local grants
- Summary of lease agreements
- Significant capital and capital facilities expenditures in future budgets (24-month projection)

Budget Overview

- Previous budget allocation by line item
- Approved current budget line-item allocations
- Year-to-date budget balance for the current budget line items
- Projected year-end budget per line item
- Proposed line-item budget allocated for the next fiscal period

Board Approved Line-Item Transfers (for the current budget)

- Overview of the total line-item transfers with a collective balance
- Narrative justification for the transferred balance

Operational Budget Increases

- Overview of the total operational line-item increases and collective balance
- Narrative justification for the increased balance request

Apparatus Replacement

- Overview of the total apparatus request with a collective balance
- Narrative justification for the new vehicle
- Finance proposal

Capital Requests and Station Improvements

- Overview of the total capital request with a collective balance
- Narrative justification for each capital item

Note: Some elected officials prefer to see a narrative justification when the account is increased or decreased by 10 percent or more.

Addendum G: Sample Calculation of Minimum Cost to Operate a Fire Department

	Capital Cost	Life Expectancy	Annual Cost	Number		Interest
Engine	250,000	20	12,500	1		750
Operation	2,500	1	2,500	1		
Station	125,000	20	6,250	1		375
Building Maintenance	1,000	1	1,000	1		
Protective Clothing	25,000	8	3,125	10		188
Health Insurance	3,500	1	3,500	1		
Workers' Compensation	1,000	1	1,000	1		varies by state
Utilities	1,600	1	1,600	1		gas, electric, phone
Communications	5,000	3	1,667	1		
Administration	1,000	1	1,000	1		postage
Training	3,000	1	3,000	1		
Fees and Licenses	500	1	500	1		
Physicals—OSHA	2,000	1	2,000	10		
Interest Cost	1,520	1	1,520	1		
Volunteer Benefits	10,000	1	10,000	1		10 per run/ average 5 people per run
Board Expenses	250	1	250	1		
Legal Expenses	1,000	1	1,000	1		
Public Education	200	1	200	1		
Equipment	2,500	1	1,000	1		
Rescue Squad	80,000	15	5,333	1		320
Insurance	2,000	1	2,000	1		
Jaws of Life	25,000	10	2,500	1		150
Station Addition	50,000	20	2,500	1		150
Total		Fire and EMS	65,945		1000	homes protected
			65.945			
		Fire Only	53,142			
			53.142			

Addendum H: Sample Statewide Firefighter Orientation Program

Module	Topic	Duration
Module 1a	Orientation	1.5 hours
Module 1b	National Incident Management System	1.5 hours
Module 2	Personal Safety/Special Hazards	2.5 hours
Module 3	Self-Contained Breathing Apparatus	2.5 hours
Module 4a	Search and Rescue	2 hours
Module 4b	Extrication	2 hours
Module 5	Hose Loads	4 hours
Module 6	Fire Streams	4 hours
Module 7	Forcible Entry	2 hours
Module 8	Ladders	4 hours
Module 9	Ventilation	3 hours
Module 10a	Apparatus Familiarization	1 hour
Module 10b	Driver Awareness Level	2 hours
Module 11	Hazardous Materials – Awareness	8 hours
Module 12	Hazardous Materials – Operations	16 hours
Module 13	EMS Awareness	2 hours

Local fire chiefs and leadership should establish rules that require members who respond to emergency medical services events to be certified in accordance with state and national training standards.

Addendum I: Sample Training Drill Calendar

January	Protective clothing/Safety
	How to get flawless execution
	Physical assessment/medical history survey—confidential/basic vitals
	Review two NIOSH LODD reports
February	Incident command
	Simulate—1 storey SFD
	Simulate—2 storey SFD
	Responsibilities of command
	Functions of command
	Command organizational positions
March	Salvage
	Communications
	Lock-out/tag-out
	Bloodborne pathogens
April	Auto extrication
	Removing a door/roof and steering wheel
	Patient packaging
May	Initial fire attack
	Protective clothing/Safety
	Pulling preconnect
	Raising ladder
	Donning and doffing SCBA
June	Search and rescue
	Protective clothing/Safety
	Donning and doffing SCBA
	Primary search
	Secondary search
	RIT
	Self-preservation
July	Ventilation
	Chain/reciprocating saw
August	Pump operations
	Tanker operations/Fold-a-tank
September	Search and rescue
	Protective clothing/Safety
	Donning and doffing SCBA
	Primary and Secondary Search Practices
	RIT and Self-preservation

October	Initial fire attack
	Protective clothing/Safety
	Pulling preconnect
	Raising ladder
	Donning and doffing SCBA
November	Pump operations
	Tanker operations/Fold-a-tank
December	Utility control
	Overhaul
	Basic fire investigation

Addendum J: Sample Experience Assessment

Experience Assessment

Unless the question specifically refers to an emergency or training experience, please note your total experience.

How many years have you been a firefighter?	1+	5+	10+
How many years have you been an officer?	1+	5+	10+
Are you a certified Firefighter I/II?	Yes	No	
Are you a certified Instructor II/III?	Yes	No	
Are you a certified Fire Officer II?	Yes	No	
Are you a certified Safety Officer?	Yes	No	
How many times have you done CPR on a real person?	1+	5+	10+
How many times have you driven an emergency vehicle?	1+	5+	10+
How many times have you pumped an engine on an emergency response?	1+	5+	10+
How many times have you dumped water into a fold-a-tank?	1+	5+	10+
How many times have you used the "jaws of life" to remove a door from a vehicle while patients were in the vehicle?	1+	5+	10+
How many times have you been inside the vehicle taking care of the patient while the door/roof was removed from the vehicle?	1+	5+	10+
How many times have you been the nozzle person on a car fire?	1+	5+	10+
How many times have you been the backup person on the hose line on a car fire?	1+	5+	10+
How many times have you been the nozzle person on a structure fire?	1+	5+	10+
How many times have you been the backup person on the hose line on a structure fire?	1+	5+	10+
How many times have you been command on an extrication?	1+	5+	10+
How many times have you been command on a brush fire?	1+	5+	10+
How many times have you been command on a house fire?	1+	5+	10+
How many times have you been assigned a sector role at an emergency event?	1+	5+	10+
How many times in the last year have you visited/trained at another fire station?	1+	5+	10+
How many times have you coached a firefighter to improve his or her personal performance in the fire department?	1+	5+	10+
How many times have you counseled a firefighter to improve his or her personal performance in the fire department?	1+	5+	10+
How many first alarm structure fires have you attended?	1+	5+	10+
How many multi-department structure fires have you attended as a firefighter?	1+	5+	10+
How many multi-department structure fires have you attended in a command role?	1+	5+	10+
How many times have you ventilated a roof?	1+	5+	10+
How many times have you been assigned a sector responsible for overhaul?	1+	5+	10+
How many times have you dealt with hazardous materials in an incident?	1+	5+	10+

Addendum K: Sample Officer Training Schedule

Month	Topic	Duration
January	Creating a Supportive and Positive Atmosphere	2 hours
February	Handling Confrontation	1.5 hours
March	Giving and Receiving Praise	1.5 hours
April	Managing Priorities/Review of Rules and Regulations	1 hour
May	Coaching For Improvement	2 hours
June	Overcoming Procrastination and Piles of Paperwork	1 hour
July	Incident Command	2 hours
August	Strategy and Tactics	2 hours
September	Handling Complaints	1 hour
October	Communications/Review of Standards of Procedure and Standards of Government	2 hour
November	Developing and Improving Our Image	1 hour
December	Leadership	1.5 hours

Addendum L: Fire Chief Checklists

Personnel Considerations

- Uniforms and acceptable clothing (on and off duty)
- Selection and appointment, including background checks and interviews
- Promotions
- Proper driving behavior, including mandatory seatbelt use
- Ethics
- Misconduct (personal and organizational)
- Diversity, equal opportunity, and nondiscrimination policies
- Health and safety policies
- Labor laws and regulations
- Involvement by spouses and children in department activities, including junior firefighter programs
- Social functions
- Friendship and respect
- Avoiding horseplay and pranks
- Honor guard policies
- Disaster operations policies, including sleeping arrangements, ice, food and power
- Guest policy for response and use of vehicles
- Criminal behavior by members, including vandalism, theft, arson, substance abuse and sexual abuse

Budgets

- Drafting a budget
- Banking
- Capital investments and savings
- Equipment purchasing and leasing
- Bonds
- Duties of the treasurer
- Fundraising
- Loans
- Investments
- Applying for grants

Legal Issues

- Workers' compensation
- Damage to member property
- Errors and omission insurance
- Death and injury procedures and liability
- Filing reports with state and federal agencies
- Occupational Safety and Health Administration requirements
- Department of Labor requirements

Station Operations

- General housekeeping
- Kitchen use, food policy, and coffee and soda machines
- Sleeping quarters
- Living at the station
- Bulletin board use
- Lounge area
- Equipment room
- Power generator
- Tools
- Apparatus fioor
- Safety issues
- Emergency response parking
- Rental of station meeting room or function area, including any risks involved
- Social and sports functions
- Alcohol and gambling policy
- Guest policy
- Exercise facilities

Station Security

- Securing the station, including security systems and distribution/return of keys
- Controlling vandalism
- Use of department tools and equipment
- Plans for natural and man-made disasters
- Insurance

Administration

- Computers (purchase, maintenance and use)
- Office furniture
- Files and security, including separate filing systems for personnel and medical records
- Incoming and outgoing mail
- Staff, both full- and part-time
- Assignment of pagers, radios, and equipment
- Licenses and approvals
  - Fire department registered with state (and information on file is current)
  - Tax-exempt status with the IRS and State
  - Nonprofit tax returns filings current
  - Radios compliant with Federal Communications Commission rules
  - Ambulances certified with state EMS agency

- Apparatus is within Department of Transportation weight guidelines
- Members have state training certifications for firefighting and EMS response
- Members have relevant law enforcement certifications (for example, to investigate arson)

■ Meetings

- Location and time
- Purpose
- Attendance
- Application of by-laws
- Agendas and action items
- Rules of order
- Minutes and records

■ Library

- Types of resources to include
- Obtaining resources
- Access to resources by members and the public

■ Training

- Training officer appointment and training
- Equipment and supplies, including audiovisuals
- Drills
- Safety at training

■ Benefits

- Workers' compensation
- Life insurance
- Errors and omission insurance
- Death and disability insurance
- Training
- Personal protective equipment
- Use of station facilities
- Relief fund
- Length of service programs
- Tax deductions

APPENDIX

D Cost Comparisons

This table shows various populations, total budget, total personnel, and a computed cost per capita (the total budget divided by population). The intent of this table is to demonstrate that combination systems can be more cost effective in growing communities prior to having a full paid staff.

Cost Comparisons National Run Survey (selected populations)

State	Population	Budget Million $	Personnel	Cost per Capita $
AL	67,000	11.0	164	164
AL	36,264	5.5	92	152
AR	15,115	1.2	39	79
AZ	50,000	9.7	85	194
CA	62,000	8.7	61	140
CA	5,000	0.4	30	80
CA	42,000	8.8	57	210
CO	27,000	3.5	73	130
CO	60,000	5.5	62	92
CT	58,860	7.2	99	122
CT	23,300	2.6	30	112
FL	20,800	3.0	39	144
FL	58,000	16.0	123	276
FL	37,540	5.8	80	155
GA	15,600	2.0	36	128
ID	39,000	4.5	50	115
IL	31,000	5.9	65	190
IL	50,000	6.0	52	120
IL	20,000	3.9	65	195
LA	54,000	11.0	203	204
MA	40,000	6.0	88	150
MD	35,000	5.3	74	151
ME	66,000	13.4	242	203
MI	65,000	13.2	117	203
MS	26,500	3.1	58	117
MS	45,000	8.4	112	187
MT	28,000	3.1	37	111
NH	17,500	2.9	32	166
NJ	14,126	1.2	17	85
NJ	24,936	4.5	49	180
NY	26,705	6.3	83	236
OH	60,690	12.0	109	198
OH	39,200	9.8	79	250
PA	40,862	7.8	68	191
SC	32,500	3.8	71	117
SC	58,000	8.9	145	153

continued

Comparisons National Run Survey (selected populations) continued

State	Population	Budget Million $	Personnel	Cost per Capita $
TN	25,435	3.3	51	130
TN	15,000	2.0	38	133
TX	32,000	1.5	27	47
TX	46,321	5.2	62	112
TX	35,000	5.0	79	143
VA	40,097	7.7	90	192
WA	44,000	9.4	83	214
WI	51,818	9.0	96	174
WY	52,000	6.5	73	125

Source: Excerpted from *Firehouse Magazine,* June 2007. Used with permission.

APPENDIX E Sample Proposed Budget

David B. Fulmer

2007 Proposed Budget

To: Town Administrator
From: D. B. Fulmer, Fire Chief
Subject: FY 2007 Budget Proposal
This should serve as the budget narrative for the FY2007 Division of Fire & EMS budget proposal.

■ Summary

The FY2007 budget proposal shows a 12.95% increase [+604,731] from the approved FY2006 or a 17.56% increase [$794,770] from the actual expenditures. The FY2006 anticipated expenditures are −4.08% [–$182,901] under the approved FY2006 budget. Significant increases and/or decreases in line items are explained in further detail below.

Salaries 50100

This account shows a 25% increase [$568,906] over the anticipated expenditures for FY2006. With that said the increase, when compared to the approved FY2006 budget, is 18.63% [$453,303]. The following increases are proposed:

1. 7% increase in non-union wages as a result of anticipated wages increases and a 3.25% COLA increases for non-union career staff and the part-time Administrative Assistant–I position. The increase from the approved FY2006 to the approved FY2007 budget is a 5.03% [$21,887].
2. 13% increase in union wages as a result of the 6% wage increase contained in the collective bargaining agreement. This is an estimation of what is likely to come out of the collective bargaining process that is still ongoing. The increase from the approved FY2006 to the approved FY2007 budget is a 7.14% [$96,402].
3. The non-union overtime line item shows a 0% increase because overtime for the Administrative Assistants has been eliminated.
4. Union overtime shows a 45% [$28,200] increase from the anticipated expenditures for FY2006. There is no change from the FY2006 approved budget and the FY2007 proposed budget. The Division is very diligent with the management of overtime funds. Expenditures from this account vary from one fiscal year to the next as a result of injuries, military deployment, and the deployment of personnel on Federal response teams such as USAR & DMAT.
5. Part-time wages shows a 66% increase [$322,714] from the anticipated expenditures for FY2006. This amount reflects a 3.25% COLA increase for part-time staff and provides the necessary funds for 52,560 personnel hours, which provides 6 part-time personnel per day, at full strength. It should be noted that this returns the funding level to staff at the levels the Division operated at in FY2005. This meets the goal of the Division to fund 6 part-time positions per day. The increase from the approved FY2006 to the approved FY2007 budget is 61% [$308,222].

6. Part-time Medicare is calculated by the finance department and is based on the part-time and paid-on-call wages. This account reflects a 73% increase.

Pensions 50200

This account shows a 13% increase [$76,285] from the anticipated expenditures from the FY2006 budget. No new career positions have been proposed in this budget proposal and therefore the increases are a result of wage increases contained in the collective bargaining agreement and anticipated wage increases for non-union career staff.

Workers Compensation 50300

This account shows a 14% increase over the anticipated expenditures for FY2006 budget. We do not anticipate any refunds through BWC in FY2007.

Employee Insurance 50400

This account shows a 19% increase from anticipated expenditures from the FY2006 approved budget. This increase is a result of the health care package negotiated and approved by the Township.

Commercial Insurance 50500

This account shows a 3% increase from the anticipated expenditures from the FY2006 budget. This reflects the projected expenditures for FY2007.

Furniture and Equipment 50600

This account shows a 0% increase from the anticipated expenditures from the FY2006 budget.

Operating Supplies 50700

This account shows a 0% increase from anticipated expenditures from the FY2006 budget. It is estimated that the account will be $4.9% [$2,000] over the approved budget as a result of increased Fire Prevention & Public Education activities conducted in FY2006. This account is used for printing expenses, mailing costs and operational costs for the Division's Bureaus.

The Division has increased 50705 fire prevention/education operations budget by 33.33% [$2,000] as a result of an increased level of public education and fire prevention education activities.

Building Repair and Maintenance 50800

This account shows a 4% increase [$1,500] from the anticipated expenditures from the FY2006 approved budget. It is estimated that this account will be 18.6% [$1,208] over the approved FY2006 budget. This account is used for routine repairs to the four fire stations.

Vehicle Maintenance and Repair 50900

This account shows a 21% increase [$31,224] from the anticipated expenditures from the FY2006 budget. This account shows a 21.65% [$31,724] unanticipated increase as a result of a mechanics position being added to the service department. It is anticipated that this account will be 3.4% [$500] over budget as a result of increased repair & maintenance costs.

The Division has proposed a −50% [$500] reduction in 50903 Supplies as a result of supplies utilized to maintain operational readiness of our fleet.

The Division is anticipating a 0% increase in 50904 Fuel for FY2007. The Division has anticipated a $5,000 surplus in this account for FY2006 as a result of a less than volatile petroleum market and a slight decrease in responses. We estimate that the same funding level should be sufficient for FY2007.

Sick Leave 51000

This account shows an increase from $0 to $19,000. The Division is anticipating the medical disability of a career employee, which is estimated to result in a $19,000 payout in FY2007.

Capital Expenditures 51100

This account shows a −8% [−$26,414] decrease from the FY2006 anticipated expenditures. The anticipated expenditures for FY2006 are expected to meet the approved FY2006 budget.

The Division is proposing a $294,081 capital improvement budget with the following projects for FY2007:

1. Debt service payment for the Aerial Ladder (2nd payment of five).
2. Debt service payment for portable radios (final payment).
3. Debt service payment (Medic 49) w/ equipment. (2nd payment of three).
4. The Division's share of the EOC (2nd payment of five).
5. Rescue tools (financed over 3 years).

Utilities 51200

This account shows a –1% [–$350] decrease from the anticipated expenditures from the FY2006 budget. It should be noted that while there is a slight decrease in the budget, certain line items have been increased or decreased based on phone costs and station connectivity costs.

County Auditor Fees 51300

This account shows an 3% increase [$1,331] from the anticipated expenditures from the FY2006 budget. FY2006 expenditures are expected to be –11.6% [–$5,331] under budget.

Publications/Subscriptions/Dues 51400

This account shows a 20% increase from the anticipated expenditures from the FY2006 budget.

The Division is proposing to reduce the funding for FY2007 by 67% [$1,000] from the approved FY2006 budget. The funding level should be adequate pending any cost increases for dues, coop fees, or licenses.

Town Hall Expense 51500

This account shows a 6% [$3,000] increase from the anticipated expenditures from the FY2006 budget, which is a result of expected costs for the Administrator charges and town hall expenses. The Division has no control over this line item, as we do not control cost associated with the Township Administrator and/or building maintenance personnel.

Computer Repair and Maintenance 51600

This account shows a 0% increase from the anticipated expenditures from the FY2006 budget. This account is utilized for replacement of computers, printers, hardware, software, and technical support.

Staff Development and Training 51800

This account shows an 11% increase [$4,500] from the anticipated expenditures from the FY2006 budget. This increase reflects the possible addition of an educational stipend for bargaining unit employees for associates and bachelors degrees, which mirrors the police department's program.

Communications Expense 51900

This account shows a 12% increase [$26,178] from the anticipated FY2006 expenditures. It is estimated that the account will be −8% [$18,700] under budget for FY2006. The Division of Fire/EMS is accountable for 40% of the Communications Center budget.

Contractual Expense 52100

This account shows a 0% increase from the approved FY2006 budget and represents an 11% [$150] increase from the anticipated FY2006 expenditures. The only expenditure from this account is hydrant maintenance fees from the Jefferson Regional Water Authority.

Employee Programs 52600

This account shows a 59% increase [$20,000] from the anticipated expenditures from the FY2006 budget. The anticipated expenditures for FY2006 are 59% [$20,000] under budget as a result of the following items:

1. Inability to reach consensus with the bargaining unit regarding mandatory physicals for career personnel.
2. Logistical issues with timely scheduling and processing of part-time candidates through the recruitment process.

That being said, the Division is requesting similar funding to that which was approved in the FY2006 budget. We are not anticipating the need for any senior staff member recruitments. Steps have been taken to streamline the hiring process and limit exposure to unnecessary expenditures for portions of the recruitment and promotion process.

Uniforms and Clothing 52700

This account shows a –12% [$3,000] decrease from the anticipated expenditures from the FY2006 budget.

Personal Protective Clothing 52800

This account shows a 0% increase from the anticipated expenditures from the FY2006 budget.

Building Repair and Maintenance Supplies 52900

This account shows a 25% increase [$1,000] from the anticipated expenditures and from the FY2006 budget. This is a result of the use of cleaning supplies and other expendable supplies for the four fire stations.

Legal Expenses 53200

This account shows a –50% [–$4,500] decrease from the anticipated expenditures of the FY2006 budget. FY2007 is a non-bargaining year but funds need to be set aside for legal counsel.

The FY2006 anticipated expenditures are –31% [−$4,000] under the approved FY2006 budget. This is a result a great deal of progress being made at the bargaining table prior to the involvement of our labor attorney, and the bargaining unit's use of a labor attorney.

CPR 53700

This account shows a 30% increase [$800] from the anticipated expenditures from the FY2006 budget. It is estimated that the anticipated expenditures for FY2006 will be –23% [–$800] under budget.

Overall the FY2007 budget proposal shows a 12.95% increase [$604,731] from the approved FY2006 or a 17.56% increase [$794,770] from the anticipated expenditures. It is anticipated that the Division will close out FY2006 with a 4.08% [$182,901] budget surplus.

Miami Township Ohio—Fire Revenue, Expenditure, and Reserve Summary

2007 Revenue and Expenditure				
Beginning Balance	$ 2,322,392	$ 2,227,514	$ 2,562,217	$ 2,975,418
Revenues:				
Property Tax	2,695,979	2,752,150	3,200,000	3,488,000
Personal Property Tax	599,306	620,218	577,000	628,930
Rollback & Homestead	806,366	814,738	440,000	479,000
Other	29,423	10,457	93,700	28,000
Donations	100	30,000		
CPR classes	2,382	2,239	930	1,000
Grants Revenue	38,698	172,800	3,500	
MVA		20,340	46,000	50,000
EMS	523,644	723,850	598,600	600,000
Total Revenues	4,695,897	5,146,792	4,959,730	5,274,930
Total Available for Appropriation	7,018,289	7,374,305	7,521,947	8,250,348
Actual Expenditures	4,737,091	4,753,219	4,485,529	5,273,161
Capital Reserves	53,684	58,869	61,000	61,000
Total Expenditures	4,790,775	4,812,088	4,546,529	5,334,161
Ending Balance Before Reserves	2,227,514	2,562,217	2,975,418	2,916,187
Reserve Balances				
EMS & MVA Revenue				
Year 2002 (EMS revenue)	100,000	100,000	100,000	100,000
Year 2003 (EMS revenue)	516,267	516,267	516,267	516,267
Year 2004 (EMS revenue)	523,349	523,349	500,000	500,000
Year 2005 (EMS revenue)		744,190	540,000	540,000
Year 2006 (EMS)			716,958	716,958
Year 2006 (MVA)			13,000	13,000
Year 2007 (EMS)				
Year 2007 (MVA)				
Total EMS & MVA Reserves	1,139,616	1,883,806	2,386,225	2,386,225
Accrued Sick Leave				
Year 1998	56,233	56,233	56,233	56,233
Year 1999	—	—	—	—
Year 2000	12,000	12,000	12,000	12,000
Year 2001	53,113	53,113	53,113	53,113
Year 2002	20,000	20,000	20,000	20,000
Year 2003				
Year 2004				
Year 2005				
Year 2006				
Year 2007				
Total Accrued Sick Leave	141,346	141,346	141,346	141,346

Miami Township Ohio—Fire Revenue, Expenditure, and Reserve Summary (continued)

Capital Reserves				
Year 1998	—	—	—	—
Year 1999	—	—	—	—
Year 2000	102,500	102,500	102,500	102,500
Year 2001	297,626	297,626	297,626	297,626
Year 2002	336,570	336,570	336,570	336,570
Year 2003	(603,916)	(603,916)	(603,916)	(603,916)
Year 2003	350,000	144,259	144,259	144,259
Year 2004	154,000	154,000		
Year 2005				
Year 2006				
Year 2007				
Total Capital Reserves	636,780	431,039	277,039	277,039
Total Reserves	1,917,742	2,456,191	2,804,610	2,804,610
Ending Balance After Reserves	$ 309,772	$ 106,026	$ 170,808	$ 111,577

2007 Fire Budget

fund	dept	account	description	2004 actual	2005 actual	2006 budget	2006 actual	2006 actual/ 2006 budget variance	2007 budget	2007 budget/ 2006 actual variance
10	510	50100	salaries	$2,302,915	$2,380,287	$2,432,212	$2,316,609	−5%	$2,885,015	25%
10	510	50200	pension	569,227	587,881	635,920	616,006	−3%	695,590	13%
10	510	50300	workers compensation	80,997	86,205	97,412	87,248	−12%	99,057	14%
10	510	50400	employee insurance	234,882	247,630	274,198	276,550	1%	328,523	19%
10	510	50500	commercial insurance	29,105	32,242	32,700	34,526	5%	35,600	9%
10	510	50600	furniture & equipment	39,567	83,749	47,000	47,000	0%	47,000	0%
10	510	50700	operating supplies	43,870	42,706	41,000	43,200	5%	43,000	0%
10	510	50800	building repairs & maint	13,368	9,569	6,500	7,708	16%	8,000	4%
10	510	50900	vehicle maintenance	142,724	151,105	146,500	147,000	0%	177,724	21%
10	510	51000	sick pay	—	16,000	—	—	0%	19,000	100%
10	510	51100	capital expenditures	711,825	435,492	320,495	320,495	0%	294,081	−8%
10	510	51200	utilities	48,185	50,068	64,300	67,750	5%	67,400	−1%
10	510	51300	county auditor fees	45,699	44,461	46,000	40,669	−13%	42,000	3%
10	510	51400	publications/subscriptions/ dues	12,507	14,700	14,750	11,475	−29%	13,750	20%
10	510	51500	town hall	50,496	53,649	53,000	53,000	0%	56,000	6%
10	510	51600	computer repair/repl/maint	16,723	39,597	24,000	27,000	11%	27,000	13%
10	510	51800	training & development	34,249	34,472	38,800	39,300	1%	43,800	11%
10	510	51900	communications	208,170	314,704	246,143	227,443	−8%	253,621	12%
10	510	52100	contractual services	1,350	1,350	1,500	1,350	−11%	1,500	11%
10	510	52600	employee programs	48,528	47,674	54,000	34,000	−59%	54,000	59%
10	510	52700	uniforms & clothing	29,333	19,373	26,000	26,000	0%	23,000	−12%
10	510	52800	personal protective clothing	51,933	51,476	45,500	45,500	0%	45,500	0%
10	510	52900	building repair & maint supplies	4,185	4,490	4,000	4,000	0%	5,000	25%
10	510	53200	legal	11,559	295	13,000	9,000	−44%	4,500	−50%
10	510	53700	cpr	5,690	4,042	3,500	2,700	−30%	3,500	30%
				$4,737,089	$4,753,217	$4,668,429	$4,485,529	−4.08%	$5,273,161	17.56%

2007 Fire Budget

fund	dept	account	description	2004 actual	2005 actual	2006 budget	2006 actual	2006 actual/ 2006 budget variance	2007 budget	2007 budget/ 2006 actual variance
10	**510**	**50100**	**Salaries and Wages**							
10	510	50101	non-union	$ 375,350	$ 362,108	$ 435,009	$ 427,681	−2%	$ 456,896	7%
10	510	50102	union	1,253,864	1,311,841	1,349,822	1,284,059	−5%	1,446,244	13%
10	510	50103	non-union overtime	13	221		—			0%
10	510	50104	union overtime	42,627	80,724	90,020	62,000	−45%	90,020	45%
10	510	50105	part-time/paid on-call	581,211	576,703	504,881	490,389	−3%	813,103	66%
10	510	50106	longevity	16,140	15,730	16,440	16,440	0%	16,550	1%
10	510	50114	part time fica & medicare	33,417	32,959	36,040	36,040	0%	62,202	73%
				2,302,915	2,380,286	2,432,212	2,316,609	−5%	2,885,015	25%
10	**510**	**50200**	**Pension**							
10	510	50201	police and fire pension	550,368	568,562	619,234	599,825	−3%	680,519	13%
10	510	50202	pers	18,859	19,319	16,686	16,181	−3%	15,071	−7%
				569,227	587,881	635,920	616,006	−3%	695,590	13%
10	**510**	**50300**	**Workers Compensation**							
10	510	50301	workers compensation	80,997	86,205	97,412	87,248	−12%	99,057	14%
				80,997	86,205	97,412	87,248	−12%	99,057	14%
10	**510**	**50400**	**Employee Insurance**							
10	510	50401	health insurance	203,601	215,347	244,259	244,259	0%	296,154	21%
10	510	50402	life insurance	2,176	2,171	2,205	2,205		2,205	0%
10	510	50403	medicare	28,232	29,546	27,734	30,086	8%	30,164	0%
10	510	50404	employee assistance	873	567					−100%
				234,882	247,631	274,198	276,550	1%	328,523	19%
10	**510**	**50500**	**Commercial Insurance**							
10	510	50501	vehicle	12,811	13,925	14,500	15,224	5%	15,700	3%
10	510	50502	building & equipment	16,294	18,317	18,200	19,302	6%	19,900	3%
				29,105	32,242	32,700	34,526	5%	35,600	3%
10	**510**	**50600**	**Furniture and Equipment**							
10	510	50601	fire suppression	12,880	20,248	18,500	18,500	0%	18,500	0%
10	510	50602	EMS	15,046	43,863	10,000	10,000	0%	10,000	0%
10	510	50603	special operations group	5,789	5,365	4,500	4,500	0%	4,500	0%
10	510	50604	station furnishings & appliances	5,822	14,273	14,000	14,000	0%	14,000	0%
10	510	50606	communications	30						
				39,567	83,749	47,000	47,000	0%	47,000	0%
10	**510**	**50700**	**Operating Supplies**							
10	510	50701	office supplies and postage	10,035	9,037	9,500	9,500	0%	10,000	5%
10	510	50702	administration operations	6,846	5,498	6,000	6,000	0%	6,000	0%
10	510	50703	fire suppression operations	5,571	4,262	4,500	4,500	0%	4,500	0%
10	510	50704	fire investigations operations	1,768	1,403	1,500	1,500	0%	1,000	−33%
10	510	50705	fire prevention/education operations	4,546	2,816	6,000	7,700	22%	8,000	4%
10	510	50706	EMS operations	14,156	18,306	12,000	12,500	4%	12,000	−4%
10	510	50707	special operations group operations	948	1,383	1,500	1,500	0%	1,500	0%
				43,870	42,705	41,000	43,200	5%	43,000	0%
10	**510**	**50800**	**Repairs**							
10	510	50801	Station 47 repairs & maint	(120)	360	500	500	0%	500	0%
10	510	50802	Station 48 repairs & maint	1,345	1,149	1,000	1,000	0%	1,000	0%
10	510	50803	Station 49 repairs & maint	1,229	877	500	1,008	50%	1,000	−1%
10	510	50804	Station 50 repairs & maint	5,587	2,445	2,500	3,000	17%	3,500	17%
10	510	50805	operations equipment repairs & maint	2,851	2,354	1,000	1,400	29%	1,000	−29%
10	510	50806	appliance repairs & maint	2,476	2,385	1,000	800	−25%	1,000	25%
				13,368	9,570	6,500	7,708	16%	8,000	4%

continued

2007 Fire Budget (continued)

fund	dept	account	description	2004 actual	2005 actual	2006 budget	2006 actual	2006 actual/ 2006 budget variance	2007 budget	2007 budget/ 2006 actual variance
10	**510**	**50900**	**Vehicle Maintenance**							
10	510	50901	parts	36,611	28,593	25,000	25,000	0%	25,000	0%
10	510	50902	labor	71,497	75,034	65,000	70,500	8%	96,724	37%
10	510	50903	supplies	1,822	536	1,500	1,500	0%	1,000	−33%
10	510	50904	Diesel fuel	32,794	46,942	55,000	50,000	−10%	55,000	10%
				142,724	151,105	146,500	147,000	0%	177,724	21%
10	**510**	**51000**	**Accrued Leave**							
10	510	51001	accrued sick leave pay	—	16,000	—	—	0%	19,000	100%
				—	16,000	—	—	0%	19,000	100%
10	**510**	**51100**	**Capital**							
10	510	51101	capital expenditures	711,825	435,492	320,495	320,495	0%	294,081	−8%
				711,825	435,492	320,495	320,495	0%	294,081	−8%
10	**510**	**51200**	**Utilities**							
10	510	51201	gas & electric	23,058	25,618	36,400	34,000	−7%	36,400	7%
10	510	51202	water	2,303	1,978	2,400	2,700	11%	3,000	11%
10	510	51203	telephone	12,986	13,415	13,500	19,050	29%	19,000	0%
10	510	51204	station connectivity service	9,838	9,057	12,000	12,000	0%	9,000	−25%
				48,185	50,068	64,300	67,750	5%	67,400	−1%
10	**510**	**51300**	**Auditor Fees**							
10	510	51301	county auditor fees	45,699	44,461	46,000	40,669	−13%	42,000	3%
				45,699	44,461	46,000	40,669	−13%	42,000	3%
10	**510**	**51400**	**Publications/Subscriptions/ Dues**							
10	510	51401	publications	826	784	800	700	−14%	800	14%
10	510	51402	subscriptions	527	682	700	625	−12%	700	12%
10	510	51403	dues	1,750	2,816	3,000	2,900	−3%	3,000	3%
10	510	51404	coop fees	9,254	10,268	10,000	7,000	−43%	9,000	29%
10	510	51405	license	150	150	250	250	0%	250	0%
				12,507	14,700	14,750	11,475	−29%	13,750	20%
10	**510**	**51500**	**Town Hall**							
10	510	51501	town hall	20,103	21,231	24,000	24,000	0%	25,000	4%
10	510	51502	administrative charges	30,393	32,417	29,000	29,000	0%	31,000	7%
				50,496	53,648	53,000	53,000	0%	56,000	6%
10	**510**	**51600**	**Computer Repair/ Replacement/Maintenance**							
10	510	51601	repair and maintenance	8,496	11,199	12,000	12,000	0%	12,000	0%
10	510	51602	hardware	300	12,033	6,000	6,000	0%	6,000	0%
10	510	51603	software	4,927	13,366	6,000	6,000	0%	6,000	0%
10	510	51607	rent	3,000	3,000		3,000	100%	3,000	0%
				16,723	39,598	24,000	27,000	11%	27,000	0%
10	**510**	**51800**	**Training**							
10	510	51801	seminars & conferences	9,983	9,515	10,000	10,500	5%	11,000	5%
10	510	51802	training classes	18,343	11,897	17,000	17,000	0%	17,000	0%
10	510	51803	training materials & supplies	3,474	5,510	5,800	5,800	0%	5,800	0%
10	510	51804	union training funds	2,449	7,550	6,000	6,000	0%	10,000	67%
				34,249	34,472	38,800	39,300	1%	43,800	11%
10	**510**	**51900**	**Communication Expenditures**							
10	510	51901	communications center	183,904	294,059	218,643	200,643	−9%	226,121	12.70%
10	510	51902	radio maint & supplies	6,293	5,009	9,000	9,000	0%	9,000	0%
10	510	51903	pager contract & maint	7,623	6,350	7,000	6,300	−11%	7,000	11%
10	510	51904	cellular service	5,436	4,228	5,500	5,500	0%	5,500	0%
10	510	51905	MCSO radio contract	4,914	5,058	6,000	6,000	0%	6,000	0%
				208,170	314,704	246,143	227,443	−8%	253,621	12%

2007 Fire Budget (continued)

fund	dept	account	description	2004 actual	2005 actual	2006 budget	2006 actual	2006 actual/ 2006 budget variance	2007 budget	2007 budget/ 2006 actual variance
10	**510**	**52100**	**Contractual Services**							
10	510	52101	jefferson county water authority	1,350	1,350	1,500	1,350	−11%	1,500	11%
				1,350	1,350	1,500	1,350	−11%	1,500	11%
10	**510**	**52600**	**Employee Programs**							
10	510	52601	emergency medical surveillance	14,629	9,869	20,000	5,000	−300%	20,000	300%
10	510	52602	employee assistance program	—	—	2,000	2,000	0%	2,000	0%
10	510	52603	employee recognition programs	9,279	5,279	7,000	7,000	0%	7,000	0%
10	510	52604	recruitment & promotion programs	24,620	2,526	25,000	20,000	−25%	25,000	25%
				48,528	47,674	54,000	34,000	−59%	54,000	59%
10	**510**	**52700**	**Uniforms & Clothing**							
10	510	52701	uniforms	17,974	15,930	20,000	20,000	0%	17,000	−15%
10	510	52702	uniform maint allowance	11,359	3,443	6,000	6,000	0%	6,000	0%
				29,333	19,373	26,000	26,000	0%	23,000	−12%
10	**510**	**52800**	**Personal Protective Clothing**							
10	510	52801	fire suppression ppe	40,587	40,029	30,000	30,000	0%	30,000	0%
10	510	52802	EMS ppe	3,130	4,492	7,000	7,000	0%	7,000	0%
10	510	52803	special operations group ppe	3,450	2,647	2,500	2,500	0%	2,500	0%
10	510	52804	scba program	4,766	4,307	6,000	6,000	0%	6,000	0%
				51,933	51,475	45,500	45,500	0%	45,500	0%
10	**510**	**52900**	**Building Repair & Maintenance Supplies**							
10	510	52901	Station 47 janitorial & maint supplies	1,000	856	1,000	1,000	0%	1,000	0%
10	510	52902	Station 48 janitorial & maint supplies	971	1,330	1,000	1,000	0%	1,500	50%
10	510	52903	Station 49 janitorial & maint supplies	957	809	1,000	1,000	0%	1,000	0%
10	510	52904	Station 50 janitorial & maint supplies	1,257	1,496	1,000	1,000	0%	1,500	50%
				4,185	4,491	4,000	4,000	0%	5,000	25%
10	**510**	**53200**	**Legal**							
10	510	53201	legal fees	11,559	295	13,000	9,000	−31%	4,500	−50%
				11,559	295	13,000	9,000	−31%	4,500	−50%
10	**510**	**53700**	**CPR Program**							
10	510	53701	cpr	5,690	4,042	3,500	2,700	−23%	3,500	30%
				5,690	4,042	3,500	2,700	−23%	3,500	30%
				$4,737,087	$4,753,217	$4,668,430	$4,485,529	−4.08%	$5,273,161	17.56%

2007 Fire Budget (continued)

Dept.	Fund	Div.	Project Name	2003 Budget	2004 Budget	2004 Accum Expend	2005 Budget	2005 Expend	2006 Budget	2006 Accum Expend	2007 Budget	2007 Accum Expend	2008 Budget	2008 Accum Expend	2009 Budget
Fire	Fire	Fire	Defibrillator Replacement	10,500											
Fire	Fire	Fire	Ladder 50 Replacement	—			97,488	545	152,230	152,230	152,230		152,230		152,230
Fire	Fire	Fire	Staff Vehicle replacement		27,357	27,357		650							
Fire	Fire	Fire	Engine 50 replacement	395,000	395,000	395,000									
Fire	Fire	Fire	Incident Management System												
Fire	Fire	Fire	Incident Command Console												
Fire	Fire	Fire	Deputy Chief Office Enhancements												
Fire	Fire	Fire	Portable Radios	—	24,000	13,651	24,000	25,065	25,065	25,065	13,651				
Fire	Fire	Fire	Ventilation Study implementation	108,046											
Fire	Fire	Fire	Pick-Up Truck	28,000											
Fire	Fire	Fire	Port-A-Count Quantitative Fit Test		8,000	7,824									
Fire	Fire	Fire	Medic 48 Replacement		140,000	129,1801									
Fire	Fire	Fire	Rescue Tool Upgrades								15,000		15,000		15,000
Fire	Fire	Fire	Heart Rate Monitor	60,000											
Fire	Fire	Fire	Medic 50 Replacement				140,000	129,668							
Fire	Fire	Fire	SCBA Cylinder Replacement	9,000											
Fire	Fire	Fire	Blazer		28,440	22,642									
Fire	Fire	Fire	Excursion		43,203	43,062		137							
Fire	Fire	Fire	Air Quality Monitors		20,000	16,652									
Fire	Fire	Fire	Thermal Imaging Cameras		36,000	578	31,725	31,995							
Fire	Fire	Fire	Knox Box Unit		6,000	5,878									
Fire	Fire	Fire	Computer Replacements		35,000	32,978		4,315							
Fire	Fire	Fire	SCBA Replacement				42,000	42,000							
Fire	Fire	Fire	SCBA Grant				172,000	172,000							
Fire	Fire	Fire	Breathing Air Compressor				52,500	52,500							
Fire	Fire	Fire	Facility Master Plan			16,634		1,177							
Fire	Fire	Fire	Reverse 911				42,639	42,639							
Fire	Fire	Fire	Video Conferencing Equipment												
Fire	Fire	Fire	Medic 49 Replacement & Equip						63,200	63,200	63,200		63,200		
Fire	Fire	Fire	EOC						50,000	50,000	50,000		50,000		50,000
Fire	Fire	Fire	Parking Lot Paving						30,000	30,000					
				$610,546	$763,000	$711,436	$602,352	$502,690	$320,495	$320,495	$294,081	$ —	$280,430	$ —	$217,230

2007 FIRE FULL-TIME STAFF

		Current	Step			salary	overtime	holiday pay	medicare	health	health	life	workers comp.	longevity	pension	total
1 Baber	j	19.84	19.84	28,966.40	28,966.40	57,932.80	3,215	3,492	937	5,214	6,518	63	2,158	540	21,977	102,048
2 Baber	m	17.71	17.71	25,856.60	25,856.60	51,713.20	3,215	3,117	842			63	1,939	300	19,735	80,924
3 basso		19.84	19.84	9,655.47	48,277.33	57,932.80	3,215	3,492	937	1,689	2,111	63	2,158	540	21,977	94,116
4 beach		17.71	17.71	25,856.60	25,856.60	51,713.20	3,215	3,117	842	5,214	6,518	63	1,939	300	19,735	92,656
5 brooks		15.03	15.03	21,943.80	21,943.80	43,887.60	3,215	2,645	721	5,214	6,518	63	1,663	1,200	16,914	82,041
6 chelmam		17.71	17.71	25,856.60	25,856.60	51,713.20	3,215	3,117	842			63	1,939	770	19,735	81,394
7 dipzinski		18.94	18.94	27,652.40	27,652.40	55,304.80	3,215	3,333	897			63	2,066	250	21,030	86,159
8 ennis		17.71	17.71	25,856.60	25,856.60	51,713.20	3,215	3,117	842	5,214	6,518	63	1,939	1,200	19,735	93,556
9 fahrney		26.31	26.31	27,362.40	27,362.40	54,724.80	3,215	4,631	907	5,214	6,518	63	2,045	1,200	21,274	99,792
10 gabbard		16.91	17.71	24,688.60	25,856.60	50,545.20	3,215	3,117	825	1,689	2,111	63	1,898	200	19,338	83,001
11 gemin		15.48	16.13	22,600.80	23,549.80	46,150.60	3,215	2,839	757	1,689	2,111	63	1,743	80	17,750	76,397
12 gilmore		17.71	17.71	8,618.87	43,094.33	51,713.20	3,215	3,117	842	5,214	6,518	63	1,939	770	19,735	93,126
13 good		17.71	17.71	25,856.60	25,856.60	51,713.20	3,215	3,117	842	3,713	4,641	63	1,939	1,200	19,735	90,178
14 harnett		17.71	17.71	25,856.60	25,856.60	51,713.20	3,215	3,117	842	5,214	6,518	63	1,939	1,200	19,735	93,556
15 johnson		17.71	17.71	25,856.60	25,856.60	51,713.20	3,215	3,117	842	5,214	6,518	63	1,939	1,200	19,735	93,556
16 krapf		15.48	16.13	33,901.20	11,774.90	45,676.10	3,215	2,839	750	1,689	2,111	63	1,726	80	17,588	75,738
17 petry		17.71	17.71	25,856.60	25,856.60	51,713.20	3,215	3,117	842	5,214	6,518	63	1,939	300	19,735	92,656
18 ruschau		16.13	16.91	27,474.77	20,573.83	48,048.60	3,215	2,976	757	1,689	2,111	63	1,810	150	18,442	79,261
19 seitz		16.91	17.71	24,688.60	25,856.60	50,545.20	3,215	3,117	825			63	1,898	200	19,338	79,201
20 shaw		14.70	15.48	35,770.00	7,533.60	43,303.60	3,215	2,724	714	1,689	2,111	63	1,642	40	16,743	72,245
21 shupert		19.84	19.84	28,966.40	28,966.40	57,932.80	3,215	3,492	937	5,214	6,518	63	2,158	980	21,977	102,488
22 shroyer		16.13	16.91	43,174.63	4,114.77	47,289.40	3,215	2,976	775	1,689	2,111	63	1,783	150	18,183	78,236
23 sweet		19.84	19.84	28,966.40	28,966.40	57,932.80	3,215	3,492	937	3,713	4,641	63	2,158	1,200	21,977	99,329
24 wenclew		16.91	17.71	24,688.60	25,856.60	50,545.20	3,215	3,117	825	5,214	6,518	63	1,898	250	19,338	90,983
25 willis		19.84	19.84	9,655.47	48,277.33	57,932.80	3,215	3,492	937	5,214	6,518	63	2,158	420	21,977	101,928
26 wilmot		17.71	17.71	25,856.60	25,856.60	51,713.20	3,215	3,117	842	5,214	6,518	63	1,939	420	19,735	92,776
27 zink	a	17.71	17.71	25,856.60	25,856.60	51,713.20	3,215	3,117	842			63	1,939	420	19,735	81,044
28 zink	c	17.71	17.71	25,856.60	25,856.60	51,713.20	3,215	3,117	842	5,214	6,518	63	1,939	910	19,735	93,266
total union						1,446,243.50	90,020	89,177	23,539	92,250	115,313	1,764	54,229	16,470	552,650	2,481,656
29 annisko		13.56	14.08	5,876.00	8,541.87	14,417.87			209				509			15,136
30 bailey		16.28	16.96	25,396.80	8,465.60	33,862.40			491	5,866	7,333	63	1,195	40	7,687	56,537
31 fulmer		41.19		28,558.40	57,116.80	85,675.20			1,242	5,866	7,333	63	3,024	250	29,130	132,583
32 hugger		15.64		21,687.47	10,843.73	32,531.20			472	4,177	5,221	63	1,148	720	7,385	51,717
33 jirka		36.05	36.05	43,740.67	31,243.33	74,984.00			1,087	5,866	7,333	63	2,647	200	25,495	117,674
34 mclean		34.01	34.67	35,370.40	36,056.80	71,427.20			1,036	5,866	7,333	63	2,521	80	24,285	112,611
35 queen		34.67	34.67	30,047.33	42,066.27	72,113.60			1,046	5,866	7,333	63	2,546	1,040	24,519	114,525
36 schmaltz		34.01	34.67	11,790.13	60,094.67	71,884.80			1,042	5,866	7,333	63	2,537	80	24,441	113,247
total non-union						456,896.27	—	—	6,625	39,374	49,217	441	16,128	80	142,940	714,031
total						$1,903,140	$90,020	$89,177	$30,164	$131,624	$164,530	$2,205	$70,357	$16,550	$695,590	$3,195,687

APPENDIX

Best Practices Models

Sample Ordinance for Dillon Rule State

■ Article I. Department of Fire and Emergency Medical Services: Establishment and Administration

Sec. 9-1. Establishment of department.
The county department of fire and emergency medical services ("the department") is hereby established. The department shall provide all fire and emergency medical services and services related to civilian protection and evacuation in disasters and emergencies. The department shall also be responsible for administration of local, state and federal emergency response, assistance and recovery programs within the county. (Ord. No. 05-03, § 1, 4-27-05)

Sec. 9-2. Composition of department.
The department shall be composed of the officials and staff of the department, and the following volunteer fire companies and volunteer rescue squads, which are an integral part of the official safety program of the county. (Ord. No. 05-03, § 1, 4-27-05)

Sec. 9-3. Responsibilities of department.
(a) The department shall be responsible for regulating and managing the provision of pre-hospital emergency patient care and services and for regulating providers of the non-emergency transportation of patients requiring medical services.

(b) The department shall be responsible for regulating and managing the provision of fire prevention, protection, investigation and suppression services, for enforcing laws relating to fire prevention, and for provision of services related to hazardous materials and similar hazards which pose a threat to life and property.

(c) The department shall also be responsible for any additional related services which are necessary for the provision of fire and emergency medical services. (Ord. No. 05-03, § 1, 4-27-05)

Sec. 9-4. Chief of fire and emergency medical services.
(a) The chief of fire and emergency medical services shall be appointed by the county administrator, shall be the director of the department, and shall provide general management of the department.

(b) The chief may delegate any and all operational authority to other officials and staff of the department. References to the chief in this chapter shall include designees.

(c) The chief shall establish and enforce departmental regulations consistent with this chapter, for the administration and operation of the department. Such regulations shall be consistent with this chapter but may establish additional and more stringent requirements applicable to the department. In no event shall any county or departmental regulations or directives be interpreted to waive requirements of federal, state and local laws and regulations, including those related to licensing.

(d) The chief shall hire and appoint and may terminate officers, staff, and volunteers of the department, including the deputies and assistants. The chief shall provide for appropriate investigation of staff and volunteer applicants and incumbents, including review of criminal and driving records. Deputies and assistants may perform any of the duties of the chief, when authorized by the chief.

(e) The chief shall provide general management of the planning, preparation and response for any disaster which occurs in the county and requires implementation of the county's emergency response plan. The chief is hereby designated by the board of supervisors and the county administrator and shall serve as the coordinator of emergency services for all purposes related to response to disasters pursuant to Title 44 of the Virginia Code.

(f) The chief, on behalf of the board of supervisors, shall have authority to enter into and take all actions necessary to implement and carry out the terms of agreements for mutual aid, disaster preparedness, and provision of services related to hazardous materials, rescue, fire suppression, investigation, medical services or other emergency response services deemed necessary in the judgment of the chief for emergency response in events exceeding the capabilities of an individual locality or government agency. The chief shall have the authority to enter into contracts on behalf of the county and to expend funds after an official disaster or emergency declaration to provide for the public safety during such events, in accordance with applicable laws and regulations. The chief shall have the authority to take all actions necessary to obtain funding and assistance from other localities and from state or federal agencies for those purposes. (Ord. No. 05-03, § 1, 4-27-05)

Sec. 9-5. Advisors to the chief.

There shall be an advisory group ("the chief's staff") composed of the two (2) highest ranking operational leaders of each of the volunteer organizations listed in section 9-2 of this chapter. The chief's staff shall be consulted by the chief prior to the issuance of any policies or regulations. (Ord. No. 05-03, § 1, 4-27-05)

Sec. 9-6. Criminal and driving record checks.

(a) Review of the criminal records of applicants for employment and volunteer status in the department shall be conducted in the interest of public welfare and safety, and review of such records of incumbents may be conducted, to determine if the past criminal conduct of any person with a criminal record would be compatible with the nature of the employment or service.

(b) Review of motor vehicle driving records of incumbents and of applicants for employment or volunteer status may be conducted in accordance with departmental regulations, to determine if the record is compatible with employment or service.

(Ord. No. 05-03, § 1, 4-27-05)

Sec. 9-7. Compliance with regulations and policies; penalties.

(a) Compliance with all regulations and directives of the chief, by the officials, staff, volunteers and entities of the department is a condition of the privilege of providing emergency medical and fire services and of participating in department functions.

(b) The chief shall have the authority to remove, suspend, or revoke the privileges of any individual or entity to operate as an EMS or fire service provider or officer in the county, for violation of regulations and policies promulgated by the chief or the medical control board, or for the purpose of protecting the public safety and providing for proper administration of the department and effective provision of services.

(c) Volunteer members may, in accordance with procedures established by department regulations, request review by the county administrator, of any disciplinary action resulting in removal, suspension or revocation of privileges.

(d) Penalties. Any violation of this chapter for which a penalty is not specified shall be a class 1 misdemeanor. Any misrepresentation made by any person to any county officer or employee in the course of obtaining or renewing a permit or in providing information for a criminal or other record investigation shall constitute a class 1 misdemeanor.

(Ord. No. 05-03, § 1, 4-27-05)

■ Article II: Volunteer Services

Sec. 9-8. Volunteer rescue squads and fire companies.

Volunteer rescue squads and fire companies may be formed, named and dissolved and shall operate in compliance with applicable statutes, provisions of this chapter and regulations, including those issued by the chief. Formation, naming and dissolution shall be effective only if approved by the board of supervisors. Rescue squads and fire companies may adopt by-laws for effecting their purposes. (Ord. No. 05-03, § 1, 4-27-05)

Sec. 9-9. Junior volunteers.

(a) Minors shall be designated "junior volunteers" and may participate in department activities in accordance with regulations of the department. Junior volunteers who have obtained certification under any applicable national or state standards, and who have written permission of parent or legal guardian, may participate in department functions in conformance with applicable laws and regulations.

(b) Only minors certified as eligible by the chief pursuant to department regulations may participate. No minor shall participate in any activity prohibited by department regulations or by regulations of the state department of labor and industry.

(c) The department shall maintain records of those minors certified. Department officers, staff and volunteers shall not permit a minor not certified pursuant to this chapter to participate in the activities of the department.

(d) The chief shall establish regulations for the certification of minors. Any minor who does not continue to meet all certification requirements or who violates department regulations, or whose removal is necessary to protect the public safety, provide for proper administration of the department or for effective provision of services, may be decertified by the chief. Department officers, staff and volunteers shall promptly notify the chief of any grounds for decertification of any participant. The chief may suspend the eligibility of any minor during investigation of the alleged grounds for decertification.

■ Article III: Fire Prevention Requirements

■ Division 1: Open Fires

Sec. 9-10. Applicability.

The restrictions on open burning and permit requirements set forth in sections 9-14 and 9-15 apply within the suburban service area as defined in the county subdivision ordinance and within other areas of the county which are zoned R-1, R-2, R-3, R-4, R-5, R-6 and RS. Section 9-16 authorizes the county administrator to ban open burning in all or part of the county in which an extraordinary condition constituting a fire hazard has been found to exist. (Ord. No. 05-03, § 1, 4-27-05)

Sec. 9-11. Definitions.

For the purposes of this article, the following words and terms shall be defined as follows:

Garbage: Solid waste from the domestic and commercial preparation, cooking and disposing of food and from the handling, storage and sale of produce.

Open burning: The burning of any matter in such a manner that the products resulting from combustion are emitted directly into the atmosphere without passing through a stack, duct, or chimney.

Open pit incinerator: A device used to burn waste for the primary purpose of reducing the volume by removing combustible matter. Such devices function by directing a curtain of air at an angle across the top of a trench or similarly enclosed space, thus reducing the amount of combustion by-products emitted into the atmosphere. The term also includes trench burners, air curtain destructors and overdraft incinerators.

Refuse: Trash, rubbish, garbage and other forms of solid or liquid waste, including, but not limited to, wastes resulting from residential, agricultural, commercial, industrial, institutional, trade, construction, land clearing, forest management and emergency operations. (Ord. No. 05-03, § 1, 4-27-05)

Sec. 9-12. Open burning in regulated area generally prohibited; exceptions.

(a) Within the regulated area, as specified in this division, open fires and open burning are generally prohibited, except as specifically permitted by this division.

(b) Open burning is permitted for bona fide firefighting instruction; for campfires or other fires used solely for recreational purposes, for outdoor noncommercial preparation of food; for safety flares or for warming of outdoor workers; provided that there shall be no burning of garbage or refuse, tires, asphaltic materials, crank case oil, impregnated wood or other rubber or petroleum based materials or toxic or hazardous materials or of containers for such materials.

(c) The open burning of garbage and refuse is prohibited, except that the open burning of land clearing refuse on the site of clearing operations resulting from commercial, industrial or residential development, the construction of roads or highways, railroad tracks, pipelines, and power or communication lines, and agricultural operations. Such burning must utilize an open pit incinerator. The burning shall be at least seven hundred fifty (750) feet from any occupied building other than a building located on the property on which the burning is conducted. Material to be burned shall consist only of brush, stumps, and other vegetative matter generated at the site and shall not include demolition or construction debris.

(d) The burning of leaves and tree, yard and garden trimmings on residential property in the regulated area is allowed only outside of the suburban service area and only during the month of April from 4:00 p.m. to 12:00 a.m.; and from November 1 through December 15 from 4:00 p.m. to 12:00 a.m. Such burning is allowed only on the premises of the private residence generating the materials.

(e) Open fires shall be attended at all times by a person eighteen (18) years of age or older to assure that the fire is controlled. (Ord. No. 05-03, § 1, 4-27-05)

Sec. 9-13. Permit requirements.

(a) Within the regulated area defined in section 9-12, any person conducting burning for recreational purposes and any person conducting burning for the purpose of disposal of land clearing refuse shall first obtain a permit from the county fire marshal.

(b) Application for any required permit shall be made to the county fire marshal at least fourteen (14) days prior to the burning, on forms prescribed by the fire marshal. The fire marshal shall establish procedures for issuance of permits and shall include in the permits restrictions and conditions determined by the fire marshal to be necessary to assure control of the fire and to minimize the impact on air quality.

(c) Permits shall be effective for a maximum period of ninety (90) days from the date of issuance. (Ord. No. 05-03, § 1, 4-27-05)

Sec. 9-14. Open fires may be prohibited in all or part of the county.

Whenever the county administrator, after consultation with appropriate agencies, declares that a drought condition exists or that forest lands, brush lands and fields have become so dry or parched or that other conditions exist so as to create an extraordinary fire hazard, the county administrator may declare that open burning is prohibited in part or all of the county. Following such a declaration it shall be unlawful for any person to burn brush, grass, leaves, trash, debris or any other flammable material or to ignite or maintain any open fire within the county or within any part of the county subject to the prohibition. The declaration of the county administrator shall remain effective until the county administrator declares the condition and the prohibition to have terminated.

When any such declaration is issued, amended or rescinded, the county administrator shall promptly post a copy of the declaration, amendment or rescission at the front door of the circuit courthouse and at each fire station in the area of the county in which the emergency has been declared. In addition, the county administrator shall publish the declaration, amendment or rescission in a newspaper of general circulation.
(Ord. No. 05-03, § 1, 4-27-05)

Sec. 9-15. Enforcement.

The provisions of this division shall be enforced by the fire marshal and the sheriff.
(Ord. No. 05-03, § 1, 4-27-05)

Sec. 9-15.1. Penalty for violation or noncompliance.

Any person violating or failing to comply with the provisions of this division shall be guilty of a Class 3 misdemeanor. Each violation or failure shall constitute a separate offense. (Ord. No. 05-03, § 1, 4-27-05)

■ Division 2: Additional Prohibitions

Sec. 9-16. Sale or discharge of fireworks.

(a) It shall be unlawful and a class 1 misdemeanor for any person to sell or expose for sale or use or discharge, in the county, any firecrackers, torpedoes, sky rockets, Roman candles, squibs or any substance intended for and commonly known as fireworks of any kind whatsoever.

(b) Nothing in subsection (a) of this section shall be construed to prevent the board of supervisors from providing for the issuance of permits, upon application in writing, for the display of fireworks by fair associations, amusement parks or by any organization or group of individuals under such terms and conditions as the board may prescribe.

(c) Subsection (a) of this section shall not apply to the use or sale of sparklers, fountains, Pharaoh's serpents, caps for pistols or to pinwheels commonly known as whirligigs or spinning jennies; provided, however, that the same may only be used, ignited or exploded on private property with the consent of the owner of such property.

■ Division 3: Fire Marshal: Enforcement

Sec. 9-18. Appointment.

There shall be a fire marshal appointed by the board of supervisors. The fire marshal shall have all the powers and duties set out in Title 27, Chapter 3 of the Virginia

Code pertaining to local fire marshals, including those powers and duties set out in the statewide fire prevention code. Deputies and assistants shall also have those powers and duties, in the absence of the fire marshal and as otherwise directed by the fire marshal, and the term "fire marshal" when used in this division shall include deputies and assistants of the fire marshal. (Ord. No. 05-03, § 1, 4-27-05)

Sec. 9-19. Deputies or assistants.
The fire marshal shall, on behalf of the board of supervisors, appoint and may terminate deputy or assistant fire marshals to assist in the functions of the department.

■ Article IV: Emergency Medical Services

■ Division 1: Medical Directors and Control Board

Sec. 9-27. Medical director and assistant medical directors.
(a) There shall be a county operational medical director ("OMD"), who shall be appointed by the chief and shall serve as chairman of the medical control board and advise the chief.

(b) The OMD shall also be responsible for:

(1) Coordinating and supervising the work of all assistant medical directors.

(2) Serving as the department's operational medical director.

(3) Recommending medical protocols to the medical control board.

(4) Advising the chief and the board of supervisors on medical issues pertaining to the provision of emergency medical services in the county.

(c) The medical director of each rescue squad or other entity included in the department shall be an assistant medical director. (Ord. No. 05-03, § 1, 4-27-05)

Sec. 9-28. Medical control board.
(a) There shall be a medical control board composed of the following members: each assistant medical director, the OMD, and the chief or designee.

(b) Medical policy shall be defined as any policy directly relating to the actual delivery of medical care to the patient. All other policies and procedures shall be operational in nature and shall not be within the purview of the medical control board.

(c) The medical control board shall establish medical policy for patient care including, but not limited to, the development of training standards, medical care procedures and protocols, medical performance standards, and general medical control policies, as well as establishing and enforcing minimum qualifications and performance standards for all assistant medical directors.

■ Division 2: Nondepartmental EMS Providers

Sec. 9-29. Permits—Required for nondepartmental providers.
(a) It shall be unlawful for any person or organization to provide emergency medical services, medical transportation, or operate vehicles for those purposes within the confines of the county for the emergency or nonemergency transportation of patients, without first being granted a permit by the board of supervisors, except when a mutual aid agreement has been approved by the chief. The board may issue such permits following a public hearing, subject to conditions established by the board.

(b) Permits shall expire one (1) year from the date of issuance unless renewed by the chief in accordance with regulations of the department and the conditions established by the board.

(c) The board of supervisors may revoke or suspend any permit upon a finding that any conditions of the permit have been violated, that federal, state or local laws or regulations have been violated or that the public health, safety or welfare is endangered by continued operation of the entity. Except in the case of a condition posing a danger to the public health or safety, no revocation or suspension action shall be taken by the board prior to notice and reasonable time being given to the permittee for correction of the condition.

(d) Nondepartmental applicants and providers shall provide to the chief any information and records requested during the course of the application or renewal process or during the effective period of any permit. (Ord. No. 05-03, § 1, 4-27-05)

Sec. 9-30. Procedure for permit applications.
Filing and content of application.

(1) Any person or organization desiring a permit shall make application to the chief on a form prescribed by the chief, who may make a recommendation to the board of supervisors regarding the application.

(2) Each application shall include an explanation of the need for the services in the area to be served by the applicant, and evidence that the applicant is trained, equipped and otherwise qualified in all respects to render first-aid, emergency and transportation services in the area indicated in the application.

(3) Each permit application shall include any other information required by the chief and the applicant's notarized certification that all requisite state permits and certifications for its vehicles, operations and personnel are current and that its operations and personnel meet all applicable current federal, state and local requirements.

(4) Each applicant shall certify that no person who is to provide services is, or shall be, under the age of eighteen (18) years. (Ord. No. 05-03, § 1, 4-27-05)

Sec. 9-31. Hearing, issuance and administration.

(a) The board of supervisors shall consider each properly filed application after holding a public hearing. Should the board find (1) that there is a need for emergency medical services within the area in question, (2) that the need will be properly served by the applicant, and (3) that the applicant and any employees or personnel of the applicant are properly trained, equipped and otherwise qualified, the board of supervisors in its sole discretion may issue a permit under such conditions as the board deems appropriate.

(b) Copies of all applications and all permits shall be retained by the chief, who shall be responsible for reviewing the permit prior to expiration and either renewing the permit or referring it to the board of supervisors for revocation or other action.

(c) The chief may suspend the permit of any permittee for a period of thirty (30) days and may recommend to the board of supervisors the continued suspension or revocation of any permit should it be found, upon investigation, that the agency is not in compliance with federal, state or local laws, regulations or conditions of the permit or that the continued operation poses a danger to public health.

(d) The board of supervisors may revoke or suspend any permit upon a finding that any conditions of the permit have been violated, that federal, state or local laws or regulations have been violated or that the public health, safety or welfare is endangered by continued operation of the entity. Except in the case of a condition posing a danger to the public health or safety, no revocation or suspension action shall be taken by the board prior to notice and reasonable time being given to the permittee for correction of the condition.

(Ord. No. 05-03, § 1, 4-27-05)

Sec. 9-32. Criminal record investigations and driving records.

Applicants for nondepartmental provider permits shall provide to the chief a copy of the complete criminal history record available from the Virginia Criminal Record Exchange and the motor vehicle driving record from the Virginia Department of Motor Vehicles, of the applicant and of all persons proposed to provide services. The chief shall determine whether the records of the individuals are compatible with the provision of emergency medical services. Permitted providers shall also provide current records upon request of the chief.

■ Division 3: Cost Reimbursement

Sec. 9-33. Service fees for emergency medical transport.

Reasonable fees shall be charged for emergency medical transport services provided by the fire and EMS department. The schedule of rates for services shall be established by resolution of the board of supervisors. (Ord. No. 05-03, § 1, 4-27-05)

Sec. 9-34. Definitions.

Basic life support (BLS), advanced life support level 1 (ALS-1), advanced life support level 2 (ALS-2) and any other transport services shall be those services as defined by applicable federal regulations, including Title 47 of the Code of Federal Regulations.

Ground transport mileage (GTM) shall be assessed based on measurement of a straight line from the location of the incident scene to a hospital or other facility where a patient is transported using the midpoint of the relevant county map grid maintained for this purpose by the geographic information system manager. (Ord. No. 05-03, § 1, 4-27-05)

Sec. 9-35. Authority of chief.

The chief is hereby authorized and directed to establish rules and regulations for the administration of the charges imposed by this division, including, but not limited to, a subscription program for county residents and payment standards for those persons who demonstrate economic hardship, as permitted by applicable law.

Sample Department Regulations

Table of Contents

Section 1: General Provisions

1.1 Purpose of the Regulations

These Regulations are issued by the Chief, Fire/EMS Department, pursuant to the Hanover County Code, Chapter 9, to govern operations of the Department and the management of all matters relating to employees and volunteers of the Department, who shall be referred to collectively as members.

1.2 Mission Statement

The mission of the Department is to serve people and protect lives and property through the provision of professional fire, rescue and emergency medical services.

1.3 Vision Statement

It is the vision of the Department to be a high performance combination emergency response and mitigation system that meets the current and future needs of the citizens of Hanover County in response to all emergency situations, accidental, natural or manmade.

1.4 Precedence of Authorities

1.4.1 All County regulations, orders, procedures, guidelines and protocols or other Departmental authorities shall supersede any volunteer fire company or volunteer rescue squad Constitution and By-Laws, general operating rules or other individual member organization rules. Where any are in conflict the County authorities shall apply.

1.4.2 All Departmental authorities shall be minimum standards unless specifically defined otherwise. When these Regulations are in conflict with any other applicable authority, the more stringent or the higher standard or requirement shall apply. In no event shall Departmental authorities be interpreted to waive or supersede federal, state or local laws or regulations.

1.4.3 In all matters, all Department operations, and the conduct of all employees and volunteers (collectively "members") shall be in conformance with federal, state and local laws and regulations.

1.4.4 References in these Regulations to the Chief shall be deemed to include designees of the Chief.

1.5 Applicability
These Regulations shall apply to all Department operations and functions. All members shall comply with these Regulations. They may be supplemented by Orders, Standard Operating Procedures, Standard Operating Guidelines and Protocols.

1.6 Enforcement
Failure to comply with these Regulations, or with Orders, Standard Operating Procedures, Standard Operating Guidelines and Protocols shall be insubordination and may result in disciplinary action including dismissal.

■ Section 2: Orders

2.1 Special Orders
The Chief or Deputy Chiefs may issue special orders for the general management of the Department. These orders shall have the standing of Regulations, shall be compiled by the Department administration, made available at the administrative office and sent to all member organizations.

2.2 General Orders
General orders may be verbal or written and may be issued by any officer of the Department to achieve the goals of the Department and to complete task or work assignments. General orders are considered valid and binding provided that they are not in conflict with these Regulations, or any applicable law or regulation. It is the duty of every member to carry out any valid general order given by a superior officer. Failure to carry out a valid order shall qualify as insubordination.

■ Section 3: Standard Operating Procedures

3.1 Definition
Standard Operating Procedures are in the nature of general orders issued in writing to provide direction for normally encountered situations to allow an individual to meet the intent of the regulations or policies of the Department or County.

3.2 Issuance
Standard Operating Procedures shall be proposed and administered by the respective Division Chiefs to govern the specific responsibilities of their Division. All SOP's require consultation with the Chief's Staff prior to issuance. The effective date of any SOP shall be the date of approval of the Chief. SOP's shall be organized by Division, shall be compiled by the Department administration, made available at the administrative office and sent to all member organizations.

■ Section 4: Standard Operation Guidelines

4.1 Definition
Standard Operating Guidelines are a management tool to provide guidance in the absence of the Divisional Officer. Guidelines are guidance and advice to all members and do not constitute an order. Officers and members shall use guidelines in their decision making process. If in the judgment of the member or officer the SOG is not applicable to their particular situation they are not bound to comply.

4.2 Issuance
Standard Operating Guidelines may be issued in writing by any level officer to their subordinates to outline expectations and to provide guidance.

■ Section 5: Medical Protocols

5.1 Regional Protocols
Hanover County conducts EMS functions through the coordination of the Old Dominion EMS Alliance. All medical treatment shall be defined and governed by the directives of the Richmond Metropolitan EMS Council's Medical Control Board and any directives issued by the Hanover County Medical Director, Assistant Medical Directors, or the Hanover County Medical Control Board. Regional Directives are published in the form of Patient Care Protocols and shall be Appendix C of these Regulations.

5.2 Hanover Medical Protocols
The County Operational Medical Director, Assistant County Medical Directors, and the Hanover Medical Control Board have the authority to issue rules, policies and directives in the form of Medical Protocols to govern the care rendered by providers within the County. Rules, policies and directives of Hanover Medical Control shall be Appendix D of these regulations.

■ Section 6: Organization

6.1 Chain-of-Command
6.1.1 The Department is organized as a paramilitary organization. Good order requires adherence to the chain-of-command. While all Chief Officers have an open door policy it is the responsibility of every

member to inform their immediate supervisor when they wish to breach the normal chain of command prior to doing so, to utilize the open door policy. This includes the use of carbon copies and blind copies on standard or electronic correspondence. The style of management of the Department shall be participatory. This style necessitates standard exceptions to chain-of-command communications. Such standard exceptions include open meetings (e.g., Chief's Staff, Steering Committee, Fire or Rescue Squad Associations, town hall meetings, or committee or task force meetings). Additionally the chain of command may be breached to report infractions of regulations, rules, SOP's, or laws, or misconduct or violation of these Regulations by superior officers.

6.1.2 The authority of the Chief pursuant to applicable statutes including those contained in the Virginia Code, Title 27, and the Hanover County Code shall be delegated as allowed automatically to the highest ranking officer on the scene of an emergency incident following the Chain of Command listed in these Regulations.

6.1.3 The Deputy Chiefs, Division Chiefs, Battalion Chiefs, District Chiefs, Assistant District Chiefs, Captains and Lieutenants and any staff or volunteer appointed by the Chief shall hereby be designated Authorized Officers of the Department and are hereby granted the authority of authorized officers as defined by Title 27 of the Virginia Code.

6.1.4 Delegated authority of the Chief shall subject all operational officers' legal authority to operate at a Fire, EMS, or other emergency incident to the discretion of the Chief. Officers may be removed or authority limited as deemed appropriate by the Chief.

6.1.5 Chain-of-Command

The following shall be the chain-of-command:

County Administrator
Chief
Deputy Chief
Division Chief
Battalion Chief
District Chief
Assistant District Chief
Line Officers
Volunteers & Staff

6.2 Contact with Elected Officials

6.2.1 Nothing in this policy shall be construed to limit a member's rights of free speech and expression. It is rather, the intent of this regulation to prescribe on-duty conduct related to contact with elected officials.

6.2.2 No member shall breach the chain-of-command to discuss departmental issues with elected officials without providing the courtesy of notifying their supervisors, except in the case of reporting wrong-doing or legal infractions. It is the policy of the Department to provide adequate opportunity for input and to air grievances, and therefore all members should utilize these structures prior to breaching the chain-of-command.

6.2.3 When contacting an elected official the individual shall ensure that they represent themselves as individuals noting that they do not speak on behalf of the Department or its recognized advisory groups. Only the Chairman of the Chief's Staff and the Presidents of the respective Associations shall have the authority to speak on behalf of the recognized advisory groups. Only those members authorized by the Chief shall have the authority to speak on behalf of the Department.

6.2.4 Political activities while on duty shall be prohibited. Uniforms shall not be worn while conducting political activities unless approved by the Chief.

6.2.5 If contacted by an elected official while on duty, the member shall meet the individual's request if it is within their scope of authority and report the contact as soon as practicable to their immediate supervisor. If the request is not within their scope of authority the member shall contact the appropriate officer and refer the request.

6.2.6 Any contact with elected officials shall meet the same standard for courteous and dignified discourse as is required for contact with any citizen. Failure to be courteous shall be deemed conduct inconsistent with service within the Department.

■ Section 7: Membership

7.1 General

Membership in the Department is a privilege that is extended on the condition of the individual's adherence to applicable rules, regulations, policies, guidelines, orders and protocols. To be an unrestricted member of the Department an individual must be a minimum of 18 years of age and a legal citizen of or legally present in the United States of America. Hanover County and the Department do not discriminate on the basis of age, race, religion or gender.

7.2 Categories of Membership

7.2.1 Volunteer Member—Fire Service Operational
7.2.2 Volunteer Member—Fire Service Non-operational
7.2.3 Volunteer Member—EMS Operational
7.2.4 Volunteer Member—EMS Non-operational
7.2.5 Volunteer Member—Officer
7.2.6 Employee
7.2.7 Employee—Officer
7.2.8 Honorary Member
7.2.9 Minors—Restricted Membership

7.3 Volunteer Member—Fire Service Operational

Members in this category fall within two subcategories; entry qualified and non-entry qualified. Entry qualified meet all of the necessary conditions to be allowed to operate fully and are allowed to enter burning structures and other hazardous environments. Non-entry qualified members are not qualified to enter burning structures or other hazardous environments. These members may operate as driver/operators or in other capacities as training and policy allow. All fire service operational members must be approved by the Chief as suitable for membership via a background check conducted by Fire Administration. All fire service operational members serve at the discretion of the Chief.

7.4 Volunteer Member—Fire Service Non-operational Fire Service

Non-operational Members are those members who serve at the discretion of the individual fire companies and include auxiliary, associate, life and other membership classifications as determined by the individual fire companies.

7.5 Volunteer Member—EMS Operational EMS

Operational Members are any members who respond to incidents and render patient care. All EMS Operational Members must be approved as eligible via a background check by Department Administration and serve at the discretion of the Chief.

7.6 Volunteer Member—EMS Non-operational

EMS Non-operational Members are those members of the individual rescue squads that do not respond to incidents or render patient care. EMS Non-operational Members serve at the discretion of the individual Rescue Squad under classes of membership defined by the squad.

7.7 Volunteer Member—Officer

Volunteer Officers are those members who are qualified and appointed by the Chief as an Officer of the Department. These Officers are within the Chain-of-Command and are authorized to give orders and exercise authority consistent with their appointed rank. All Officers of the Department serve at the discretion of the Chief. This category of membership does not apply to corporate officers of the individual fire company or rescue squad. Corporate Officers shall not be considered Officers of the Department. Corporate Officers shall serve in accordance with individual constitution and by-law requirements and serve at the discretion of the individual organization. Only Departmental Officers shall hold the delegated authority of the Chief as allowed pursuant to the Virginia Code. Selection and appointment practices of volunteer officers shall be defined by standard operating procedures.

7.8 Employee

Members of the Department including administrative staff who are full or part-time employees of the County of Hanover are categorized in this manner. Employees may be uniformed or non-uniformed and are governed by all Departmental as well as all County policies and procedures.

7.9 Employee—Officer

Only uniformed employees of the Department who are promoted or appointed to rank may serve as Officers of the Department. Delegated authority of the Chief pursuant to the Virginia Code is extended to these officers following the Department's Chain-of-Command. Employee Officers are governed by all Departmental as well as County policies and procedures.

7.10 Honorary

Citizens, retired employees or volunteers may be granted honorary membership by the Chief to recognize outstanding contributions to the Department. Individual fire companies and rescue squads may appoint honorary members to their organizations as allowed by their individual constitutions and by-laws.

7.11 Minors—Restricted Membership

7.11.1 Purpose

The purpose of this policy is to govern participation of minor restricted volunteer members, 16 and 17 years of age. This policy also governs all adult members regarding responsibility for supervision of members 16 and 17 years of age. Membership is allowed in the junior program of the Department at age 16.

7.11.2 General
Volunteer members 16 and 17 years of age shall adhere to all Hanover County policies and procedures. Their participation will also be governed by the Virginia Department of Fire Programs, and Department of Labor and Industry regulations, and other applicable laws and regulations pertaining to minors in the Commonwealth of Virginia. Each volunteer company shall appoint a junior advisor who shall manage the program.

7.11.3 Special Rules Relating to Minors
These rules shall govern the conduct of minors and of adults. Adult members shall be responsible for enforcing these rules.

7.11.3.1. No one under the age of eighteen is permitted in any Department building after 10:00 p.m. Sunday through Thursday and 11:00 p.m. Friday and Saturday, while public schools are in session. No one under the age of eighteen is permitted in any Department building after 11:00 p.m. Sunday through Thursday and 12:00 midnight during holidays and other times when public schools are not in session. If an individual who is under the age of eighteen is on a call and does not return to a building until after the above stated curfew the individual must prepare and leave the premises within fifteen (15) minutes.

7.11.3.2. No one under the age of eighteen is permitted in any Department building at any time without the presence and supervision of two responsible adults. If the minor's parent is on premises, the two responsible adults are not required at that time. All adult members including officers and non-officers shall be responsible for properly supervising any individuals under the age of eighteen in their presence.

7.11.3.3. Failure to provide appropriate supervision for members under the age of eighteen is considered conduct inconsistent with service in the Department.

7.11.3.4. It shall be the responsibility of all adult members to enforce this Policy. Failure to comply shall be deemed insubordination and shall subject the member to discipline up to and including termination.

7.11.3.5. Members 16 and 17 years of age may not participate in any firefighting or support activities unless certified to NFPA 1001, Level 1. This includes responding in emergency vehicles, pulling hose lines, raising ladders, or operating equipment unless training to meet the requirements of NFPA 1001, Level 1. Training must be an approved class with all the proper documentation complete and on file with the Hanover County Training Division and Virginia Department of Fire Programs.

7.11.3.6. Members 16 and 17 years of age may participate in activities of the Department including response to emergency calls once certified under the National Fire Protection Association Standard 1001, Firefighter Level 1.

7.11.3.7. Members 16 and 17 years of age shall not serve in a command role on an emergency scene.

7.11.3.8. Members 16 and 17 years of age shall adhere to the laws of the Commonwealth of Virginia and are prohibited from the use of alcohol and tobacco products in the station or during any function related to Hanover County Fire/EMS.

7.11.3.9. Members age 16 and 17 shall adhere to the laws of the Commonwealth of Virginia and be enrolled in high school as required. All members must maintain a grade of "C" in each class taken, as evidenced by copies of grade reports submitted in accordance with prescribed procedures. Failure to maintain a "C" grade shall result in suspension from the Department and continue until the next scholastic grading period or receipt by Department administration of an interim report to provide proof the grade has been improved.

7.11.3.10. EMS members 16 and 17 years of age shall not serve as attendant in charge on an ambulance.

7.11.3.11. Member age 16 and 17 years of age shall not drive Fire/EMS apparatus.

7.11.3.12. Members age 16 and 17 years of age shall be prohibited from responding to emergency scenes in personal vehicles and must obey all traffic laws at all times and are prohibited from responding to stations using warning devices in personal vehicles.

7.11.3.13. No waiver of these requirements shall be allowed. Written parental permission must be obtained and filed with Department administration on forms provided by the Department, prior to participation in any Department activities.

■ Section 8: Conduct

8.1 Employees

8.1.1 Applicability of Hanover County Human Resources/Personnel Policy Employees of the Department shall be governed by the Hanover County Personnel Policies, and by all other policies applicable to County employees. In no event shall those policies or these Regulations be deemed to be a waiver of requirements of federal, state or local laws and regulations.

8.2 Volunteers

8.2.1 Volunteer Hanover County Human Resources/Personnel Policies Volunteers of the Department shall be governed by the following standards of conduct included in the County Personnel Policies (references to "employee" and "employment" and similar terms shall be deemed to refer to "volunteer," "volunteer service," or similar, for purposes of these Regulations):

8.2.1.1. Equal Opportunity Employment/Non Discrimination:
Hanover County is an Equal Opportunity Employer. It is committed to the maintenance and promotion of the policy of nondiscrimination by incorporating sound merit principles in all aspects of personnel management affecting its employees and applicants. Personnel management shall be free from such prohibited personnel practices as discrimination, sexual harassment, or any other conduct inconsistent with sound merit principles. Is shall provide equal employment opportunity to all employees and all applicants without regard to race, color religion, national origin, political affiliation, disability, sex or age, except where such is bona fide occupational qualification. The adoption of this policy by the Board of Supervisors is a reaffirmation of adherence to and promotion of the policy of nondiscrimination. Any person employed by the County of Hanover who fails to comply with this policy is subject to the County disciplinary procedures.

8.2.1.2. Conflict of Interests Act:
Employees shall comply with the State and Local Government Conflict of Interests Act. Questions concerning interpretation or the application of the Act should be directed to the department head.

8.2.1.3. Prohibition Against Harassment:
It is the policy of Hanover County that all employees have a right to work in an environment free of discrimination, which includes freedom from harassment whether that harassment is based on sex, age, race, national origin, religion, sexual orientation, marital status, disability, or membership in other protected groups. The County prohibits harassment of its employees in any form by supervisors, co-workers, customers, or suppliers. Such conduct may result in disciplinary action up to and including dismissal of the employee who harasses others or the supervisor or department head who tolerates such conduct. Persons who are not employees who engage in offensive and/or harassing behaviors or language will be asked by the Supervisor or Department Head to leave the premises. No Supervisor shall threaten or insinuate either or implicitly that any employees' submission to or rejection of sexual advances will in any way influence any personal decision regarding that employee's employment, performance appraisal, wages advancement, assigned duties, shifts, or any other condition of employment or career development. Other harassing conduct in the workplace, whether physical or verbal, committed by supervisors or others is also prohibited. This includes: slurs, jokes or degrading comments concerning sex, age, race, national origin, religion, sexual orientation, marital status, disability, or membership in other protected groups; repeated offensive sexual flirtation, advances, or propositions; continual or repeated abuse of a sexual nature; graphic verbal comments about an individual's body; and the display in the workplace of sexually suggestive objects or pictures.

8.2.1.4. Employee Dress:
During work hours, employees are considered to be representatives of the County and are required to dress and groom themselves in a manner that portrays a professional image. Department heads are to determine appropriate appearance and apparel. Any manner of dress or personal hygiene that is disruptive to the work of the Department or to those being served, shall be considered inappropriate. Employees issued uniforms or protective clothing are required to wear them while on duty and are expected to maintain them in a neat, clean and operational state. Position descriptions will specify those jobs where there are special grooming and/or dress requirements imposed to ensure the safety of the employee.

8.2.1.5. Confidentiality:
Employees having access to personal information or data in the course of providing County services to clients shall maintain the confidentiality of that information and shall release that information only in accordance with the Virginia Privacy Protection Act and any other regulations which are applicable to specific programs. Failure to adhere to those requirements and to maintain the confidentiality of personal information may result in disciplinary action, including termination of employment or volun-

teer status. Non-members may observe operations only in accordance with procedures established by the Chief, and shall not have access to personal information.

8.3 General Conduct—All Members

8.3.1 Knowledge of Regulations

Every member is required to establish and maintain a working knowledge of these Regulations, Orders, Standard Operating Procedures, Standard Operating Guidelines and Protocols. In the event of improper action or breach of discipline, it will be presumed that the member was familiar with the authority in question. Violation of any authority may be grounds for disciplinary action, including termination of employment or volunteer status.

8.3.2 Obedience to Laws and Regulations

All members shall observe and obey all laws and ordinances, these Regulations, Orders, Standard Operating Procedures, Standard Operating Guidelines and Protocols as well as orders of superior officers.

8.3.3 Chain of Command; Requests for Interpretation

The unbroken line of authority extends from the County Administrator to the Chief, and through a single subordinate at each level of command, down to the level of execution. The chain of command is the proper route for vertical communication among Department members. Requests for interpretation of these Regulations shall be directed through the chain of command to the superior officer, who may consult with the Chief, or with the designees of the Chief.

8.3.4 Performance of Duty

All members shall perform their duties as required or directed by law, SOP, Departmental rule, Operating Manual, General Order, or by order of a superior officer. All lawful duties required by competent authority shall be performed promptly as directed, notwithstanding the general assignment of duties and responsibilities.

8.3.5 Human Relations

Every member is expected to perform his/her duties in an efficient, courteous, and orderly manner, employing patience and good judgment at all times. All members shall refrain from harsh, profane, or insolent language or acts, and shall be courteous and civil in their dealings with others. Members shall not use racial or ethnic slurs. Members shall conduct themselves in a manner which will best accomplish the Department's mission, and in accordance with these Regulations. They shall use respectful, civil forms of address to all persons. Displays of bias toward any person are prohibited. Failure to act in keeping with the Department's standards as set forth in these Regulations, shall be considered conduct inconsistent with service in the Department, and may result in disciplinary action, including termination.

8.3.6 Cooperation/Coordination

Members shall coordinate their efforts with all other members of the Department and County and other public and private agencies, with the objective of ensuring maximum achievement and continuity of purpose through teamwork. All members are charged with the responsibility of fostering and maintaining a high degree of cooperation both within the Department and with all other agencies.

8.3.7 Reporting Violations

Any member who has knowledge of other members, individually or collectively, who are either knowingly or unintentionally violating any laws, these Regulations, SOPs, or who are disobeying orders, shall bring any and all facts pertaining to the matter to the attention of a superior officer. The superior officer contacted shall then take appropriate action. A member may bypass the official chain of command and directly advise the Chief in writing of the violation(s). A member's failure to report violations shall be deemed as complicity in the act and that member may be punished as a party to the infraction.

8.3.8 Accepting Gifts, Gratuities; Conflicts of Interest

Members shall not knowingly accept services, tickets to amusement places, or other material benefits at a discounted rate or at a rate lower than offered to the general public while on duty or by virtue of their position with the Department, unless previously approved by the Chief, or necessary for the performance of duties. All members are expected to be knowledgeable of conflict of interest laws, or to inquire as to the application of those laws, and are required to act in a manner that does not create an appearance of a conflict of interest. Any questions related to this subject should be addressed to Department administration.

8.3.9 Firearms

Firearms shall not be carried by on-duty personnel or on any Department equipment. Firearms shall not be allowed in or stored in any Department facility unless approved by the Chief. Firearms on the person of a sworn officer of the Hanover County Fire Marshal's Office, Hanover Sheriff's Department, Virginia State Police, ATF, or FBI are allowed on/in any Department facility or equipment at the discretion of the officer.

8.3.10 Gambling
Members shall not indulge in any gambling while on duty or while wearing the Department uniform or any material with identification of the Department.

8.3.11 Intoxicants—Purchase and Consumption
Members shall not consume intoxicants while off-duty to the extent that evidence of such consumption is apparent when reporting for duty or to the extent that the ability to perform on duty is impaired. In addition, wearing the Department uniform, or parts thereof, or any clothing bearing the Department's identification while consuming or purchasing intoxicants is prohibited.

8.3.12 Intoxicants on Department Premises
Members shall not bring onto, store or possess any intoxicants on Department premises, except for approved fund raising or other functions approved by the Chief and consistent with requirements of the Alcoholic Beverage Control Board. Intoxicants shall not be transported in any Department vehicle (County or volunteer-owned).

8.3.13 Intoxication
Members shall not report for duty under the influence of any intoxicant or consume or possess any intoxicant while on duty.

8.3.14 Malingering
A member shall be absent from duty because of sickness only when suffering from an illness or injury which would prevent the proper performance of duty, or as otherwise provided for in the County's sick leave policy. He/she shall not feign sickness or disability, or attempt to deceive a supervisor concerning his/her physical or medical condition. Notification of the use of sick leave will be made to the employee's supervisor prior to his/her scheduled reporting time as required in Department SOPs.

8.3.15 Obligation to Duty
Uniformed members of the Department are always subject to duty, although periodically relieved from its routine performance. They shall at all times respond to the lawful orders of superior officers. Proper action consistent with the member's training must be taken whenever required. Uniformed employees assigned to special duties are not relieved from taking proper action outside the scope of their specialized assignment when necessary.

8.3.16 Reporting to Duty
Members shall report for duty at the time and place required by assignment or orders and shall be properly uniformed, equipped, and prepared to assume duty. They shall give their undivided attention to orders, instructions, and any other information that may be disseminated.

8.3.17 Truthfulness
All members shall supply complete and truthful information when questioned by any authority in the chain of command about matters related to the performance of the employee's official duty and/or fitness to perform such duty. Members reporting infractions or submitting complaints shall report only information that they know to be factual. Misrepresentation of facts and spreading of rumors is forbidden and shall be deemed conduct inconsistent with participation as a member of the Department.

8.3.18 Unbecoming Conduct
Members shall conduct themselves at all times, both on and off duty, in such a manner as to reflect most favorably on the Department. Conduct shall be deemed to fail to meet that standard if it brings the Department into disrepute or reflects discredit upon the member as a member of the Department or impairs the operation or efficiency of the Department or member.

8.3.19 Insubordination
All members must comply with any legal order given by a superior officer as defined by the Department chain-of-command. Failure to comply will be deemed insubordination.

8.3.20 Personal Accountability
Each member shall be held personally accountable for their actions. It is incumbent upon each member to fulfill their respective duties and advance the goals, mission, vision, and values of the Department. Respect for and fulfillment of each individual's responsibility is a basic condition of membership within the Department.

8.4 Honor Code

All members shall subscribe to the following commitments throughout their tenure within the Department:

- I will not lie.
- I will not cheat.
- I will not steal.
- I will not tolerate those who do.
- I will not conceal facts to evade discipline or influence an investigation.
- I will not coerce others to violate the Honor Code.
- I will not knowingly falsely accuse others.
- I will act ethically and with compassion at all times.

- I will protect the rights of others.
- I will maintain the confidentiality of personal information of those to whom I provide service.

8.5 Department Values

8.5.1 Honor/Integrity
All members will be held to high standards and expected to behave in a manner that represents a strict adherence to moral and ethical values.

8.5.2 Respect
All members will be expected to display self-respect. They will treat others as they wish to be treated. They will hold great regard for all they serve and protect their dignity.

8.5.3 Responsibility
All members have a personal obligation to honor their commitment to their respective organizations and to the Department. They shall make decisions using good judgment and common sense, keeping safety as a first priority.

8.5.4 Accountability
All members will be held accountable for their actions. Members have a responsibility to our mission to help others. Members must respect the leaders and the rules that govern the system.

8.5.5 Professionalism
All members will be expected to operate within the boundaries of professional standards. This includes, but is not limited to, appropriate public behavior, neat personal appearance and promptness.

8.5.6 Quality
Members of the Department will always seek to provide the highest quality of service possible in all their endeavors and continuously strive to improve the quality of the entire system.

Section 9: Rights and Responsibilities

9.1 Disciplinary Procedures

As noted above, all employees are subject to the Hanover County Personnel Policies, including those related to discipline and the Grievance Procedure. Members, including volunteers, shall be subject to discipline including termination.

9.2 Volunteer Review Procedure

To provide volunteer members access to a review of issues, the following procedure shall be used, when informal attempts to resolve issues are unsuccessful. This review procedure is provided to volunteers, for grievances as defined.

9.2.1 Definition.
A grievance is a complaint or dispute by a member relating to his or her membership, including but not necessarily limited to:

9.2.1.1. Disciplinary actions, including disciplinary demotions, suspensions, and dismissals provided that such dismissals result from formal discipline or unsatisfactory performance.

9.2.1.2. The application of policies, procedures, rules, and regulations, and the application of ordinances and statutes.

9.2.1.3. Acts of retaliation taken as the result of utilization of this grievance procedure or the participation in the formal grievance (under this grievance procedure) of another member.

9.2.1.4. Acts of retaliation because the member has complied with any law of the United States or of the Commonwealth of Virginia, has reported any violation of such law to a governmental authority, or has sought any change in law before the United States Congress or the General Assembly of Virginia, or has reported an incident of fraud, abuse, or gross mismanagement.

9.2.1.5. Complaints of discrimination on the basis of race, color, creed, religion, political affiliation, age, disability, national origin, or sex.

9.3 County Management Rights and Prerogatives:

The County reserves to itself the exclusive right to manage the affairs and operations of County government. Accordingly, complaints involving the following management rights and prerogatives are not grievable:

9.3.1 Establishment and revision of volunteer position classification, or general benefits.

9.3.2 Work activity accepted by the member as a condition of membership, or work activity which may reasonably be expected to be a part of the member's duties.

9.3.3 The contents of ordinances, statutes, or established policies, procedures, rules, and regulations.

9.3.4 The methods, means, and personnel by which work activities are to be carried on, including but not necessarily limited to:

9.3.4.1. The provision of equipment, tools, and facilities necessary to accomplish tasks.

9.3.4.2. The scheduling and distribution of personnel resources.

9.3.4.3 Training.

9.3.5 The acceptance of membership, promotion, appointments to new position, transfer, assignment, and retention of members in positions within the County service.

9.3.6 The relief of members from duties, or taking action as may be necessary to carry out the duties of the Department in emergencies.

9.3.7 Direction and evaluation of the work of members.

9.4 Coverage of Personnel

All non-probationary volunteer members of the Department may use this procedure for grievances as defined.

9.5 Operation of the Volunteer Review Procedure

9.5.1 Step 1—Supervisor Level:

A member who believes he or she has a grievance shall discuss the grievance informally with the immediate supervisor within twenty calendar days after the occurrence of the event giving rise to the grievance. If the grievant alleges discrimination or retaliation by the immediate supervisor, however, the grievance may be initiated with the next level supervisor. If the issue is not resolved after an informal discussion, the grievant shall complete the Step 1 section of the Volunteer Review Procedure form. Grievability shall be determined by the County Administrator (or designee), in consultation with the Chief. Resolution of the grievance by the immediate supervisor shall be communicated, in writing, to the grievant within ten calendar days after the date of the discussion. At Step 1, the only persons who may normally be present in the meetings are the grievant, the supervisor, and appropriate witnesses. Witnesses shall only be present while actually presenting testimony.

9.5.2 Step 2—Chief Level:

If the grievant is not satisfied with and does not accept the Step 1 response, or if the immediate supervisor fails to respond within the required time period, and the grievant wishes to advance to Step 2 of this procedure, the grievant complete the Step 2 section of the Volunteer Review Procedure form and shall file the completed request form with the Chief within ten calendar days of receipt of the supervisor's response or the deadline for that response, whichever occurs first. The grievant shall specify the issues and state the relief that he or she expects to gain through the use of this procedure. If there is more than one outstanding grievance, the Chief may combine them to be heard in the same Step 2 meeting or consider them in separate Step 2 meetings. The Chief or designee shall promptly meet with the grievant. The Chief or designee shall render a written response to the grievant within ten calendar days following receipt of the request form. At Step 2, the only persons who may normally be present in the meetings are the grievant, the Chief, the Division Chief charged with overseeing volunteer services and appropriate witnesses. Witnesses shall only be present while actually presenting testimony.

9.5.3 Step 3—County Administrator Level:

If the grievant is not satisfied with and does not accept the Step 2 written response, or if the Chief or designee fails to respond within the required time period, and the grievant wishes to advance to Step 3 of this procedure, the grievant shall complete the Step 3 section of the Volunteer Review Procedure form indicating the intent to advance the grievance to Step 3. The form shall be filed by the grievant with the office of the Chief within ten calendar days following receipt of the Step 2 response or the deadline for that response, whichever occurs first. At the discretion of the County Administrator (or designee), all of the grievant's outstanding grievances can be combined to be heard at the same Step 3 meeting, or considered in separate Step 3 meetings. If the County Administrator or designee determines (or has previously determined) that the complaint is grievable, the Administrator or designee shall meet with the grievant together with a representative of the Department, appropriate witnesses for each side, and such other persons as the County Administrator or designee deems necessary and appropriate. Witnesses shall be present only while actually providing testimony. If the grievant is represented by legal counsel, the County likewise has the option of being represented by counsel. The County Administrator or designee shall render a written response resolving the grievance within ten calendar days following receipt of the Grievance form from the Chief.

9.6 Grievability

Grievability and access are determined by the County Administrator or designee. Only after the County Administrator or designee has determined that a complaint is grievable and that the grievant is eligible to file a grievance may a grievance be advanced through Step 3 of this procedure. When the question of grievability arises at the (Step 2) level, or whenever the question of access to this procedure arises, the grievant or the department head may request a ruling on grievability or a ruling on access, as the case may be, by the County Administrator. The County Administrator or designee shall render a decision within ten calendar days of receipt of the request, and shall send a copy of the decision to the grievant and the Chief.

9.7 Compliance

The County Administrator or designee shall determine compliance issues. The County Administrator or designee may require a clear written explanation of the basis for a request for just cause extensions or exceptions.

9.8 Procedures for and Conduct of Volunteer Review Procedure Hearings

Except as otherwise noted, the following rules apply to all levels of grievance hearings.

9.8.1 Time intervals may be extended by mutual consent of the parties.

9.8.2 When a deadline falls on a Saturday, Sunday, or County holiday, the next calendar day that is not a Saturday, Sunday, or County holiday shall be considered the deadline.

9.8.3 As far as practical, all grievance meetings and hearings shall be held during normal County working hours.

9.8.4 County employees who are necessary participants at grievance hearings shall not lose pay for time necessarily lost from their jobs and will not be charged leave because of attendance at such hearings.

9.8.5 The use of recording devices or a court reporter is not permitted at hearings.

9.8.6 Hearings are not intended to be conducted like proceedings in court and the rules of evidence do not necessarily apply.

9.8.7 The grievant shall present evidence first.

9.8.8 Both the grievant and the County may call appropriate witnesses. All witnesses, including the grievant, shall be subject to examination and cross-examination.

9.8.9 Witnesses shall be present only while actually giving testimony.

9.8.10 The grievant shall not be entitled to recover more than that which the grievant has lost.

9.8.11 When a grievant has obtained partial relief at one level of his grievance procedure but decides to appeal to the next higher level, the filing of a request form for appeal to the next higher level shall constitute rejection of, and relinquishment of any claim to, any and all relief granted at the previous level.

9.8.12 Each party shall bear the costs and expenses of legal counsel or representative, if any.

Sample Standard Operating Procedures

Disciplinary Investigations

Note: The Commonwealth of Virginia has codified rules regarding disciplinary investigations in the form of the Fire Fighter's Bill of Rights. This policy is specifically designed to comply with that law.

Purpose

Members of Hanover Fire EMS shall observe all departmental policies, directives, orders and applicable regulations and laws. It is the responsibility of every officer to ensure compliance of members. This policy is intended to provide guidance to officers in the fulfillment of that responsibility and to outline a fair process for disciplinary action. In the case of employees, where this policy is in conflict with Hanover County Human Resources Policies, County Policy shall be used.

Authority

All officers appointed by the Chief shall have the authority to institute necessary disciplinary action to ensure that subordinates operate in compliance with all departmental policies, orders, and applicable regulations and law. This authority includes the ability to enforce immediate removal from duty of a member pending an investigation when warranted. Any disciplinary action that contemplates dismissal, demotion or suspension for punitive reasons must be approved by the Chief or a Deputy Chief prior to implementation.

Procedure

Informal Counseling Sessions

All Officers are encouraged to counsel their subordinates on a routine and frequent basis as a method to continually improve performance. Informal counseling sessions shall be documented as a note to file and a copy shall be provided to the Department's personnel specialist. Informal counseling sessions and notes to file are not limited to critical sessions. Good performance should also be noted in the Department file. Notes to file shall be used as background information by supervisors to support performance ratings.

Formal Counseling Sessions

Whenever a member is repeatedly counseled on a issue for improvement or if the issue is of serious nature, the session shall be documented in memorandum form. A copy of the record shall be forwarded to the Department Personnel Specialist who shall forward a copy to the Human Resources Department for inclusion in the member's personnel record.

Investigations

An Officer may begin an informal investigation into any complaint observed or reported against a subordinate. Upon determination that the complaint may warrant suspension or demotion for punitive reasons, or dismissal the investigation shall be designated as a formal investigation and the following procedure shall be applied.

Step 1

The individual to be investigated shall be notified in writing of the charge and investigation to include:

a. Nature of charge(s)
b. The name, rank and assignment of the investigating officer
c. Listing of the interrogator and any individual(s) to be present during the interrogation.
d. The date, location and time of the interrogation
e. The identity of the accusor(s)

Step 2

The following procedures shall be utilized for the interrogation:

a. The interrogation shall be conducted during normal business hours at the individual's assigned station or at Fire Administration. Interrogations involving volunteers may be held during evening or weekend hours at the request of the individual charged.
b. Interrogations shall be limited to 1 hour in duration and may be repeated as necessary to fully investigate the charges.
c. No offensive language shall be used nor shall any inducements be offered.
d. If a recording is made a copy shall be provided to the individual charged at no cost to the individual upon request.

All members are required to answer any question forthrightly at the time asked. Refusal to answer questions in any disciplinary investigation shall constitute insubordination and may be punished accordingly. If at any time a matter is determined to be of a possible criminal nature the matter shall be reported to the proper law enforcement agency for investigation. The individual charged shall be placed on paid administrative leave for the duration of the criminal investigation and the departmental administrative investigation shall be suspended until any criminal investigation is completed. At such time the administrative investigation shall be resumed.

Advisory
All members are advised that criminal and administrative investigations are separate matters and different standards, procedures and rights apply. This policy in no way applies to criminal investigations with the exception that the member shall be placed on administrative leave (paid for employees) for the duration of any criminal investigation.

Orientation Program

Purpose
The purpose of this course is to provide the student with the knowledge and skills needed to perform as a productive team member of the emergency service community. The student will be educated on the safety considerations needed while performing tasks on the emergency scene. The information provided will satisfy the training requirements of the new member training requirements of Hanover Fire EMS.

Scope
The information contained in this course is designed to provide a basic understanding of Fire EMS operations to new members of the Department. In addition, the subjects outlined in this course can provide an opportunity for existing members to sharpen or enhance their emergency scene skills. This course can also benefit persons wanting to gain a basic understanding of emergency operations.

Policy
New member

1. Once a new member is accepted into the department, he or she will be scheduled for the Department orientation program. The first night is mandatory. The topics covered will be department orientation and infectious disease control.
2. The County-level course runs Monday through Thursday for two consecutive weeks. Classes begin at 18:30 each night and end at 21:30. In the event a student cannot attend the classes due to scheduling conflicts, they must notify the Fire EMS Training Academy instructors and schedule a make up time. They must successfully complete the written and practical test at the Fire EMS Training Academy to finish the course.
3. All members must attend the first session. All members must attend a test session conducted by Fire Academy staff to complete the course. Test sessions are the last night of the course schedule. The course "re-starts" 4 weeks from the conclusion of the last course.
4. All new members entering Hanover Fire EMS must complete this course before being allowed to ride any emergency apparatus under emergency conditions.

Volunteer Officer Selection Process
Note: This is a new policy with very stringent training requirements—it was acknowledged that a grace period would be necessary to allow volunteers time to attend necessary training to be fully compliant.

Purpose
The purpose of this standard is to identify the appointment process and the minimum training standards for volunteer chief officers and rescue squad captains for the department according to the approved volunteer work plan.

Policy
I. Each Fire/EMS company will provide a letter of recommendation to the Hanover County Fire/EMS Chief with potential candidates for the positions of **District Chief, Assistant Chief(s), and Rescue Squad Captain.**

II. Each candidate shall provide copies of required certifications to the HR Analyst of the Human Resources and Volunteer Services Division for verification and scheduling of interviews.

III. Each qualified candidate will then have an interview with the Hanover County Fire and EMS Chief.

IV. The Hanover County Fire and EMS Chief will appoint operational chief officers and squad captains to serve for a period of one (1) year from the time of appointment. The following shall serve as the minimum requirements for each position: **District Chief —Full compliance date July 1, 2010**

a. Minimum of 2 years of Hanover County service or equivalent
b. 21 years of age or older
c. Firefighter Level I (VDFP, NFPA 1001)
d. Firefighter Level II (VDFP, NFPA 1001)
e. Hazardous Materials Operations (VDFP)
f. Mayday-Firefighter Down (VDFP)
g. National Incident Management Systems (ICS, IMS)
h. Leading the Attack (HFEMS)
i. EMT- Basic (VAOME)

j. EVOC
k. CPR (American Heart or equivalent)
l. MCI I
m. MCI II
n. Shaping the Future (NFA)
o. Fire Officer I (VDFP)
p. Fire Officer II (VDFP)
q. HIPAA Medical/Medical Class
r. Hanover County Harassment/Diversity Training

Assistant District Chief—Full compliance date July 1, 2010

a. Minimum of 2 years of Hanover County service or equivalent
b. 21 years of age or older
c. Firefighter Level I (VDFP, NFPA 1001)
d. Firefighter Level II (VDFP, NFPA 1001)
e. Hazardous Materials Operations (VDFP)
f. Mayday-Firefighter Down (VDFP)
g. National Incident Management Systems (ICS, IMS)
h. Leading the Attack (HFEMS)
i. EMT-Basic (VAOME)
j. EVOC
k. CPR (American Heart or equivalent)
l. MCI I
m. MCI II
n. Shaping the Future (NFA)
o. Fire Officer I (VDFP)
p. HIPAA Medical/Medical Class
q. Hanover County Harassment/Diversity Training

Squad Captain—Full compliance date July 1, 2010

a. Minimum of 2 years of Hanover County service or equivalent
b. 21 years of age or older
c. EMT-Basic (VAOEMS)
d. Hazardous Materials Awareness (VDFP)
e. CPR (American Heart or equivalent)
f. National Incident Management Systems (ICS, IMS)
g. MCI Level I
h. MCI Level II
i. EVOC
j. EMS Officer I (HFEMS—EMS Track)
k. EMS Officer II (HFEMS—EMS Track)
l. HIPAA Medical/Legal Class
m. Hanover County Harassment/Diversity Training

Battalion Chief

a. 21 years of age or older
b. Minimum of 5 years as an active volunteer
c. Firefighter I (VDFP, NFPA 1001)
d. Firefighter II (VDFP, NFPA 1001)
e. Emergency Medical Technician with current CPR
f. Hazardous Materials—Operations (VDFP)
g. National Incident Management Systems (ICS, IMS)
h. EVOC
i. Fire Officer I (VDFP)
j. Fire Officer II (VDFP)
k. MCI I
l. MCI II
m. Managing Company Tactical Operations
n. Shaping The Future (NFA)
o. Managing EMS Systems (HFEMS)
p. Mayday—Firefighter Down (VDFP)
q. HIPAA Medical/Legal Class
r. Hanover County Harassment/Diversity Training
s. 45 Educational Points

V. All personnel shall provide and maintain copies of required certification with the HR Analyst of the Human Resources and Volunteer Services Division at all times. Failure to provide copies of certifications will result in the determination that the required training is incomplete.

VI. Each Member of the department is responsible for maintaining their own recertification requirements.

Apparatus Equipment Changes

■ Roseville Fire Department Standards

SUBJECT: Apparatus Equipment Changes
INDEX: 104.08
DATE DRAFTED: December 8, 2002
DATE EFFECTIVE: March 15, 2003
DATE REVISED:

1. PURPOSE: The purpose of this Standard is to serve as a guideline for fire department personnel to implement any change involving equipment carried on fire apparatus, including what equipment should be carried, where it should be stored and on which apparatus. The purpose of this Standard is also to maintain uniformity of equipment carried on apparatus wherever possible.
2. RESPONSIBILITY: It is the responsibility of each employee to understand and use this Standard whenever applicable, such as before changing the location of, removing, adding or altering any piece of equipment carried on any fire apparatus.
3. SCOPE: This Standard shall apply to all employees of the Roseville Fire Department.

GENERAL: Below is the procedure to be followed for effecting any change involving fire apparatus equipment, including proposed changes initiated by firefighters. No piece of equipment shall be moved, added or altered on any fire apparatus without prior authorization obtained by complying with the procedure noted below.

A.) Submit your written request to Fire Administration, describing in detail what you are requesting. Provide rationale and/or justification for the change.

B.) Fire Administration will evaluate the request and may grant, deny or refer the request to the Equipment Committee for consideration and recommendation.

C.) Based on the nature of the equipment change, before implementation it may be required to provide department training or to simply notify the membership of the equipment change.

Incident Management Systems

Roseville Fire Department Standards

SUBJECT: INCIDENT MANAGEMENT SYSTEM
INDEX: 106.01
DATE DRAFTED: May 25, 2000
DATE EFFECTIVE: December 1, 2000
DATE REVISED: April 16, 2001

PURPOSE OF INCIDENT MANAGEMENT

1. Fix the responsibility for command on a specific individual through a standard identification system, depending on the arrival sequence of members, companies, and chief officers.
2. Ensure that a strong, direct, and visible Command will be established from the onset of the incident.
3. Establish an effective incident organization defining the activities and responsibilities assigned to the Incident Commander (IC) and to other individuals operating within the Incident Management System (IMS).
4. Provide a system to process information to support incident management, planning, and decision making.
5. Provide a system for the orderly transfer of Command to subsequent arriving officers.

RESPONSIBILITIES OF THE INCIDENT COMMANDER

1. The IC is responsible for the completion of the tactical priorities. The tactical priorities are:
   a. Remove endangered occupants and treat the injured.
   b. Stabilize the incident and provide for life safety.
   c. Conserve property.
   d. Provide for the safety, accountability, and welfare of personnel. (This priority is ongoing throughout the incident.)

FUNCTIONS OF COMMAND

1. Assume and announce Command and establish an effective operating position (Command Post location), on a mobile radio if possible.
2. Rapidly evaluate the situation (size up).
3. Initiate, maintain, and control the communications process.
4. Identify the overall strategy, develop an incident action plan, and assign companies and personnel consistent with plans and established departmental Standards.
5. Develop an effective Incident Management Organization.
6. Provide tactical objectives.
7. Review, evaluate, and revise (as needed) the incident action plan.
8. Provide for the continuity, transfer, and termination of command.

INCIDENT PRIORITIES

1. Life Safety:
   a. Responding fire department personnel.
   b. Other emergency workers.
   c. Occupants and victims.
   d. Bystanders.
2. Incident Stabilization:
   a. Actions that minimize the impact to surrounding areas.
   b. Actions that prevent further expansion of the incident.
3. Property Conservation
   a. Working smarter to reduce damage.
   b. Salvage operations.
4. Continuity of Community
   a. Recovery of business and industry.
   b. Returning to pre-incident conditions.
   c. Reducing the impact the incident has on the community.

ESTABLISHING COMMAND

1. The first arriving officer or crew leader to arrive at the scene shall assume command of the incident and initiate whatever parts of the IMS that are needed to effectively manage the incident scene.
2. The initial IC shall remain in command until command is transferred or the incident is stabilized and terminated.

INITIAL RADIO REPORT

1. The first arriving fire department unit activates the command process by giving an initial radio report.
2. This "windshield report" is what the officer sees as they approach the scene before the officer leaves the apparatus.
3. The rolling size-up may not be entirely accurate, and may be revised as the officer assesses the emergency.
4. Includes (as a minimum) the following information:
   a. Number of stories (examples: single story, story and a half, two story, three story, multi-story).

b. General construction-type of building (wood-frame, masonry, steel, bow string).
c. Occupancy of the structure (examples: residential, multi-family residential commercial, industrial, institutional).
d. Conditions present (nothing showing, smoke showing, working fire, people outside).

5. A description of the action to be taken.
6. Declaration of the incident strategy.
7. Any obvious safety concerns.
8. Assumption, identification, and location of command (street name or business name).
9. Request for (or release of) additional resources as required.

INCIDENT MANAGEMENT MODES

1. Investigating Mode:

a. The first arriving unit sees no obvious problem.
b. The Incident Commander will be investigating the reason for the call.

2. Fast Attack Mode:

a. The situation requires immediate action to stabilize.
b. The incident requires the Incident Commander's assistance and direct involvement in the attack ("Command Working").
c. The Incident Commander goes with the crew to provide the appropriate level of supervision.
d. Examples of Fast Attack Mode situations include:
   - Offensive fire attacks (especially in marginal situations).
   - Critical life safety situations (i.e., rescue) which must be achieved in a compressed time.
   - Any incident where the safety and welfare of firefighters is a major concern.
   - Obvious working incidents what require further investigation by the Incident Commander.

e. The fast attack mode should not last more than a few minutes and will end with one of the following:
   - The incident is stabilized.
   - The situation is not stabilized and the Incident Commander must withdraw to the exterior and establish a command post. (At some point the Incident Commander must decide whether or not to withdraw the remainder of the crew, based on the crew's capabilities and experience, safety issues, and the ability to communicate with the crew. No crew shall remain in a hazardous area without radio communications capabilities.)
   - Command is "assumed" or "transferred" to a new Incident Commander.

   ASSUMED COMMAND: If the IC is still a "working command" (in Fast Attack Mode) when the next officer arrives, the IC may not be able to make face-to-face contact with the next arriving officer. In this instance, next arriving officer should make radio contact with the IC, advise that he is on the scene, obtain a progress report, and "assume" the command at the street level. When this happens, the new IC advises the old IC that Command has been assumed, and the new IC gives the old IC a designated sector name to operate as (for example, Attack Sector, Interior Sector, etc.).

   TRANSFERRED COMMAND: The IC gives the next arriving officer a face-to-face report of conditions (or by radio if face-to-face is not possible). The Command may then be "transferred" to another on-scene officer. When command is transferred, the new IC is briefed with more detail than when command is "assumed", and includes:
   – Incident conditions (fire location, extent, haz-mat, number of victims, etc.)
   – The Incident Action Plan.
   – Progress toward completion of the tactical objectives.
   – Safety considerations.
   – Deployment and assignment of operating companies and personnel.
   – Appraisal of the need for additional resources.
   – A review of the tactical worksheet with the new IC.
   – The new IC gives the old IC an assignment and a designated sector name.

f. When a Chief Officer arrives on the scene at the same time as the first arriving apparatus, the Chief Officer should assume Command of the incident.
g. The first arriving Chief Officer should become the Incident Commander using the "assumption" or "transfer" of command pro-

cedure, depending on what mode command is operating in when the first chief arrives.

h. Additional arriving Chief Officers should report directly to the Command Post for assignments.

i. Later arriving, higher-ranking Chief Officers may choose to assume Command, or assume adviser positions.

j. When Command is assumed or transferred, dispatch should be informed who the new IC is (officer number).

3. Command Mode: Certain incidents, by virtue of their size, complexity, or potential for rapid expansion, require immediate strong, direct, overall Command. In such cases, the first arriving officer will initially assume an exterior, safe, and effective Command position and maintain that position until relieved by a higher ranking officer. A tactical worksheet shall be initiated and utilized to assist in managing this type of incident.

   a. The IC may appoint another member of the company to be the company officer and give the company an assignment.

   b. The IC may elect to assign the crew members to perform staff functions to assist command.

   c. The first arriving officer assuming Command has a choice of modes and degrees of personal involvement in the tactical activities, but continues to be fully responsible for the Command functions.

   d. The IC operating in the Command Mode shall utilize a Command Worksheet to document and organize the incident.

   e. Sector officers may benefit from utilizing a Sector Worksheet to document and organize sector activities.

   f. Sector officers operating in Command Mode shall utilize command vests where appropriate to do so (Especially: Command, Operations, Safety, and Staging).

INCIDENT ACTION PLAN:

1. The Incident Commander is the primary developer of the incident action plan.
2. On small incidents, the action plan will be organized completely by the IC and may not need to be written down.
3. On more complex incidents, the action plan will be a written document developed by the IC, with staff assistance.
4. Action plans must be flexible and continually assessed.
5. All incident action plans (written or unwritten) should include:
   a. Strategy goals.
   b. Tactical objectives.
   c. Establish priorities.
   d. Resource needs and assignments of crews.
   e. Having a "Plan A" and a "Plan B".
   f. Anticipated outcomes.
   g. Set timelines for progress.
   h. Consider unanticipated outcomes (two working fires at the same time).
6. The incident action plan defines where and when resources will be assigned to the incident to control the situation.
7. The plan is the basis for developing a Command organization, assigning all resources, and establishing tactical objectives.

STRATEGIC GOALS:

1. Before strategic goals are established, the ATTACK MODE (Offensive, Defensive, or Marginal) is set.
   a. OFFENSIVE: Conduct interior firefighting and victim search. Interior victims are deemed to be savable. Structural integrity intact and fire has not extended to beyond room of origin. Primary goals are to save savable lives, extinguish the fire, and save savable property.
   b. DEFENSIVE: Conduct exterior firefighting. Interior victims are deemed to be unsavable. Structural integrity is NOT intact and fire involvement is significant. Primary goals are to save exposure lives and save exposure properties.
   c. MARGINAL: Conduct limited interior firefighting and victim search. Fire and structural conditions are deteriorating. Primary goals are to save savable lives, attempt to stop fire progression and to limit time of firefighter exposure within the structure.
2. Strategic Goals establish what needs to be done to achieve the incident priorities (Life Safety, Incident Stabilization, Property Conservation, Community Continuity).
3. A sample list of strategic goals include:
   - Rescue
   - Protect exposures
   - Confine the fire (or haz-mat)

- Extinguish the fire
- Ventilation
- Salvage
- Overhaul

TACTICAL OBJECTIVES:

1. Tactical objectives direct operational activities.
2. The accumulated achievement of tactical objectives should accomplish the strategic goals.
3. Examples of tactical objectives would be:
   - Search the second floor for victims.
   - Advance interior hose lines to extinguish the fire.
   - Vertically ventilate.

TASK ASSIGNMENTS

1. Task assignments refer to the specific activities that are accomplished by company personnel.
2. Task assignments are the details of the actual work to be done.
3. The accumulated achievement of task assignments should accomplish tactical objectives.
4. Examples of task assignments would be:
   - Two firefighters take the interior stairs to the second floor and conduct a search of the bedrooms.
   - Three firefighters take a $1^3/_4$ attack line and advance it to the basement and extinguish the fire.

Four firefighters throw an extension ladder to the roof, cut a 4×4 hole over the fire.

GEOGRAPHIC SECTORING

In order to provide a point of reference on the scene of incidents, the incident (structure) will be broken into geographic quadrants (four parts). "Side 1" will be the street-side (address side) of the structure or incident location. This is most often where the IC will be located.

From "Side 1" the incident will then be divided in a clockwise direction with the remaining quadrants being labeled as "Side 2", "Side 3", and "Side 4."

EXAMPLE:

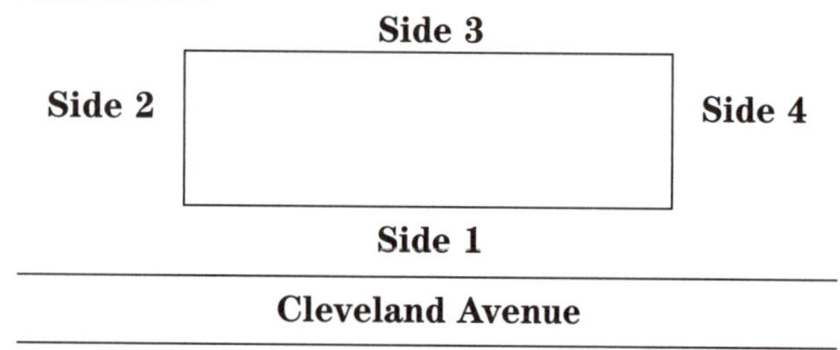

The structure can also be broken down further in the following manner.

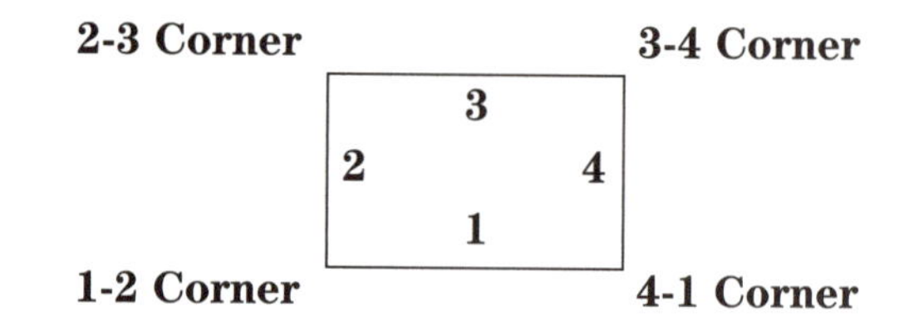

Cleveland Avenue

Exposures can also be identified based on which side of the structure they are on. Using the example above, if the exposure is on the left, it would be identified as the "Side 2 Exposure".

For multi-story structures, it may be necessary to identify each floor that personnel will be working on. Each floor will be called by the floor number, **STARTING WITH THE GROUND FLOOR ON THE STREET SIDE (address side) OF THE STRUCTURE.**

EXAMPLE:

Floor 6
Floor 5
Floor 4
Floor 3
Floor 2
Floor 1

Ground level at the street-side.

If the structure has a basement, the geographic designation will be "Basement." If the structure has multi-level basements, it may be necessary to identify each basement level that personnel will be working on. In this case, each basement level is a "Basement." Starting with the first subterranean floor on the street side (address side) of the structure as "Basement 1."

Ground level at the street-side

Basement 1
Basement 2
Basement 3
Basement 4
Basement 5
Basement 6

Examples of radio designations for geographic sector officers include:

- Side One
- Side Two
- Side Three
- Side Four
- Third Floor

- Basement
- Roof
- Attic

FUNCTIONAL SECTORING

Functional sectors are specialized sectors assigned to perform specific tasks or activities which do not necessarily coincide with geographic sectors. The IC will pick and choose the sector names needed, using the combination of geographic and functional sectors that best apply. This combination can be different for each alarm.

Examples of functional (task oriented) sectors include:

- Attack
- IRIT (Initial Rapid Intervention Team)
- RIT (Rapid Intervention Team)
- Ventilation
- Search
- Extrication
- Back-up
- Operations
- Safety
- Staging
- Water Supply
- Medical
- Rehab
- Public Information

SAFETY

The safety, accountability, and welfare of firefighters are the top priorities for the Incident Commander. A separate Roseville Fire Department Standard exists to address our operational policies for incident safety. The Safety Standard should be used, in conjunction with this Standard when operating at the scene of an incident.

COMMUNICATIONS

1. With the exception of units calling on-scene, Command should be the **ONLY** unit on the emergency incident scene that communicates with dispatch.
2. With the exception of acknowledging units arriving on-scene, all radio communications from dispatch should be directed to Command.
3. Each working sector on the fireground should be assigned a functional or geographic name by the IC. That assigned name should be well understood and well communicated by the IC to the officer. Officer numbers should not be used to identify personnel assigned to perform a duty.
4. Each working sector officer should be informed whom they are to report to, so they know who to call with progress reports or requests for additional resources.
   a. Benchmarks for progress include:
      - **ALL CLEAR**: Indicates that all victims are clear from the hazard zone.
      - **UNDER CONTROL**: Indicates the fire is controlled.
      - **LOSS STOPPED**: Indicates that property conservation is complete.
5. Radio orders and progress reports should be repeated by the receiver to ensure that what was said, is what was heard and understood.
6. When calling an unassigned crew, via radio, call the crew by their vehicle number (example: "Engine 11 from Command").
7. When the IC wants to reach the operator of the rig, call the rig number, followed by the term operator (example: "Engine 11 operator from Command").
8. When a crew has been giving an assignment, under our IMS, the crew should be called by their functional assignment or geographic location (example: "Roof Sector from Command" or "Interior Team 1 from Command").

COMMAND STRUCTURE (Fireground Organization)

The Incident Commander carries the responsibility for everything that happens on a fireground. For very small incidents, this should be a manageable task. However, as an incident escalates and personnel are deployed into hazardous environments, the span of control for the IC can exceed their capabilities. The IC can become overwhelmed and overloaded with information management, assigning companies, filling out and updating the tactical worksheets, planning, forecasting, requesting additional resources, talking on the radio, and fulfilling all the other functions of command. Thus, the IC should, as staffing will permit, delegate duties to other officers (or firefighters). Common roles that should be given consideration to filling include:

1. INCIDENT COMMANDER

   The IC is the person responsible for managing the incident on the strategic level. The IC establishes the overall operational plan, develops an effective organizational structure, allocates resources, makes assignments to carry out the attack plan, manages information to develop and revise decisions, and continually attempts to achieve the basic command objectives.

The IC should establish and operate from a stationary command post as soon as possible. The command post should offer the IC a relatively quiet vantage point, with lighting, radio equipment, and sufficient space to manipulate written reference material.

2. SENIOR ADVISOR
The Senior Advisor shall be a chief officer who serves as an assistant to the IC. The Senior Advisor assists with the development and documentation of the operational plans, organizational structure, resource allocation, communications, and can serve as a liaison with other agencies.

3. COMMAND AIDE
An Aide is a person assigned to assist the IC. During large operations, Sector Officers may also have aides to assist them. Aides assist by managing information and communications. Aides keep track of assignments, locations and the progress of companies, assist with tactical worksheets, use reference materials, and refer to pre-fire plans. The Command Aide may also provide reconnaissance and operational details for the IC (be his eyes and ears). Each working incident should have a Command Aide.

4. OPERATIONS
The Operations officer is responsible for the direct management of all incident tactical activities, the tactical priorities, and the safety and welfare of the personnel working in the Operations section. The Operations officer communicates, face-to-face and by radio with the tactical groups, divisions, and sectors.

5. SAFETY
The Safety officer's job is to anticipate and deal with unsafe or hazardous conditions. Safety's function is to monitor the condition of personnel, observe safe practices, and ensure proper use of all equipment, especially personal protective equipment. The Safety officer is responsible for evaluating structural safety, ongoing monitoring of toxic or explosive conditions, and assisting in the management of any special situations that expose firefighters to hazards. The Safety officer ensures incident accountability through use of the PAR system, ensures a rapid intervention team is staffed and equipped, ensures the rehabilitation of firefighters, and addresses any concerns with regard to freelancing. The Safety officer is a highly mobile position on a fire incident.

6. STAGING
The Staging officer's job is to assist the IC with establishing a location (and staffing a location) for apparatus and personnel staging. Refer to Staging Standard (108.02).

7. PUBLIC INFORMATION
The Public Information Officer (PIO) establishes an effective link with the media and provides a place for the media to assemble (away from the IC). The PIO should be the sole source of information for media and should be briefed with information from the IC to ensure the accuracy of the report. The PIO should escort the media on an incident scene, ensuring that member of the media are able to obtain video and photographs from a safe distance.

8. INITIAL RAPID INTERVENTION TEAM (IRIT)
To ensure compliance with the OSHA 2-in/2-out mandate, an Initial Rapid Intervention Team (IRIT), consisting of a minimum of two persons from the first arriving companies, shall be established and maintained. The IRIT shall remain in a state of ready deployment until relieved by a Rapid Intervention Team (RIT). Refer to the Rapid Intervention Teams Standard (107.03) for details.

9. RAPID INTERVENTION TEAM (RIT)
To ensure compliance with the OSHA 2-in/2-out mandate, a Rapid Intervention team (RIT), consisting of a minimum of two, but preferably four, persons shall be established and maintained from secondary arriving companies. The RIT shall relieve the IRIT from "stand-by" duties. Refer to the Rapid Intervention Teams Standard (107.03) for details.

SUMMARY

■ THE FOUR STEPS TO INITIALIZE COMMAND

STEP ONE: GIVE A ROLLING SIZE-UP

1. A windshield report of what the officer sees as they approach the scene before the officer leaves the apparatus.
2. The rolling size-up may not be entirely accurate, and may be revised as the officer assesses the emergency.

3. Includes (as a minimum) the following information:
   - Number of stories (examples: single story, story and a half, two story, three story, multi-story).
   - General construction-type of building (wood-frame, masonry, steel, bow string).
   - Occupancy of the structure (examples: residential, multi-family residential commercial, industrial, institutional).
   - Conditions present (nothing showing, smoke showing, working fire, people outside).

STEP TWO: ESTABLISH COMMAND

1. Establish command on every incident.
2. This will be accomplished by stating the following after the size-up:
   a. Officer number who will be in command.
   b. Name of the command being established (The "default" will be the street name for command. If you use the business name instead of the street name, state that on the radio).
   c. State your operational mode. Command is "investigating", "fast attack mode", or the fixed command post location.

STEP THREE: MAKE AN INCIDENT ACTION PLAN

1. This is the overall direction of the incident.
2. It includes the strategy, tactics, priorities, resource needs, and outcomes.
3. It includes a "Plan A" and a "Plan B".

FOUR: COMMUNICATE

1. Give assignments to your crew.
2. Give assignments to other incoming crews.
3. Give assignments to mutual aid/auto-response companies.
4. Assign people to IMS roles.
5. Ask for progress reports from sector officers.
6. Give periodic updates to dispatch.

EXAMPLES: ARRIVING RADIO TRAFFIC

- From Engine 11:
  - Dispatch, Engine 11 is on the scene
  - We have a single story, wood-frame, residential structure, with smoke showing
  - 101 will have Dale Street Command in fast attack mode
- From Engine 31:
  - Dispatch, Engine 31 is on the scene
  - We have a multi-story, brick, apartment house, with nothing showing and people outside
  - 933 will have Dale Street Command, investigating
- From 910
  - Dispatch, 910 is on the scene
  - We have a 100 by 400, single story, brick, warehouse, with fire and smoke showing from Side 2
  - 910 establishing Medtronics Command at intersection of the Cleveland and Lydia

Sample Job Description—Firefighter

City of Roseville Fire Department Standard

SUBJECT: Position Description—Firefighter
INDEX: 101:01.14
DATE DRAFTED: July 15, 1996
DATE EFFECTIVE: July 15, 1996
DATE LAST REVISION:

Reporting relationship: Captain, District Chief, Deputy Chief, Chief

Guidance/Autonomy: Works under general supervision with majority of time spent as team member. To act within limits of the Fire Department policy, procedures, and applicable by-laws. Acts as a crew officer when assigned.

Scope:
Number of Subordinates: None
Number of Incumbents: 84 (Average)

Primary objective of position:
Expected Results: Provides emergency services to the City of Roseville, and community members. Includes minimizing loss of life and property from fires, natural disasters, life-threatening situations, and other hazards. Administers emergency medical treatment, provides property inspection, and assists other emergency agencies/departments through fire suppression, rescue, administration of first aid, and other associated activities.

Functional Information:

General Activity: Time: 5%

1. Responds to emergency calls through use of personal paging system.
   - Provides self transportation within 5 minutes (maximum) to designated station for emergency response.
   - Dons appropriate personal protective equipment within 2 minute (maximum), including but not limited to, helmet/eye shield, hood, boots, gloves, protective coat, protective trousers, self contained breathing apparatus (SCBA), personal alert safety system (PASS), eye protection, hearing protection, life safety harness, and biological exposure control.
   - Mounts, dismounts, and operates around emergency response apparatus.
   - Assists in vehicle and crowd control in emergency situations.

Fire Activity: Time 45%

2. Responds to fire calls, and provides extinguishment or fire control as part of a team (company) under the supervision of a fire officer.
   - Selects, carries, lifts, drags, or pushes attack/supply lines (hoses) and nozzle as necessary. Connects and disconnects hose couplings to establish water supply, sprinkler/standpipe connection, and position attack line for suppression of fire. includes moving hose lines up and down ladders/stairways, and entry into structures depending on type of fire. Uses hose line, or fire extinguisher to apply a stream of water or agent onto fire.
   - Positions and climbs ground/aerial ladders, while carrying fire fighting tools/equipment, to gain access to upper levels of building or to assist individuals from burning structures.
   - Raise or lower fire fighting tools/equipment, or victims using ropes, and rope rescue systems.
   - Perform forcible entry, or create openings in building for access and ventilation, using ax, crow bar, power saw, or other power equipment.
   - Protects property from water and smoke damage by use of positive pressure ventilation equipment, smoke ejectors, or placement of waterproof salvage covers.
   - Performs overhaul by removal of debris, pulling apart burned materials, and exposing fire by opening ceilings, walls, and floors.
   - Participates in initial arson investigation by preserving evidence and noting the location and condition of related objects upon entering into a burning building.

Emergency Activity: Time: 10%

3. Responds to medical emergency calls, performs rescue and first aid as needed.
   - Removes injured persons from immediate hazards by use of carries, drags, and stretcher.
   - Administers first aid and CPR to injured persons.
   - Determines nature and extent of illness and injury. Establishes priority for required emergency medical care.
   - Extricates trapped individuals, using power tools, rescue equipment, or other appliances in accordance with prescribed techniques.

Maintenance Activities: Time: 10%

4. Maintains fire apparatus, equipment, and facilities.

- Operating equipment in an efficient and safe manner.
- Performs assigned duties in maintaining apparatus, equipment, building, quarters and grounds.
- Inspects and tests equipment for proper operation as needed for emergency calls.

Training/Drill Activities: Time: 20%

5. Attends regular and assigned training sessions/drills, to maintain and upgrade firefighter skills.
- Actively participates in drills, work details, demonstrations and courses in fire fighting/emergency response techniques.
- Demonstrates proficiency in techniques and procedures during drills.

Driver/Operator Activities: Time: 5%

6. Response as a driver/operator of fire apparatus under supervision of a fire officer.
- Drives and operates emergency response vehicles and equipment.
- Operates equipment in an efficient and safe manner.

Community Education Activities: Time: 5%

7. Provides community relations on fire prevention and emergency response.
- Provides fire safety programs for commercial or industrial employees. Demonstrates safety equipment and techniques.
- Provides public safety activities through open houses, tours, displays, demonstrations, and talks in public places.
- Provides fire safety programs for commercial or industrial employees. Demonstrates safety equipment and techniques.
- Participates in pre-fire planning of structures or target areas.
- Participates in pre-fire planning of structures or target areas.

Environment:

The environmental factors described in this section are representative of those an employee encounters while performing the essential functions of this position.

A. Operates both as a member of a team and independently at incidents of uncertain duration.

B. Spends excessive time outside exposed to the elements.

C. Tolerate extreme fluctuations in temperature while performing duties. Must perform physically demanding work in hot (up to 400 degree Fahrenheit), humid (up to 100 %) atmospheres while wearing equipment that significantly impairs body-cooling mechanisms.

D. Experience frequent transition from hot to cold and from humid to dry atmospheres.

E. Work in wet, icy, muddy areas, and uneven terrain.

F. Perform a variety of tasks on slippery, hazardous surfaces such as on rooftops or from ladders.

G. Work in areas where sustaining traumatic or thermal injuries is possible.

H. Faces exposure to carcinogenic dusts such as asbestos, toxic substances such as hydrogen cyanide, corrosives, carbon monoxide, or organic solvents either through inhalation or skin contact.

I. Faces exposure to infectious biological agents such as hepatitis B or HIV.

J. Wears personal protective equipment that weighs approximately 50 pounds while performing fire fighting tasks.

K. Perform physically demanding work while wearing positive pressure breathing equipment with resistance to exhalation and a flow rate specified by current SCBA manufacture.

L. Perform complex tasks during life-threatening emergencies.

M. Work for long periods of time, requiring sustained physical activity and intense concentration.

N. Faces life or death decisions during emergency conditions.

O. Be exposed to grotesque sights and smells associated with major trauma and burn victims.

P. Make rapid transitions from rest to near maximal exertion without warm-up periods.

Q. Operate in environments of high noise, poor visibility, limited mobility, at heights, and in enclosed or confined spaces.

R. Use manual and power tools in the performance of duties.

S. Rely on senses of sight, hearing, smell, and touch to help determine the nature of the emergency, maintain personal safety, and make critical decisions in a confused, chaotic, and potentially life-threatening environment throughout the duration of operation.

T. Encounter smoke filled environments, and a variety of physical hazards, damaged structures, moving mechanical equipment, electrical equipment, radiant energy, and possible exposure to explosives.

Worker Requirements:

To perform this position successfully, an individual must be able to perform each of the essential job functions, however, reasonable accommodations may be made in accordance with the Americans With Disabilities Act for individuals with disabilities. The requirements listed below are representative of the knowledge, skills and/or abilities required.

Skills Involved:

- Operates hand and/or foot controlled equipment/tools, such as, but not limited to, fire apparatus, pike poles/hooks, ax, forcible entry pry-bars/tools, chain saws, circular saws, gas powered fans, atmospheric monitoring equipment, air bags, hydraulic spreader/shears/rams, hand saws, wet vacuum, sledge hammer, and other cutting, prying, or striking tools. Uses fine motor skills/dexterity to operate radios, phones, fire apparatus controls (includes valves, switches, levers), and tieing various knots.
- Knowledge of fire behavior, fire chemistry, building construction, wiring practices, fire codes, hydraulic principles, and the characteristics of various fuels.

Minimum Educational Qualifications:

- High school graduate or GED equivalent (see below).
  - Minimum of 18 years of age.
  - Possession of a valid class C driver's license.

GED Requirements:

- Mathematical Development (GED level 2): Must be able to multiply, divide, use fractions and read graphs.
- Language development (GED level 3): Must be capable of reading fire protection textbooks, write reports with proper grammar and speak correctly in public.
- Reasoning development (GED level 4): Must be able to interpret instructions and use logic to solve complex problems.

Schedules and Other Conditions:

- On call—carries a pager, responds to emergency calls per standard operating procedures all hours other than normal working hours or vacations at a frequency specified in department by-laws. Follows established departmental by-laws, safety requirements, and operating procedures.

Physical Demands:

- Each applicant/member must be able to meet the physical requirements outlined in NFPA 1582 (Medical requirements for fire fighters), perform the tasks outlined in NFPA 1001 (Fire fighter professional qualifications), and meet the physical requirements outlined in the attached Physical Specification Worksheet.

Mental Abilities:

- Must be able to work as part of a team under stress caused by emergencies, danger or criticism.
- Must not be claustrophobic or afraid of heights.

Personal Attributes:

- Present a positive, constructive image and attitude in the performance of their duties.

Judgment Demands:

- Decisions regarding the safety of firefighter's team members, own self, and civilians.

NOTE: The requirements of this position description may not be fully inclusive and members/applicants are responsible for other functions as assigned. In addition, the Roseville Fire Department reserves the right to change or modify the duties outlined in this position description at any time.

Attachment: Position Physical Specification Form

ROSEVILLE FIRE DEPARTMENT POSITION SPECIFICATION WORKSHEET

Identifying Information

Position title: Firefighter

Note: In terms of time that may be spent at an emergency incident: *Occasionally* equals 1% to 33%; *Frequently* equals 34% to 66%; and *Continuously* equals 67% to 100%.

1. Employee's job requires:

	Activity	Not at all	Occasionally	Frequently	Continuously
a.	Sit		X		
b.	Stand				X
c.	Walk			X	

2. Employee's job requires:

	Activity	Not at all	Occasionally	Frequently	Continuously
a.	Bend/Stoop		X		
b.	Squat		X		
c.	Crawl		X		
d.	Climb/Height		X		
e.	Reach above shoulder level		X		
f.	Crouch		X		
g.	Kneel		X		
h.	Balance		X		
i.	Push/Pull		X		

3. Employee's job requires he/she carry:

	Activity	Not at all	Occasionally	Frequently	Continuously
a.	Up to 10 lbs.			X	
b.	11 – 24 lbs.			X	
c.	25 – 34 lbs.			X	
d.	35 – 50 lbs.			X	
e.	51 – 74 lbs.		X		
f.	75 – 100 lbs.		X		

4. Employee's job requires he/she lift:

	Activity	Not at all	Occasionally	Frequently	Continuously
a.	Up to 10 lbs.			X	
b.	11 – 24 lbs.			X	
c.	25 – 34 lbs.			X	
d.	35 – 50 lbs.			X	
e.	51 – 74 lbs.		X		
f.	75 – 100 lbs.		X		

5. Job requires that employee use feet for repetitive movements (as in operating foot controls):

Right				Left				Both			
Yes	X	**No**		**Yes**	X	**No**		**Yes**	X	**No**	

6. Job requires employee use hands for repetitive action such as:

	Simple Grasping		Firm Grasping		Fine Manipulating	
a. Right	X Yes	No	X Yes	No	X Yes	No
b. Left	X Yes	No	X Yes	No	X Yes	No

7. Employee's job requires:

	Yes	No	Comments
a. Working on unprotected heights	X		
b. Being around moving machinery	X		
c. Exposure to marked changes in temperature and humidity	X		
d. Driving automotive equipment	X		
e. Exposure to dust, fumes, and gases	X		

Comments: ______________________________

____

____

____

____

Employer representative signature: ______________

Title: ______________

Date completed: ______________

INDEX

Page numbers followed by *f*, *t*, or *n* denote figures, tables, and footnotes, respectively.

O

P